Sustainable Development Goals Series

World leaders adopted Sustainable Development Goals (SDGs) as part of the 2030 Agenda for Sustainable Development. Providing in-depth knowledge, this series fosters comprehensive research on these global targets to end poverty, fight inequality and injustice, and tackle climate change. The sustainability of our planet is currently a major concern for the global community and has been a central theme for a number of major global initiatives in recent years. Perceiving a dire need for concrete benchmarks toward sustainable development, the United Nations and world leaders formulated the targets that make up the seventeen goals. The SDGs call for action by all countries to promote prosperity while protecting Earth and its life support systems. This series on the Sustainable Development Goals aims to provide a comprehensive platform for scientific, teaching and research communities working on various global issues in the field of geography, earth sciences, environmental science, social sciences, engineering, policy, planning, and human geosciences in order to contribute knowledge towards achieving the current 17 Sustainable Development Goals.

This Series is organized into eighteen subseries: one based around each of the seventeen Sustainable Development Goals, and an eighteenth subseries, "Connecting the Goals," which serves as a home for volumes addressing multiple goals or studying the SDGs as a whole. Each subseries is guided by an expert Subseries Advisor.

Contributions are welcome from scientists, policy makers and researchers working in fields related to any of the SDGs. If you are interested in contributing to the series, please contact the Publisher: Zachary Romano [Zachary.Romano@springer.com].

More information about this series at http://www.springer.com/series/15486

Godwell Nhamo • Lazarus Chapungu
Editors

The Increasing Risk of Floods and Tornadoes in Southern Africa

Springer

Editors
Godwell Nhamo
Institute for Corporate Citizenship
University of South Africa
Pretoria, South Africa

Lazarus Chapungu
Institute for Corporate Citizenship
University of South Africa
Pretoria, South Africa

ISSN 2523-3084 ISSN 2523-3092 (electronic)
Sustainable Development Goals Series
ISBN 978-3-030-74191-4 ISBN 978-3-030-74192-1 (eBook)
https://doi.org/10.1007/978-3-030-74192-1

This Springer imprint is published by the registered company Springer Nature Switzerland AG
The registered company address is: Gewerbestrasse 11, 6330 Cham, Switzerland

Peer Review Process

This book is blind peer-reviewed. Apart from being the international norm, this double-blind peer review process is mandatory for South African-based authors in order to fulfil the requirements of the Department of Higher Education and Training's (DHET) policy for recognised research outputs for subsidy purposes. The authors invested their time to incorporate observations from the blind peer review process, an aspect that enhanced the quality of the product. However, the authors take full responsibility for any liabilities associated with their work, including plagiarism.

Preface

The 2019/2020 and 2020/2021 rain seasons in southern Africa witnessed unprecedented flooding. Part of the extreme flooding resulted from four tropical cyclones that made landfall in March and April 2019, one in December 2020 and another in February 2021. The tropical cyclones were, namely, Idai, Kenneth, Chalane and Eloise in that order. However, there were also floods not associated with cyclone events that took place, particularly in South Africa and eSwatini. Apart from floods, southern Africa has been experiencing an increasing intensity of tornadoes (also known as twisters in other regions). Tornadoes are known to be intense, rare and more localised events, which create large amounts of damage and destruction. To this end, there has always been the desire and need to increase the understanding of the causes, nature and impact of these natural hazards. One definition by Edwards (2012) highlights a tornado as a violently rotating column of air, in contact with the ground, either pendant from or underneath a cumuliform cloud and often (but not always) visible as a funnel cloud. As such, many eye witness account describe a tornado as a snake in the sky. Given the difficulty associated with predicting tornado occurrence, efforts have been made to capture and understand this weather phenomenon using digital beamforming, or radar imaging, which is often used to observe tornadoes. However, such advanced instrumentation remains a major drawback in southern Africa, although advancement is taking place. Tornadoes can leave devastating impacts on human settlements, infrastructure, agriculture and other economic sectors. In some instances, tornadoes cause loss of life and ruin livelihoods. As identified in the book, the intensity and destructive nature of both tornadoes and floods is increasing. As there is limited documentation and case studies profiling both floods and tornado events in southern Africa, this book is dedicated to doing just that, especially in the era of the Sendai Framework on Disaster Risk Reduction and the 2030 Agenda for Sustainable Development. The phenomena under investigation have resulted in financial, technical, human and other resources being diverted from the developmental agenda for emergency use. To this end, the book advocates for southern African countries to continue learning more about floods and tornadoes in order to prepare and put in place effective early warning systems at all spatial administrative levels.

Pretoria, South Africa — Godwell Nhamo
Pretoria, South Africa — Lazarus Chapugu

Acknowledgements

We, the editors, Prof. Godwell Nhamo and Dr. Lazarus Chapungu, wish to thank all the blind peer reviewers for their invaluable inputs during the writing and publishing process of this book. We thank Springer for taking on board this book project and for the quality control processes. We also wish to thank our families for their ongoing support of our work. The book project was coordinated through the Exxaro Chair in Business and Climate Change at the University of South Africa (UNISA). The Exxaro Chair is a research Chair funded by the Exxaro Resources (Pty) Ltd Chairman's Fund and hosted by the Institute for Corporate Citizenship at UNISA. The Exxaro Chair was established in 2008 and is now in its fourth term running (2018–2022).

The editors further extend their gratitude to the following people who took part during the fieldwork: Ms. Foxi Gcobisa, Mr. Khuselo Jokazi, Ms. Sbonginkosi Ndlovu, Ms. Anele Mayekiso, Mr. Mxolisi Mtyobile, Mr. Simbarashe Nyika, Dr. James Ayuk, Dr. Saul Ngarava, Mr. Sbusiso Hoho, Mr. Sibulele Bongoza, Mr. Lwandiso Mdiya, Dr. David Chikodzi, Dr. Kaitano Dube, Ms. Asanda Madikizela, Ms. Nompe Gladys Ntombela, Ms. Busisiwe Faith Magagula, Ms. Nthabiseng Mashula, and Ms. Hlengiwe Precious Kunene. We also thank all the respondents, be it from the focus group discussions, household questionnaire survey, interviews and/or those that cooperated with us during field observations. Mr. Humbulani Masha is also recognised for his work of transcribing the interviews.

Contents

About the Contributors

Andrea Ajmar has a PhD and is an expert in the field of spatial data science, geodatabase management and spatial data infrastructure (SDI) development and implementation. He has solid expertise in environmental management, gained during working experiences both for the public and private sector. He was responsible for the realization of UN WFP SDI, a dedicated infrastructure for emergency management. He currently holds the position of fixed-term researcher at the Politecnico di Torino, with teaching positions in master's degrees and doctoral courses. He is part of the laboratory called SDG11Lab, which aims to create a spatial data infrastructure to support research activities.

Simone Balbo specializes in geospatial data processing, satellite remotely-sensed data analysis and webGIS development in the context of emergency management. He holds a PhD in environmental protection and management (Politecnico di Torino, 2012). He has been collaborating with ITHACA since 2009 working on: (1) the Copernicus EMS—Rapid Mapping in the production site and as GIS tool developer; (2) GeoNode (geonode.org) based projects as code developer; (3) the Copernicus In-Situ component (insitu.copernicus.eu) as consultant to the EEA on geospatial data; and (4) extreme rainfall detection through satellite data analysis as code developer (erds.ithacaweb.org).

Piero Boccardo is a full professor in the field of surveying and mapping at the Politecnico di Torino, Interuniversity Department of Regional and Urban Studies and Planning, since 2001. Piero is director of ITHACA (Information Technology for Humanitarian Assistance, Cooperation and Action) since 2006, a joint project between Politecnico di Torino, SiTI and WFP (World Food Programme) for the distribution of IT products and services in support of emergency management operations. Author of more than 100 publications, he has attended more than 75 National and International scientific congresses as speaker and chairman.

Petronella Chaminuka is the principal economist and senior manager of the Economic Analysis Unit of the Agricultural Research Council in South Africa. The Unit specialises in economic impact assessment. She holds a PhD from the Environmental and Natural Resource Economics Group of Wageningen University, Netherlands, and an MSc in agricultural economics from the University of Zimbabwe. A research specialist in agricultural, environmental and rural development research, she has more than 20 years' international experience working as a lecturer, researcher and facilitator in rural development processes. She has authored numerous articles in peer-reviewed journals, book chapters and popular media.

Lazarus Chapungu (co-editor) is a postdoctoral research fellow in the Institute for Corporate Citizenship, UNISA. He has worked as a senior lecturer and research chair in the Department of Physics, Geography and Environmental Science at Great Zimbabwe University. He also serves as a chapter scientist in the 6th Assessment Report of the Intergovernmental Panel on Climate Change (IPCC)(AR6), Working Group III, Chapter 2. His research interests are in climate change, sustainable development and biodiversity (lchapungu@gmail.com).

David Chikodzi is a postdoctoral fellow with the Exxaro Chair in Business and Climate Change at the University of South Africa. His research interests are in climate change, water resources management, tourism, sustainable development and application of Earth observation technologies for societal benefit. He has worked for over 10 years in academia at Great Zimbabwe University (Zimbabwe) and has published over 30 journal articles, one co-authored book titled 'Counting the Cost of COVID-19 on the Global Tourism Industry' (2020) and several book chapters. Dr. Chikodzi has also taken part in several local and international funded research projects across southern Africa and has previously worked as a research scientist at the Scientific and Industrial Research and Development Centre in Zimbabwe. Dr. Chikodzi is an Alumnus of the ISIbalo Africa Young Statisticians and also the Zimbabwe Young Academy of Sciences.

Sosten Chiotha is a professor and specialist in environmental sustainability, public health and climate change. He is a graduate of the University of Malawi (B.Ed.), London School of Hygiene and Tropical Medicine (MSc in medical parasitology) and the University of Maryland, USA (PhD in marine, estuarine and environmental science). He served in various portfolios at the University of Malawi, including lecturing in the biology department and as university research coordinator. He is currently the regional director of LEAD (Leadership for Environment and Development).

Felistus Patience Chipungu is an experienced Root and Tuber Crops (RTC) breeder in the Department of Agricultural Research Services, the Ministry of Agriculture, Irrigation and Water Development in Malawi. In 2008, Felistus obtained her PhD in plant breeding and crop improvement from the University of Malawi. Felistus has developed and released several varieties of sweet potato and cassava over the years, as well as contributing to the release of other crops such as potato, tomato and bananas which are widely grown by farmers and contributing to the improved livelihoods of many Malawians. Felistus joined the International Potato Centre (CIP) in 2014 as a sweet potato crop specialist based in Chipata, Zambia, before moving to the CIP office at Bvumbwe Agricultural Research Station.

Kaitano Dube is an ecotourism management lecturer at the Vaal University of Technology, South Africa. He is one of the African leading tourism geographers researching in the area of tourism, climate change, sustainability and green aviation. He has published in high-impact international peer-reviewed journals with his work receiving global attention. His work has received extensive media coverage including in National Geographic, Wunderground, Atlasobscura and AFP. He has granted a number of international television and radio interviews. Dr. Dube holds a PhD and MSc from the University of South Africa. He graduated with a BSc Hons from Midlands State University in Gweru, Zimbabwe. He holds several other qualifications from UNISA, the Vaal University of Technology and University of the Witwatersrand Business School in Johannesburg, South Africa. Dr. Dube is an executive member of the Tourism Educators Association of South Africa and Tourism Sector Human Resource Development Governance and Institutional Coordination Forum hosted by South Africa's National Department of Tourism.

Miriam Dalitso Kalanda-Joshua is a senior lecturer in geography at the University of Malawi. She holds a PhD in geography with a focus on water governance from the School of Geography and Environment at the University of Southampton, United Kingdom. Miriam has over 16 years of research and work experience with communities, government, higher education and private sector institutions in areas of environmental management, climate change adaptation, indigenous knowledge systems and agriculture. Her published work mainly focuses on climate change adaptation and indigenous knowledge systems.

Ruth Kalinga-Chirwa is an environmental scientist and obtained her MSc in environmental sciences in 2013 from the University of Malawi, Chancellor College. She is a qualified secondary school teacher with over 20 years of experience in government schools. Her research interests include the assessment of flood and drought impacts on rural communities, climate change adaptation and mitigation, and sustainable urban and rural livelihoods. She is an activist championing education for girls with deprived backgrounds through her reusable sanitary pads project.

Donald Reuben Kamdonyo holds a Master of Science degree in climatology and applied meteorology and a PhD in change management. He retired after working for 38 years for the national meteorological service, 12 of which he was the director. He is now the senior associate of his consulting firm, Kamdonyo & Associates, a climate change and change management firm.

Raymond Kasei holds a PhD in hydrology and natural resources from the University of Bonn-Germany. He is currently a senior lecturer in the Department of Climate Change and Food Security of the University for Development Studies, Ghana. Kasei's research experience ranges from understanding hydrological processes, water allocation practices, ecosystem services, water storage options for climate change adaptation, transboundary water management and dam/irrigation operations in Africa. Kasei acted as a guest editor in the journal of the Physics and Chemistry of the Earth. Kasei is currently a party lead negotiator for Africa on agriculture and capacity building at the United Nations Framework on Climate Change Conventions (UNFCCC).

Hlengiwe Precious Kunene is currently a postgraduate research assistant for the Exxaro Chair in Business and Climate Change at the University of South Africa. She holds a Master of Science (MSc) degree in geography, honours and an undergraduate degree in geography and environmental management from the University of KwaZulu Natal, South Africa. Her research interests are in climate change, sustainable development and the use of earth observation technologies including GIS and Remote Sensing software for societal benefit. Hlengiwe is also an active member of the Golden Key International Honour Society and Friends in Disaster Risk Reduction (FIDRR).

Patrick Likongwe is a PhD student in environmental sciences at Rhodes University, South Africa. His research is focusing on the use of ecosystem services to help mitigate the effects of climate change and build resilience in urban areas through use of green infrastructure. Since 2012, his interest in climate change mitigation and adaptation grew when he joined the Lake Chilwa Basin Climate Change Adaptation Programme (2010–2017) in Malawi, which he later managed as the Program Manager in its final 2 years. He was one of the editors of 'Socio-Ecological Resilience to Climate Change in a Fragile Ecosystem: The Case of the Lake Chilwa Basin, Malawi', a book published in 2018 by Routledge Earthscan Studies in Natural Resource Management.

Asanda Madikizela manages the administration aspects within the UNISA Institute for Corporate Citizenship. She holds an MPhil in development practice from the University of Pretoria and a master's in business leadership degree from UNISA Graduate School of Business Leadership. Within the Institute, she is responsible for overseeing an effective, efficient management and development of human resources, the implementation of UNISA policies aimed at improving governance, evaluating and reporting on compliance. She is also a teaching assistant for the Sustainability and Greed module offered by the Institute. Her interests are in the nexus between sustainable development, leadership and social justice.

Nthivhiseni Mashula holds a master's degree in climate change, agriculture and food security (environmental management), obtained from the National University of Ireland Galway. She is currently a postgraduate research assistant for Exxaro Chair and Climate Change at the University of South Africa. Her research interests are in climate adaptation, sustainable agriculture/climate smart agriculture and food security.

Paola Mazzoglio is a PhD student at Politecnico di Torino. She worked for 2 years at ITHACA, developing an early warning system for the monitoring and forecasting of exceptional rainfall events (ERDS—Extreme Rainfall Detection System). She was also involved in disaster mapping activities in the framework of an international project for the provision of geospatial information derived from satellite data. Before ITHACA, Paola worked for 8 months as Junior Research Fellow at Politecnico di Torino and also collaborated with WaterView SRL (a start-up based in Torino) as a data scientist, developing a rainfall measurement methodology from photographic images.

Enrico Borgogno Mondino received his PhD in 'Geodesy and Geomatics' by Politecnico di Milano in 2004. He is associate professor of geomatics at the Department of Agricultural, Forest and Food Sciences (DISAFA)—the University of Torino, lecturer in BSc, MS, post-graduate masters and PhD courses since 2002, member of the Directive Council of the Italian Association of Remote Sensing, and vice president of the Scientific Council and member of the Directive Council of the Italian Confederation of Scientific Associations for Territorial and Environmental Information (ASITA). He is the author of over 100 papers in National and International Scientific Proceedings, Journals and Books.

Maurice Monjerezi is an associate professor in the Department of Chemistry at Chancellor College, a constituent college of the University of Malawi. His current research activities are in the application of chemical principles and water quality assessments to understand how changes in climate, hydro-biogeochemical and anthropogenic pressures influence ground- and surface water resources. He applies a combination of field- and laboratory studies, statistics as well as geochemical models to study these processes. He was awarded the 2014 African Union-The World Academy of Sciences (AU-TWAS) Young Scientists National Award in the Earth and Life Sciences category.

Simbarashe Mukonzo is an insurance and reinsurance expert involved in structuring insurance policies and risk products for large corporates, as well as infrastructure development projects in sub-Saharan Africa. Simba is the CEO of Mukoma Financial Services (MUKFIN), a pan-African reinsurance broking firm based in Johannesburg which provides risk advice and intermediary services across the continent. He holds a Master of Business Administration degree and a Bachelor of Commerce Honours degree in Insurance and Risk Management from the National University of Science and Technology in Zimbabwe. Simba is a fellow of the Insurance Institute of South Africa (IISA) having attained the Programme in Advanced Insurance Practice and the Programme in Risk Management, both with UNISA, as well as a Marketing Diploma from the Institute of Marketing Management (IMM). Simba is also involved in providing skills development training across the Continent through local insurance institutes.

Abbyssinia Mushunje is a full professor in the Department of Agricultural Economics and Extension at the University of Fort Hare, South Africa. The department focusses on issues that deals with the economics of agriculture. He holds a PhD in agricultural economics (University of Fort Hare, South Africa) and an MSc in agricultural economics (University of Fort Hare, South Africa). Abbyssinia has supervised many PhD and master's students. He specialises in rural livelihoods, natural resources economics, natural resource management and land reform. He has authored numerous articles in peer-reviewed journals, book chapters and technical reports.

Evance Mwathunga is a development geographer, urban and regional planner and is presently a lecturer in Development Geography in the Department of Geography and Earth Sciences at Chancellor College, University of Malawi. He holds a PhD in urban and regional planning obtained in 2014 from the University of Stellenbosch, South Africa. His research interests include environmental and spatial analysis, urban planning and development and climate change adaptation. He currently serves as a Commissioner of the Malawi National Planning Commission. Between 2015 and 2017, Evance was the head of the Department of Geography and Earth Sciences.

Chalo Ng'ambi is general manager of Emeritus Resseguros, SA Mozambique, the only resident reinsurance company in Mozambique. He holds a Master of Business Administration degree from MANCOSA; he is an Associate of Chartered Insurance Institute (ACII) UK and also an Associate of the Insurance Institute of South Africa. Besides, he holds an advance diploma in Applied Insurance Studies from the Malta Insurance Training Centre in Malta. He is the coordinator of the Insurance Institute of Zimbabwe professional training program.

Saul Ngarava is a postdoctoral fellow in the Risk and Vulnerability Science Centre (RVSC) at the University of Fort Hare, South Africa. The Centre specialises in global and climate change issues. He holds a PhD in agricultural economics (University of Fort Hare, South Africa), MSc in agricultural economics (University of Fort Hare, South Africa) and MBA (National University of Science and Technology in Bulawayo, Zimbabwe). Saul has 3 years of research experience on aspects of agricultural economics, climate change and rural development. He has authored numerous articles in peer-reviewed journals, book chapters and technical reports.

Nonkululeko Ngcobo attended her primary school years in Nhlangakazi Primary School and finished senior primary in Lyndhurt Primary. She attended her high school years in Pinetown Girls' High School and thereafter did an environmental management and geography degree at the University of KwaZulu-Natal in Pietermaritzburg. Nonkululeko is currently completing her honours degree in BSc Environmental Management at the University of KwaZulu-Natal whilst entering the corporate industry as she is self-employed.

Cosmo Ngongondo is a professor of Earth sciences in the Department of Geography and Earth Sciences, University of Malawi, Chancellor College, and is the current dean of the Faculty of Science. He holds a PhD in hydrology from the University of Oslo, Norway. He is the current chief editor of the Malawi Journal of Science and Technology (MJST) and also serves as a guest editor for the Journals: Physics and Chemistry of the Earth (JPCE) and Hydrology Research. His research work in hydrological and climatic sciences was awarded the 2013 African Union-The World Academy of Sciences (AU-TWAS) Young Scientists National Award in the Earth and Life Sciences category.

Godwell Nhamo editor-in-chief and accounting officer, is a full professor and Exxaro Chair in Business and Climate Change at the University of South Africa (UNISA). He undertakes research in the fields of climate change and governance, sustainable tourism, green economy and sustainable development. Prof Nhamo has conceptualised and completed 11 book projects (eight edited and three co-authored). The most recent being Counting the Cost of COVID-19 on the Global Tourism Industry (2020); Scaling up SDGs Implementation: Emerging cases from the state, development and private sectors published by Springer (2020); SDGs and Institutions of Higher Education published by Springer (2020); and SDG 7: Ensure Access to Affordable, Reliable, Sustainable and Morden Energy by Emerald (2020). Prof Nhamo has also published over 95 journal articles. Since 2013, Prof Nhamo has graduated 11 PhDs and hosted 10 postdoctoral fellows. Professor Nhamo sits in a number of both international and national boards and has received several awards and recognitions for his outstanding work both locally and internationally. Finally, Prof Nhamo was one of the four-member African Union High-Level Panel that drafted the Green Innovation Framework for the continent (nhamog@unisa.ac.za).

Nompe Ntombela holds an MA and is currently a researcher and a secondary lecture for Sustainability and Greed at the Institute for Corporate Citizenship at the University of South Africa (UNISA). Ms. Ntombela research interest adopts an interdisciplinary approach that looks at the relationship between people, the environment and interaction with technology. From time to time, she focuses on migration issues including women migration and environmental issues that influence migration.

Simbarashe Nyika is a lecturer at Walter Sisulu University, South Africa, and an IBM Consultant in Industry 4.0. He holds a master's degree, and worked briefly at Solusi University (Zimbabwe), Bethel College (South Africa), and currently serves as a lecturer in Computer Science, at Walter Sisulu University, also in South Africa. He holds a number of IBM badges and has trained students in South Africa and Botswana Universities offering IBM badge programs in partnership with IBM Africa. His areas of interest are in artificial intelligence. He has presented on Artificial Intelligence Applications and Emerging Technologies at selected international conferences. He currently lectures computer science courses, at Walter Sisulu University in South Africa. Simbarashe Nyika has been with Walter Sisulu University since 2005.

Francesca Perez holds a master's degree and a PhD in environmental engineering from Politecnico di Torino, Italy. She has a long-lasting experience in the processing of satellite multispectral imagery for thematic information extraction in support of environmental analysis. She is currently a senior researcher and remote sensing specialist at ITHACA, where she is doing research in the field of satellite remote sensing data exploitation for monitoring and responding to natural disasters. Main research fields: support to natural disaster early warning and early impact phases; identification and analysis of long-term vegetation trends, land cover, land cover changes, land degradation processes; natural disaster hazard, vulnerability, risk modelling.

Nedson Phophiwa, PhD is a visiting research fellow at the Centre for Learning on Evaluation and Results (CLEAR) based at the University of Witwatersrand. He holds a PhD in economic history from the University of Kwazulu-Natal, South Africa. He has over a decade of experience in academic and socio-economic policy research.

Guy Schumann received his PhD in geography in 2008 and is an international expert in modelling and remote sensing of floods. He is active in the private R&D sector, both in Europe and in the USA, and holds a research affiliation with the University of Bristol and the University of Colorado Boulder. Guy's research and development interests focus around flood risk, from local to the global scale, using flood modelling, prediction and remote sensing technologies. He is regularly involved in international innovation and challenge programs, and works with many organizations to make better use of scientific innovation for improving flood disaster response.

Désirèe Sehlapelo has a master's degree in environmental and sustainable development from the University of Glasgow. She has a local government/municipal experience as well as an experience of academia having worked as a director of planning and economic development in the Waterberg District Municipality as well as having been a senior lecturer at the University of South Africa. She became a visiting scholar at Georgia Institute of Technology (US) focusing on science technology and innovations studies. Desiree is an impeccable natural resource economist with over 20 years of environmental, socio-economic and educational experience having worked on both public and private sector research and projects.

Soul Shava holds a PhD in environmental education from Rhodes University, South Africa, and is currently an associate professor in environmental education (Education for Sustainable Development) in the Department of Science and Technology Education at the University of South Africa (UNISA). Soul's main research interests are in environmental sustainability education processes in southern Africa, Indigenous Knowledge (particularly their representation in knowledge generation processes at the interface of modern institutions and local communities and their application in community development contexts) and decoloniality. He also has research interests in community-based natural resources management (CBNRM), socio-ecological resilience, climate change education, green economy, intangible cultural heritage, sustainable agriculture, and the use and conservation of traditional agrobiodiversity and indigenous food plants to achieve food security and resilience to climate change impacts.

Shelton Siwedza is an insurance and risk management practitioner, who has an in-depth knowledge and experience in short-term insurance and reinsurance underwriting in sub-Saharan Africa's commercial and large corporate insurance market space, with an emphasis on property, agriculture, mining, oil and energy risks. Shelton has a great interest in the fields of enterprise risk management, business continuity planning and climate change risks. He holds a master's in Business Leadership (UNISA-SBL), a Diploma in Marketing Management and a BBA (Marketing) from the IMM's Graduate School of Marketing. He is a fellow of the Insurance Institute of South Africa (FIISA), an associate of the Institute of Risk Management of South Africa (AIRMSA) and a Certified Risk Management Practitioner (CRM Prac) with the same institute. He is currently pursuing a Master of Commerce in Risk Management with North-West University.

Leocadia Zhou is the director of the Risk and Vulnerability Science Centre (RVSC) at the University of Fort Hare, South Africa. The Centre specialises in global and climate change issues. She holds a PhD in geography and environmental science from the University of Fort Hare, South Africa. Leocadia specialises in capacity development through strengthening individual and institutional capacity in various climate change-related issues. She was awarded the Vice-Chancellor Excellent Community Engagement Award at the University of Fort Hare in 2016. She has authored numerous articles in peer-reviewed journals and book chapters.

Part I

Introduction and Background

Check for updates

1 Contextualising the Increasing Risk of Floods and Tornadoes

Lazarus Chapungu and Godwell Nhamo

Abstract

Natural hazard-related disasters have been progressively increasing across the globe over the last several decades. Notably, the frequency of floods and tornadoes has accelerated leading to the proliferation of related calamities over time. Although some significant amount of work has been done to understand flood risks on communities around the world, a lot more still needs to be researched given the evolving magnitude and frequency of the flood events. For tornadoes, the science still remains veiled in obscurity and their risks still need further scientific inquiry. It is in view of this that this chapter discusses the increasing risk of floods and tornadoes in different geographic and temporal contexts with a view to understand the consequences of increasing risks associated with both phenomena to society and the environment. The chapter presents this background and puts into context the various studies contained in this book. Generally, the chapter shows that floods and tornadoes are monstrous hazards that are seriously threatening societal progress and economic development around the globe. They are among the driving forces behind the slow progress towards the achievement of the 2030 Agenda for Sustainable Development and its associated sustainable development goals (SDGs). The need to build community preparedness and response capacity is also highlighted.

Keywords

Floods · Tornadoes · SDGs · Risk · Southern Africa

1.1 Introduction and Background

Natural hazards have increased in magnitude and frequency over time, inflicting a plethora of socio-economic problems on societies across the world (IPCC, 2004). Floods and tornadoes (also known as twisters in other regions) have contributed immensely to the deluge of calamities befalling communities the world over. The frequency of both phenomena has increased in the recent past and the character of each has evolved, owing partly to human-induced changes in the climate system and possibly to natural cycles occurring in the earth's atmospheric system. Climate-induced hazards such as floods, droughts, tropical cyclones, and tornadoes are cooperatively accountable for the superfluity of

L. Chapungu · G. Nhamo (✉)
Institute for Corporate Citizenship, University of South Africa, Pretoria, South Africa
e-mail: nhamog@unisa.ac.za

G. Nhamo, L. Chapungu (eds.), *The Increasing Risk of Floods and Tornadoes in Southern Africa*, Sustainable Development Goals Series, https://doi.org/10.1007/978-3-030-74192-1_1

disaster fatalities and economic losses globally (UNISDR, 2015). Remarkably, the frequency of floods has increased more rapidly than all other disasters leading to the triplication of major flood-related catastrophes since the 1980s (ADB, 2013). See (2019) observed that floods have affected more people than any other natural disaster, and since 2015, floods were responsible for 47% of disaster-related losses across the globe. In 2018 alone, about 34.2 million people were affected by flooding (Louvain, 2019). Exposure and sensitivity to flooding have increased over the years owing to the increased frequency and severity of flood events. The increased vulnerability to flooding can be attributed to the growing trend of urbanisation, which alters land use and exposes more people to natural hazards (GFDRR, 2015). Communities residing within river basins and closer to large water bodies have been greatly affected (Hallegatte, 2011). The poor and vulnerable communities have been the most affected.

Unlike floods, tornadoes' impacts on humanity and influence on disaster fatalities have not been adequately studied. However, their increase in frequency and reported fatalities are a cause for concern. Scientific observations show that there has been a foreseeable increase in tornado risk over the past few decades associated with a rise in the number of tornadoes in large tornado outbreaks, a shift in the time of the year they occur and changes in jet stream behaviour that favour severe thunderstorms (Blamey et al., 2017). Scientists generally agree that the mechanisms responsible for the generation of tornadoes are not yet well understood. However, they are known to be a result of a combination of wind shear and atmospheric instability (Denton, 2018). They feed on warm, moist air from strong winds that change direction with altitude. The increase in atmospheric greenhouse gas (GHG) concentration adds more energy to the climate system and possibly exacerbates wind shear which contributes to the generation of tornadoes. Like many other disasters, floods and tornadoes are increasing their influence on humanity, posing some risks and related spill-over effects that have affected the achievement of personal and policy level goals, specifically the 2030 Agenda for Sustainable Development and the associated interlinked 17 SDGs (United Nations, 2015).

This chapter discusses the increasing risks of floods and tornadoes putting into context the various geographic and contextual themes covered by the chapters in this book. This includes modelling, early warning systems, partnerships, impact on insurance sector, and urban and rural development, among other themes. The next section looks at the increasing risk of floods, from a global to a Southern African perspective.

1.2 The Increasing Risk of Floods

The radiative effects of anthropogenic activities have altered the climate system and will continue to do so in the future. This will lead to global warming that results in climate change. Climate change may, in turn, lead to the intensification of the global water cycle with a resultant increase in flood risk (Cubasch, 2001). Climate change and flooding are frequently featuring among the top ten of global risks, and these two disasters remain deeply connected. Climate change is predicted to be the driver of sea level rise and more extreme weather events. This modifies the frequency, severity and duration of hydro-meteorological hazards (Allen et al., 2018). Consequently, water levels in several reservoirs are expected to change and the frequency and severity of flood hazards are expected to rise. Recent years have seen an increase in the frequency and severity of flood disasters, and they have been regarded as the world's most costly natural hazard (Henstra et al., 2019).

Although there are different types of floods, Gaume et al. (2009) observed that flash floods are among the most dangerous of them all. Flash floods are "fast surface flows with high peak discharge values, often limited in their spatial extent" (Gaume et al., 2009, p. 1) and commonly result from heavy rainfall events (Hill et al., 2010). To this end, flash floods are consequently referred to as pluvial flash floods (Acosta-Coll et al., 2018). Coastal flooding is another type of floods that often occur in coastal areas (Riama et al., 2021). It is mainly a culmination of storm waves, inland flooding, shallow coastal flooding as well as tsunami waves (Adaku 2020; NOAA 2021). Jonkman and Dawson (2012) argued that coastal flooding results in some of the greatest losses economically, socially and physically. There is a deluge of physical factors that work in a synergistic manner to influence the incidence of flooding and affect coastal communities. Land subsidence, sea level rise, flood runoff, heavy rainfall and storms can combine their influence on coastal flooding and expand the areas that are susceptible (Hofmann & Schüttrumpf, 2019). This has been the case for most Indonesian coastal cities (Chaussard et al., 2013). Farhan and Lim (2010) state that such situations complicate management of islands flooding.

The increase in flooding events has already cost the world billions of dollars and the situation going into the future is expected to get worse (Allen et al., 2018; Berardeli, 2021). The number of fatalities associated with flood events is expected to double by the year 2030. About 132 million are predicted to be highly susceptible to riverine flooding by 2030 and increase from 65 million who were affected in 2010. On the other hand, coastal flooding is predicted to affect 15 million individuals, a rise from seven million in 2010 (Kuzma & Luo, 2020). Figure 1.1 shows the projections for populations and urban assets expected to be exposed to coastal flooding by 2080.

As shown in Fig. 1.1, there is an expected sharp increase in exposure to coastal flood risk between 2030 and 2080. It is most likely that most of the flood victims will be located in vulnerable regions such as Asia and Africa (Allen et al., 2018). Vulnerability in Southern Africa is mainly determined by the lack of effective early warning systems (EWS), limited access to information and communication technologies (ICT) and general lack of capacity to take early responsive action in the aftermath of a flood event (United Nations Development Programme (UNDP), 2018). Countries in coastal regions are also expected to be among the highly affected.

The increase in flood risk is threatening communities and their economies (Smith & Ward, 1998). Property destruction in urban areas by riverine floods is also projected to rise significantly

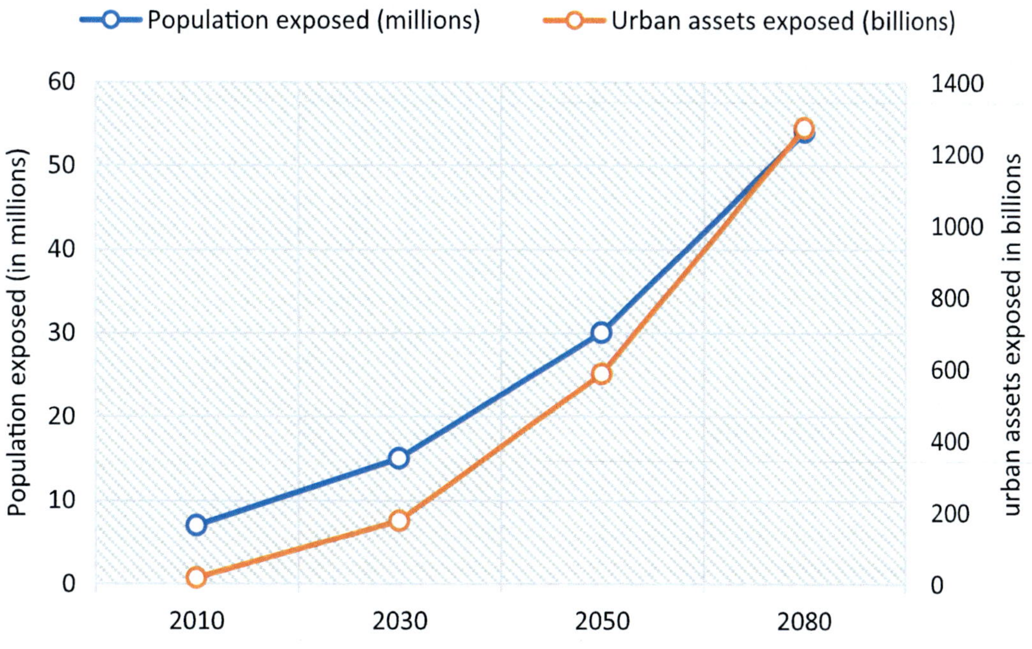

Fig. 1.1 Expected annual damage from coastal floods up to 2080. (Source: Authors, data from World Resources Institute, 2021)

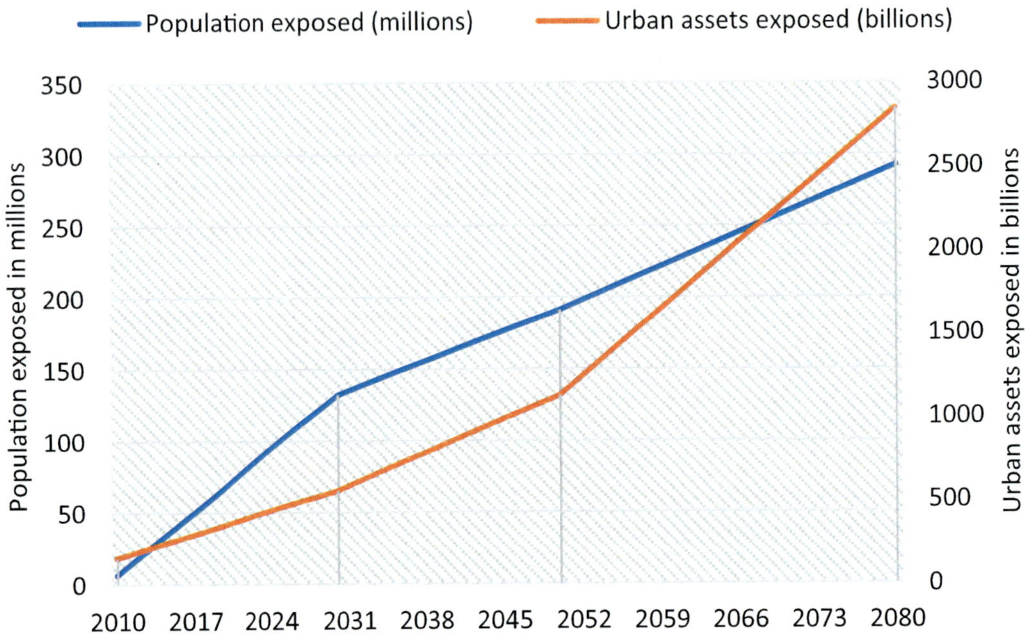

Fig. 1.2 Expected annual damage from riverine flooding. (Source: Authors, data from World Resources Institute 2021)

from $157 to $535 billion per annum (Kuzma & Luo, 2020). Figure 1.2 presents a picture expected from riverine floods.

On the other hand, there is a predicted tenfold increase in costs related to coastal storm surges and sea level rise—from US$17 billion to US$177 billion per annum (ibid). At a time when the world is grappling with the vice of the COVID-19 pandemic, which is already bullying global economies, it is evident that flood fortification needs to be prioritised by governments and policy makers, especially in Southern Africa where the physical environment is already fragile and communities highly vulnerable to a plethora of vagaries driven by climate change.

1.2.1 Why Is Flooding Risk Increasing

The increase in flooding risk is triggered by a plethora of factors, some of which are natural, while others are driven by anthropogenic activities (Gai et al., 2017). Heavy rainfall and storms naturally increase the risk of flooding. However, they are a result of human activities such as the release of GHGs into the atmosphere, leading to global warming and consequently climate change. Some social and economic factors such as population increase and development along rivers and close to the coasts contribute to the increase in flooding risks. Land subsidence driven by overdrawing of water also contributes to the increase in flooding risk. Areas highly vulnerable to flooding are affected by a combination of these factors. In India, Bangladesh and Indonesia, for example, there are significantly large communities impacted by both coastal and riverine floods on an annual basis. Kuzma and Luo (2020) noted that the three countries annually constitute 44% of the world's population impacted by riverine floods and 58% of the world's population affected by coastal flooding.

Zhang et al. (2011) postulated that climate change has been and will continue to influence rainfall patterns and coastal storm surges in various areas across the globe. This increases vulnerability of many people to flooding risks (Choi et al., 2015). In Puerto Rico, for example, scientists predict that climate change will be a key driver of increased flood risk. It is expected that, if all factors remain constant, by 2030, the population affected by riverine floods will double. Damage to urban properties is projected to increase to US$340 million (Kuzma & Luo, 2020).

Socio-economic changes also contribute significantly to the increase in the occurrence of flooding (Ahmad & Afzal, 2020). The world's population continues to increase and, consequently, demand on resources also increases resulting in massive concentration of people in small areas that are resource rich and have opportunities for survival. Thus, urban development in flood plains has occurred and this will increase the risk of riverine and coastal flooding. It is projected that by 2030 about 614,000 individuals will be exposed to the risk of riverine flooding annually due to urban development near rivers (Kuzma & Luo, 2020). An estimated US$1.6 billion will be lost annually. Coastal flooding, on the other hand, is predicted to affect 15,500 more people and cause losses to the tune of US$1.1 billion by 2030. About 87% of the flooding is attributed to the high rate of urbanisation (Ibid.).

Subsidence is propounded by several scholars as one of the key factors influencing increased risk of flooding (Gai et al., 2017). Coastal cities are sinking mainly because of overexploitation of groundwater. This is predicted to add a further two million individuals to the list of those who are going to be at risk of coastal flooding. Subsidence-driven flooding is high in the USA where the damages are predicted to cost more than US$16 billion by 2030 (Kuzma & Luo, 2020).

1.2.2 Flood Preparedness

Flood preparedness and EWS are among the best available options to mitigate against the calamities associated with flooding everywhere, including in Southern Africa. The concept of flood preparedness implicates the development of knowledge and capacities in expectation of a flood hazard. The knowledge and capacities should help in the fortification of communities against the impending hazards (Chapungu, 2020; Mabuku et al., 2018). Adequate preparation by institutions and individual households saves lives and reduces economic losses and consequently promotes societal development. This necessitates the need for perpetual calls for perpetual systems adjustments at institutional level and public behaviour in the context of forthcoming floods. The commitment to precautionary measures is key to resilience building among individuals and organisations (Schlör et al., 2018).

Alexander (2015) noted that there is a rapid transition of the disaster preparedness field. There is a shift from focusing on post-disaster responses to a system of pre-disaster planning with a deliberate emphasis on preparedness (Keating et al., 2017; Suryani et al., 2018). The rationale for this shift is the realisation of the fact that, as time progresses, financial returns from preparedness has the potential to offset costs of disaster response and recovery (Tim et al., 2016). However, the major stumbling block to the development of preparedness systems is that preparedness remains politically less attractive and mobilising resources for preparedness is a daunting task (Tim et al., 2016). As a result, in some countries, especially in Southern Africa, the level of disaster preparedness is still very low, explaining why the region remains highly vulnerable to natural disasters.

Doong et al. (2012) observed the need to urgently develop EWS for flooding, especially in coastal flooding in order to mitigate the associated casualties and property damage. Riama et al. (2021) added that the EWS should be understood by communities for effective mitigation efforts. Furthermore, Fakhruddin (2015) reiterated the need to support EWS with early action to effectively prevent flood hazards from turning into flood disasters. Well-coordinated and collaborative action through strategic partnerships is key in converting flood early warnings into effective and efficient implementation by all people exposed to a flood disaster and in all vulnerable regions (Perera et al., 2020. Such action will protect thousands of people and their livelihoods, reduce susceptibility and reinforce pliability. Executing EWS can be a daunting task as this is dependent on the manner in which the community consents to the information and acclimatise to it in their lifestyles. To achieve the best results, EWS must use approaches that are community driven. Thus, EWS should take into consideration the community's values and methods of execution (The International Federation of Red Cross and Red Crescent Societies (IFRC), 2009).

Execution of EWS must be preceded by the measurement of public acceptance to determine its success before it is operationalised. Several factors should be taken into consideration since behaviour during flood events is influenced by their preparedness and various traits including perceived benefits, risk awareness or even denial (Lopez et al., 2017). New EWS have the propensity for triggering the emergence of new problems related to how the community understands and acts upon the information provided. Accordingly, assessment of public acceptance leads to understanding of the information, and it will be received efficiently and effectively as part of the EWS in line with the main tenets, that is, precisely, timely and consistently (Sattele et al., 2015).

1.3 Tornadoes: Trends and Distribution

Tornadoes are helical atmospheric rotations of air, which are connected to a cumulonimbus or a towering cumulus cloud that make contact with the earth's surface (TORRO, 2021; Bluestein, 2013). However, the definition has evolved over time and the taxonomy has been updated regularly by the American Meteorological Society (AMS) (Agree, 2014). The evolution of the definition and updating of taxonomy is evidence that the atmospheric mechanisms behind their formation and prognosis remain veiled in obscurity (Kimambo, 2018). Consequently, modelling of tornado activity is not adequately robust to assertively ascertain future trends. However, there are indications of increasing tornado activity over time across the globe. The predicted high frequency of severe weather conditions the world over might generate more severe tornadoes.

Tornadoes occur mainly between 20 and 60° north and south of the equator (Kimambo, 2018). They have been recorded in all continents except Antarctica where the environmental conditions are not favourable for convective storms to develop. Most of the damages associated with tornado occurrence are found in the Northern Hemisphere with the most, and the strongest and most violent tornadoes occurring in the USA. In the USA, tornado occurrences are mainly concentrated in the central part commonly known as Tornado Alley (Agree, 2014).

Canada ranks second in terms of tornado frequency with most of them occurring in Ontario and the Prairie provinces. The world over, other areas that experience highly frequent, strong and violent tornadoes include Bangladesh, Philippines, Argentina, New Zealand, northern Mexico, South Africa, Europe, Uruguay and far eastern Asia (Agree, 2014). In the last few decades, the USA recorded 1274 tornadoes per year while Canada recorded an average of 100 per annum. The United Kingdom has the highest per area per year occurrences of tornadoes although they are usually weak (Snow, 2020).

1.3.1 The Meteorology of Tornadoes

Although the meteorology of tornadoes remains open for further scientific inquiry, it has been established that tornadoes are principally formed when the environmental conditions comprise a deep layer of mid-atmospheric dry air above moist surface, high surface temperatures, vertical wind shear, steep temperature gradient, steep moisture gradient and atmospheric instability (Ramseyer et al., 2016). The occurrence of a tornado is preceded by a thunderstorm, which is driven by unstable warm air under a cold air mass. The process of thunderstorm development is punctuated by upward and downward atmospheric motion which causes lightning and thunder. The thunderstorms can be classified into four types, namely, supercell, multicell line, multicell cluster and single cell storm (Blamey et al., 2017). The multicell line thunderstorm is the strongest and can lead to small tornadoes. The supercell storm is the most violent with a high propensity for producing tornadoes, according to the National Weather Services (NWS, 2009). The NWS observed that tornadoes occur most frequently in spring compared to winter. The recurrent transition during autumn and spring stimulates the development of extra-tropical cyclones and frontal systems that propel enormous convective storms. Some scientists also observed that land falling tropical cyclones have the propensity for masking tornadoes. This occurs when the tornadoes are focused in the right poleward section of the cyclone. Sid (2002) noted that tornadoes can also be produced consequent to eyewall meso-vortices, which continuously occurs until landfall.

However, conditions that favour the development of a tornado can occur at any time of the year. It is important though to note that tornado development is highly reliant on the time of the day because of solar radiation. Globally, the multitude of tornadoes occur in the late afternoon between 15:00 and 19:00. However, destructive tornadoes can occur at any time of the day even during the evening (Roger, 2006). Lastly, the magnitude of tornadoes is generally computed using the Enhanced Fujita Scale which measures tornado intensity from EF0 to EF5 by the amount of damage to the human environment as well as wind speed. The measurements are made post-tornado dissipation and the damage trail is assessed by experts (Ibid.).

1.3.2 The Trends of Tornadoes

In Southern Africa, tornadoes are a frequent phenomenon with most of them located in South Africa (Blamey et al., 2017; Kimambo, 2018; Kantamaneni et al., 2017). Recently, South Africa has been experiencing tornadoes almost every year, with most of them occurring in the eastern parts of the country. The tornadoes have been responsible for potential to severe damages and loss of lives (Reason, 2017). Figure 1.3 shows the trend of tornadoes in South Africa between 1905 and 2020 and the USA between 1950 and 2010.

The history of tornadoes in South Africa indicates that a significant number of the tornadoes range from light to severe on the Fujita scale (F Scale) (Bluestein, 2013). The trend, however, shows that the frequency of light category tornadoes is higher than the most severe ones.

1.3.3 Tornadoes and Climate Change

There is still paucity of scientific evidence of the link between tornadoes and climate change (Elsner et al., 2015). This is mainly because for most countries, current data on tornadoes is erratic (Long et al., 2018). In most cases, tornado assessment is dependent on observer explanations and aftershock destruction evaluation, instead of computable variable data. More so, it is a daunting task to determine trends in tornado incidences over a short period of time. For example, the US incidence data dates back only to the 1950s (Elsner et al., 2015). Trends can only be identified now since more records are becoming available due to the fact that communities affected by tornadoes across the globe are increasing, contributing to increased eyewitness accounts. In addition, technological advancements in the form of advanced radar also help in the detection of tornado hazards that have not been identified some decades back (Blamey et al., 2017). Another barrier in determining the influence of climate change of tornado activity is that tornadoes are geographically too small to be well simulated by climate models (Long et al., 2018). Only some of the conditions contributing to the formation of thunderstorms that spawn tornadoes can be simulated by climate models. Several scientific observations have established that high magnitude thunderstorms are associated with warmer conditions, indicating the possibility of a changing climate to influence tornado activity. However, it remains a daunting task to confidently ascertain the influence of climate change on tornado activity unless the observational trends, physics and related processes of tornado development are improved.

Furthermore, the physics associated with tornadoes remains obscured and scientists are still working to establish authentic facts about how

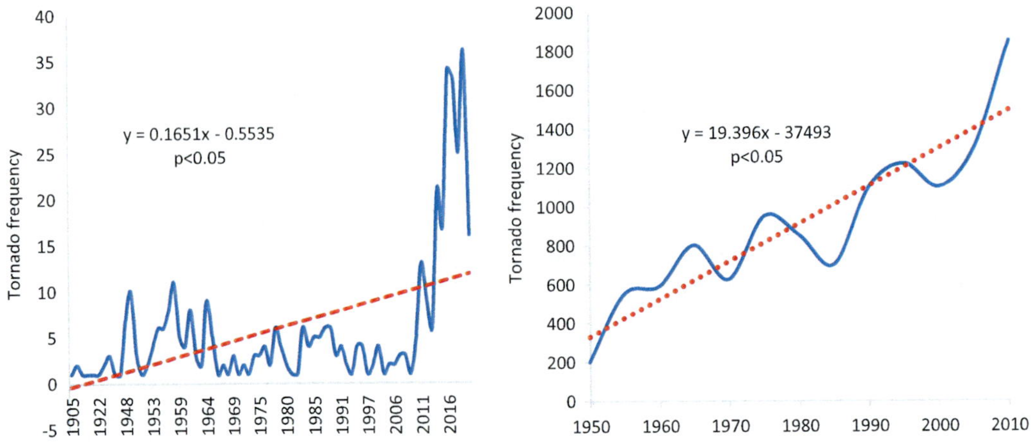

Fig. 1.3 Tornado frequency (left) in South Africa between 1905 and 2020 and (right) in the USA between 1950 and 2010. (Source: Authors, data from the South African Weather Service)

the key processes for tornadoes—atmospheric instability and wind shear—will respond to atmospheric temperature changes. However, there is a high probability of a warmer and more humid world to be associated with instability and, consequently, tornado occurrences (Snow, 2020). Studies in the USA show a relationship between tornado frequency and temperature patterns in the USA, a trend which contributes in indicating the possibility that climate change may influence tornado activity (CCES, 2021). In the USA, temperatures have increased since the 1950s. The increase is associated with an increase in tornado frequency. Although cause-effect relationships have not been ascertained, there are greater chances that temperature is playing a significant role in influencing tornado occurrence trends. However, CCES (2021) argues that it is possible that a warmer world will eliminate chances for wind shear, a key building block in the formation of tornadoes.

From the little that is known thus far, the changing climate is possibly shifting the timing of tornadoes as well as their numbers and magnitude across the globe. Thus, the current increase in global temperatures, the variability of weather patterns and several other indications of a changing climate point to the fact that tornadoes are likely to continue changing in character and posing a serious meteorological hazard to communities around the world. So far, homes have been destroyed, people have died and infrastructures have been vandalised, and the globe waits for more to come. Appropriate measures should be put in place to ensure that vulnerable areas are cushioned from tornado aftermaths.

1.3.4 Risks Posed by Tornadoes

The reported increases in risks of tornadoes justify the need to investigate their effects on societies. Tornadoes across the world have been associated with massive socio-economic losses. Every year, the USA counts human fatalities and damage to property which amount to massive economic losses. For example, an estimated cost of US$8.5 billion was incurred in the first half of 2016 (CCES, 2021). The greatest impacts emerge from tornado outbreaks, which are sequences of tornadoes occurring in close succession. The number of tornadoes during outbreaks has increased since 1954. The strong winds and debris carried by the wind pose a significant threat to human life and property from tornado occurrences. Long et al. (2018) postulate that, while discrete tornadoes may distress relatively small area compared to tropical storms, they pose a monstrous threat to homes, people and communities at large that are comparable to damages caused by some of the tropical storms.

However, in the USA, the death toll from tornadoes has decreased owing to the increasing capacity among forecasters to detect dangerous weather and warn communities (CCES, 2021). However, tornadoes continue to drain billions of US dollars through property damage every year, especially in the USA where more than a thousand occur annually. In the USA, the most economically damaging tornado occurred in 2011 when seven tornadoes and severe weather outbreaks each resulted in more than US$1 billion in damages. The total damage amounted to US$28 billion (Kimambo, 2018). In South Africa, damages associated with tornado occurrences amounted to US$800 million between 1990 and 2014 (DEA, 2016). In 2019, a tornado hit South Africa's coastal province of KwaZulu-Natal leaving in its wake two deaths and the destruction of homes in the eMpolweni area of New Hanover. Some homes were reduced to rubble, some people were thrown off roofs and several people were left homeless (Shaba, 2019).

Kimambo (2018) indicated that cold season tornadoes have become a common feature in the US Tornado Alley region. These tornadoes have the propensity for posing deleterious societal risks because they catch people off guard due to the infrequency of tornadoes during winter (Fike, 1993).

Table 1.1 Tornado events with outstanding characteristics

Tornado characteristic	Parameter	Date	Length of record
Highest mortality tornado	Approximately 1300 individuals	26 April 1989	1873–present
Longest lasting—greatest distance travelled (single tornado)	352.4 km/3.5 h duration	18 March 1925	
Biggest (numerical) tornado outbreak	207	27 April 2011	1787–present
Widest tornado (maximum diameter)	4184 m in width	31 May 2011	
Calendar month with greatest number of tornadoes	543 tornadoes	5 May 2003	1950–present
Highest recorded tornadic wind speed	5 May 2003	3 May 1999	1996—present
Longest tornado transport	359 km	11 April 1991	1871–present

Source: Kimambo (2018)

1.3.5 The Changing Character of Tornadoes

The nature and character of tornadoes have evolved over time, probably signifying the potency of climate change in regulating atmospheric thermodynamics that influence the proliferation of tornado formation processes. Thus, it is expected that the form and character of tornadoes will continue to change with the progressively changing climate, specifically the increased warming of the earth-atmospheric system (Snow, 2020). Consequently, the risks associated with tornadoes also evolve over time in terms of frequency, severity and casualties. Table 1.1 shows some isolated tornado events that partly indicate changes over time in terms of size, frequency and wind speed, among other characteristics.

Tornado characteristics have changed in many aspects over time and the changes have exposed communities to greater risks. In South Africa, for example, the number of tornadoes has increased from around three per annum in 1905 to over 38 per annum in 2020 (de Villers, 2019). This indicates an increase in the number of individuals and geographic regions exposed to tornado risk. The seasonality of tornadoes has also changed; for example, Long et al. (2018) observed that in the USA, tornadoes are starting earlier in the year and their frequency in the colder months has also increased. However, there has been a notable decline in summer tornadoes.

1.3.6 Building Resilience to Tornadoes

It is imperative for communities to build resilience to impacts associated with tornado occurrences. This can be achieved by adopting more stringent building codes in tornado-prone areas (de Villers, 2019). Tornado chasers argue that it is possible to develop risk-informed performance criteria for houses and other infrastructures exposed to a broad spectrum of natural hazards. The criteria can be matched to societal goals and building attributes can be identified and parametrised to support the risk-informed decision framework. This will enable enhanced community resilience and sustainability.

Furthermore, there is a need for the continued support for research on severe weather forecasts. This enables the capacity to predict future tornado risks and appropriate fortification measures taken before disaster strikes. The need to heed warnings from emergency alert systems cannot be overemphasised. It helps in building resilience through the institution of appropriate mechanisms and frameworks that contribute in warding off impacts associated with tornado occurrences (Roper, 2016). The next section details the materials and methods used in the book volume.

1.4 Materials and Methods Used in the Book

The chapters contained in this book are born out of a mix of research methodologies. The studies were mainly conducted in South Africa, Malawi and Zimbabwe. The countries are among those that are mostly affected by floods and tornadoes. The mixed methods research approach was adopted as the key strategy of inquiry for this book. It is an approach to inquiry that purposefully combines qualitative and quantitative procedures in research (Shorten & Smith, 2017). The method was considered ideal as it provides cross-disciplinary rigour and triangulation in addressing research questions. Shorten et al. (2014) state that the mixed methods approach promotes "purposeful data integration, which enables researchers to pursue a more panoramic view of their research landscape, viewing phenomena from different viewpoints and through diverse research lenses" (p. 74).

The purposive sampling method was the main sampling approach used in selecting study subjects and sites. This was accomplished by applying researcher knowledge of the population and areas under investigation. The use of this approach was considered desirable given that it is among the most efficient sampling methods in terms of time and costs (Saunders et al., 2012). More so, it is considered the best applicable method in circumstances of inadequate sources of primary data (Black, 2010). Saunders et al. (2012) reiterate that purposive sampling contributes significantly in "exploring anthropological situations where the discovery of meaning can benefit from an intuitive approach" (p. 2).

Some authors made use of primary data collected from different areas of study through repeated field observations, household questionnaire surveys and key informant interviews. The data was collected from areas affected by phenomena under investigation. The methods stated above complemented each other to safeguard authenticity of findings and conclusions drawn from the studies. While some of the questionnaire surveys were administered physically, multi-pronged online questionnaires hosted on QuestionPro were also used.

In addition, the case study approach was also used by authors of this book. This was a deliberate choice in order to have a deeper understanding of the concepts under investigation as it provides both scope and insight into the phenomena under study. Earth observation systems, specifically geographical information systems (GIS), were also used in some of the chapters of this book. This involved analysis and interpretation of remotely sensed imagery to obtain useful information.

Secondary data sources and archival data were also key in determining the findings and conclusions of this book. Researchers gained insights about phenomena under investigation by using available secondary data sources, including annual reports, agency reports and other authoritative documents.

1.5 Book Outline

This book comprises six parts, all focused on unravelling the increasing risk of floods and tornadoes in general drawing from country-specific experiences and contexts. The first part has an introductory chapter, which puts its context the themes covered in the book. The second part has four chapters focusing on flood modelling, EWS and partnerships. The third part of the book, which comprises five chapters, investigates urban flooding and insurance matters. Part four interrogates a plethora of flooding challenges bedevilling rural settlements. It comprises four chapters, each exploring a different theme focusing on rural flooding. Part five of this volume comprises two chapters, which focus on the increasing risk of tornadoes. Part six of the book is made up of the single concluding chapter focusing on the policy space in the context of the emerging key findings.

References

Acosta-Coll, M., Ballester-Merelo, F., Martinez-Peiró, M., & de la Hoz-Franco, E. (2018). Real-time early warning system design for pluvial flash floods—A review. *Sensors, 18*, 2255.

Adaku, J. E. (2020). The impact of flooding on Nigeria's sustainable development goals (SDGs). *Ecosystem Health and Sustainability, 6*, 1. https://doi.org/10.1080/20964129.2020.1791735.

ADB, A. D. (2013). *The rise of natural disasters in Asia and the Pacific: Learning from ADB's experience*. Metro Manila: ADB.

Agree, E. (2014). A revised tornado definition and changes in tornado taxonomy. *Weather and Forecasting, 29*, 1256–1258. https://doi.org/10.1175/WAF-D-14-00058.1.

Ahmad, D., & Afzal, M. (2020). Flood hazards and factors influencing household flood perception and mitigation strategies in Pakistan. *Environmental Science and Pollution Research, 27*, 15375–15387. https://doi.org/10.1007/s11356-020-08057-z.

Alexander, D. E. (2015). Evaluation of civil protection programmes, with a case study from Mexico. *Disaster Prevention and Management, 24*(2), 263–283.

Allen, M. R., Dube, O. P., Solecki, W., Aragón-Durand, F., Cramer, W., Humphreys, S., Kainuma, M., Kala, J., Mahowald, N., Mulugetta, Y., Perez, R., Wairiu, M., & Zickfeld, K. (2018). Framing and context. In: *Global Warming of 1.5°C. An IPCC Special Report on the impacts of global warming of 1.5°C above pre-industrial levels and related global greenhouse gas emission pathways, in the context of strengthening the global response to the threat of climate change, sustainable development, and efforts to eradicate poverty* [Masson-Delmotte, V., P. Zhai, H.-O. Pörtner, D. Roberts, J. Skea, P.R. Shukla, A. Pirani, W. Moufouma-Okia, C. Péan, R. Pidcock, S. Connors, J.B.R. Matthews, Y. Chen, X. Zhou, M.I. Gomis, E. Lonnoy, T. Maycock, M. Tignor, and T. Waterfield (eds.)]. In Press.

Berardeli. (2021). Climate change is responsible for billions of dollars in floods costs, study says. *CBS News*. Retrieved February 17, 2021, from cbsnews.com/news/climate-change-flood-costs-study/

Black, K. (2010). *Business statistics: Contemporary decision making* (6th ed.). Wiley.

Blamey, R. C., Middleton, C., Lennard, C., & Reason, C. J. C. (2017). A climatology of potential severe convective environments across South Africa. *Climate Dynamics, 49*, 2161–2178. https://doi.org/10.1007/s00382-016-3434-7.

Bluestein, H. (2013). *Severe convective storms and tornadoes*. Oklahoma: Springer.

Chapungu, L. (2020, May). *Mitigating the impact of cyclone disasters: Lessons from Cyclone Idai*. South African Institute of International Affairs, Policy briefing, Climate Change and Migration.

Chaussard, E., Amelung, F., Abidin, H., & Hong, S. H. (2013). Sinking cities in Indonesia: ALOS PALSAR detects rapid subsidence due to groundwater and gas extraction. *Remote Sensing of Environment, 128*, 150–161.

Choi, W., Nauth, K., Choi, J., & Becker, S. (2015). Urbanization and Rainfall–Runoff Relationships in the Milwaukee River Basin. Professional Geographer, pp. Article in Press.

Centre for Climate and Energy Solutions (CCES). (2021). *Tornadoes and climate change*. https://www.c2es.org/content/tornadoes-and-climate-change/

Cubasch, U. (2001). Ch 9: Climate change 2001: The scientific basis. In J. T. Houghton et al. (Eds.). Cambridge: Cambridge Univ. Press.

de Villers, J. (2019). Tornadoes in South Africa are likely to grow more common-as thunderstoms grow more severe, Business insider, SA, Nov 13 2019.

DEA. (2016). *Climate information and early Warning Systems for Supporting the disaster risk reduction and management Sector in South Africa under future Climates*. Pretoria. Retrieved from https://www.environment.gov.za/sites/default/files/reports/ltas-book2of7_climateinformationandearlywarningsystemsforsupportingtheDRR.pdf

Denton, S. (2018). *Breakdown: Why wind shear is necessary for tornado development*. Retrieved February

15, 2021, from https://www.wmcactionnews5.com/2018/10/16/breakdown-why-wind-shear-is-necessary-tornado-development/#:~:text=Wind%20shear%20is%20usually%20caused%20by%20thunderstorms.&text=Tornadoes%20form%20in%20very%20strong,higher%20levels%20of%20the%20atmosphere

Doong, D.-J., Chuang, L., Wu, L.-C., Fan, Y.-M., Kao, C., & Wang, J.-H. (2012). Development of an operational coastal flooding early warning system. *Natural Hazards and Earth System Sciences, 12*, 379–390.

Elsner, J. B., Elsner, S. C., & Jagger, T. H. (2015). The increasing efficiency of tornado days in the United States. *Climate Dynamics, 45*, 651–659. https://doi.org/10.1007/s00382-014-2277-3.

Fakhruddin, S. (2015). Risk communications for coastal inundation forecasting to the community. *Journal of Psychology and Psychotherapy, 5*, 203.

Farhan, A. R., & Lim, S. (2010). Integrated Coastal Zone Management Towards Indonesia Global Ocean Observing System (INA-GOOS): Review and recommendation. *Ocean and Coastal Management, 53*, 421–427.

Fike, P. C. (1993). *A climatology of nocturnal severe local storms outbreaks*. In: 17th Conference on Severe Local Storms, St. American Meteorological Society.

Gai, L., Baartman, J. E. M., Mendoza-Carranza, M., Wang, F., Ritsema, C. J., & Geissen, V. (2017). A framework approach for unravelling the impact of multiple factors influencing flooding. *Journal of Flood Risk Management, 11*(2), 111–126. https://doi.org/10.1111/jfr3.12310.

Gaume, E., Bain, V., Bernardara, P., Newinger, O., Barbuc, M., Bateman, A., Blaškoviˇcová, L., Blöschl, G., Borga, M., & Dumitrescu, A. (2009). A compilation of data on European flash floods. *Journal of Hydrology, 367*, 70–78.

GFDRR, G. F. (2015). *Bringing resilience to scale*. GFDRR.

Hallegatte, S. (2011). *How economic growth and rational decisions can make disaster losses grow faster than wealth*. Washington, DC: World Bank.

Henstra, D., Thistlethwaite, J., Brown, C., & Scott, D. (2019). Flood risk management and shared responsibility: Exploring Canadian public attitudes and expectations. *Journal of Flood Risk Management, 12*(1), e12346.

Hill, C., Verjee, F., & Barrett, C. (2010). *Flash flood early warning system reference guide*. Boulder, CO: University Corporation for Atmospheric.

Hofmann, J., & Schüttrumpf, H. (2019). Risk-based early warning system for pluvial flash floods: Approaches and foundations. *Geosciences, 2019*(9), 127. https://doi.org/10.3390/geosciences9030127.

IPCC. (2004). *Climate change 2014: Synthesis report. Contribution of Working Groups I, II and III to the Fifth Assessment Report of the Intergovernmental Panel on Climate Change*. Geneva: Intergovenmental Panel on Climate Change.

De Villiers, J. (2019, November 13). *Tornadoes in South Africa are likely to grow more common—As thunderstorms grow more severe*. Retrieved February 15, 2021, from https://www.businessinsider.co.za/tornado-south-africa-kwa-zulu-natal-severe-thunderstorms-stellenbosch-university-witwatersrand-university-2019-11

Jonkman, S. N., & Dawson, R. J. (2012). Issues and challenges in flood risk management—Editorial for special issue on flood risk management. *Water, 4*, 785–792.

Kantamaneni, K., Alrashed, I., & Phillips, M. (2017). Cost vs. safety: A novel design for tornado proof. *HBRC Journal, 13*(2), 223–232. https://doi.org/10.1016/J.HBRCJ.2015.05.004.

Keating, A., Campbell, K., Szoenyi, M., Mcquistan, C., Nash, D., & Burer, M. (2017). Development and testing of a community flood resilience measurement tool. *Natural Hazards and Earth System Science, 17*, 77–101.

Kimambo, O. N. (2018). Tornadoes in South Africa: An overview with a case study. *Researchgate*. https://doi.org/10.20944/preprints201811.0547.v1.

Kuzma, S., & Luo, T. (2020). *The city fix*. Retrieved from https://thecityfix.com/blog/number-people-affected-floods-will-double-2030-samantha-kuzma-tianyi-luo/

Lopez, G. M., Di Baldassarre, G., & Seibert, J. (2017). Impact of social preparedness on flood early warning systems. *Water Resources Research, 53*, 522–534.

Long, A. J., Stoy, P. C., & Gerken, T. (2018). Tornado seasonality in the southeastern United States. *Weather and Climate Extremes, 20*(2018), 81–91.

Louvain. (2019). *Natural disasters 2018; Centre for Research on the Epidemiology of Disasters CRED*. Geneva, Switzerland: USAID.

Mabuku, M. P., Senzanje, A., Mudhara, M., Jewitt, G., & Mulwafu, W. (2018). Rural households' flood preparedness and social determinants in Mwandi district of Zambia and eastern Zambezi region of Namibia. *International Journal of Disaster Risk Reduction, 28*, 284–297.

National Oceanic and Atmospheric Administration. *Coastal Services Center, Coastal Inundation Toolkit*. Retrieved February 14, 2021., from http://www.csc.noaa.gov

NWS. (2009). *15 January 2009: Lake Champlain Sea Smoke, Steam Devils, and Waterspout: Chapters IV and V*. National Oceanic and Atmospheric Administration.

Perera, D., Agnihotri, J., Seidou, O., & Djalante, R. (2020). Identifying societal challenges in flood early warning systems. *International Journal of Disaster Risk Reduction, 51*, 101794.

Ramseyer, C., Holliday, L., & Floyd, R. (2016). Enhanced residential building code for tornado safety. *Journal of Performance of Constructed Facilities, 30*(4), 2016.

Reason, C. (2017). Climate of Southern Africa. *Climate Science*. https://doi.org/10.1093/acrefore/9780190228620.013.513.

Riama, N. F., Sari, R. F., Rahmayanti, H., Sulistya, W., & Nurrahmat, M. H. (2021). The level of public acceptance to the development of a coastal flooding early warning system in Jakarta. *Sustainability, 13*, 566. https://doi.org/10.3390/su13020566.

Roger, E. (2006). *The Online Tornado FAQ*. Storm Prediction Center, National Oceanic and Atmospheric Administration.

Roper, L. D. (2016). *Extreme Weather: Tornadoes*. Retrieved from http://www.roperld.com/personal/roperldavid.htm

Sattele, M., Brundl, M., & Straub, D. (2015). Reliability and effectiveness of early warning systems for natural hazards: Concept and application to debris flow warning. *Reliability Engineering and System Safety, 142*, 192–202.

Saunders, M., Lewis, P., & Thornhill, A. (2012). *Research methods for business students* (6th ed.). Pearson Education.

Schlör, H., Venghaus, S., & Hake, J. (2018). The FEW-Nexus city index—Measuring urban resilience. *Applied Energy, 210*, 382–392.

See, L. (2019). A review of citizen science and crowdsourcing in applications of pluvial flooding. *Frontiers in Earth Science, 7*(44). https://doi.org/10.3389/feart.2019.00044.

Shaba, A. (2019). *Deaths, destruction as tornado hits South African Province*. Retrieved from https://www.africanews.com/2019/11/13/deaths-destruction-as-tornado-hits-south-african-province/

Shorten, A., & Smith, J. (2017). Mixed methods research: Expanding the evidence base. *Evidence-Based Nursing, 20*, 74–75.

Shorten, A., Shorten, B., & Kennedy, H. P. (2014). Complexities of choice after prior cesarean: A narrative analysis. *Birth, 41*, 178–184. https://doi.org/10.1111/birt.12082.

Sid, P. (2002, May 11). Science News Online. *The Weekly Magazine of Science News*. Retrieved from http://www.sciencenews.org/articles/20020511/bob9.asp

Smith, K., & Ward, R. (1998). *Floods: Physical processes and human impacts*. Chichester, England: Wiley.

Snow, J. (2020, October 29). *Tornado. Encyclopedia Britannica*. Retrieved February 16, 2021, from https://www.britannica.com/science/tornado

Suryani, R. S., Shiwaku, K., Munadi, K., & Shaw, R. (2018). Enhancing community resilience towards disaster: The contrib uting factors of school-community collaborative network in the tsunami a ffected area in Aceh. *International Journal of Disaster Risk Reduction, 29*, 3–12.

Tim, P., Michel, H., Tabea, K., & Florian, R. (2016) International civil protection adapting to new challenges. Center for Security Studies (CSS), ETH Zurich. https://doi.org/10.3929/ethz-a-010819210.

TORRO. (2021). *Severe storm definitions & whirlwind classification*. London: The Tornado & Storm Research Organisation.

The International Federation of Red Cross and Red Crescent Societies (IFRC). (2009). *World Disaster Report 2009, Focus on Early Warning, Early Action*. Paris, France: IFRC.

UNISDR. (2015). *Making development sustainable: The future of disaster risk management. Global assessment report on disaster risk reduction*. Geneva: UNISDR.

United Nations. (2015). *Transforming our world: The 2030 agenda for sustainable development*. New York: United Nations.

United Nations Development Programme (UNDP). (2018). *Five approaches to build functional early warning systems*.

World Resources Institute (2021) Water. Flooding Could Worsen Impacts of COVID-19 and Other Infectious Diseases, https://www.wri.org/news/release-new-datashows-millions-people-trillions-property-risk-flooding-infrastructure

Zhang, S., Hua, D., Meng, X., & Zhang, Y. (2011). Climate change and its driving effect on the runoff in the "Three-River Headwaters" region. *Journal of Geographical Sciences, 21*(6), 963.

Part II

Flood Modelling, Early Warning and Partnerships

2 Satellite-Based Approaches in the Detection and Monitoring of Selected Hydrometeorological Disasters

Paola Mazzoglio, Andrea Ajmar, Guy J. P. Schumann, Simone Balbo, Piero Boccardo, Francesca Perez, and Enrico Borgogno-Mondino

Abstract

Earth observation satellite systems play an important role in the provision of a wide range of information, especially in data-scarce regions. This role becomes extremely relevant during a disaster and when direct access to the affected area is difficult. This chapter provides an overview of the state of the art remote sensing techniques and tools for the detection and monitoring of hydrometeorological disasters, focusing on tropical cyclones Idai and Kenneth. In a perspective of disaster preparedness, rainfall measurements provided by satellites can enable decision-makers to take urgent measures in the pre-event phase and can be used as input for early warning systems. Data acquired from satellite missions are used for a set of different tools developed to map flood extent, while several satellite-based emergency mapping mechanisms provide timely post-event information by taking advantage of observations provided by satellites (including commercial platforms providing images at very high geometric resolution). Analysis of vegetation dynamics derived from time-series of multispectral imagery is used to assess changes, such as the detection of cyclone-damaged areas, identification of critical conditions in vegetation health/productivity, and land cover change mapping.

Keywords

Cyclone · Flood · Hurricane · Rainfall · Remote sensing · Satellite

P. Mazzoglio (✉)
Department of Environment, Land and Infrastructure Engineering, Politecnico di Torino, Turin, Italy
e-mail: paola.mazzoglio@polito.it

A. Ajmar · P. Boccardo
Interuniversity Department of Regional and Urban Studies and Planning, Politecnico di Torino, Torino, Italy
e-mail: andrea.ajmar@polito.it; piero.boccardo@polito.it

G. J. P. Schumann
School of Geographical Sciences, University of Bristol, Bristol, UK

Dartmouth Flood Observatory (DFO), University of Colorado Boulder, Boulder, CO, USA

S. Balbo · F. Perez
ITHACA - Information Technology for Humanitarian Assistance, Cooperation and Action, Torino, Italy
e-mail: simone.balbo@ithaca.polito.it; francesca.perez@ithaca.polito.it

E. Borgogno-Mondino
Department of Agricultural, Forest and Food Sciences, Università degli Studi di Torino, Grugliasco, Italy
e-mail: enrico.borgogno@unito.it

G. Nhamo, L. Chapungu (eds.), *The Increasing Risk of Floods and Tornadoes in Southern Africa*, Sustainable Development Goals Series, https://doi.org/10.1007/978-3-030-74192-1_2

2.1 Introduction

In recent years, hurricanes and tropical storms have been particularly devastating globally, with rainfall and inundation footprints far exceeding records and national response capabilities. On a global scale, the frequency and severity of floods have increased, causing massive damage to societies and economic systems, jeopardizing infrastructure, transportation, and communication networks, and reducing crop yields. In addition to the high number of people killed or displaced from their homes, floods affect human health for long periods. According to the Emergency Events Database (EMDAT), flood events are the natural disasters that have affected the greatest number of people in the twenty-first century. Single disastrous flood events can affect and displace many hundreds of thousands of people, cause several hundreds of deaths, and cost billions of USD in damages. The 2019 South-West Indian Ocean cyclone season, with Cyclones Idai in March and Kenneth in April and numerous others, affected large areas in Mozambique and neighboring countries and devastated entire nations. Complete recovery from these disasters will most likely be impossible for most people in these regions (Schumann, 2019). Considering the high vulnerability of people living in that area, a detailed study that investigates both the formation and the motion of tropical cyclones and heavy rainfall events is necessary to contribute to the process of reducing the number of people affected by water-related disasters. This objective is achievable through the integration of climate change measures into national strategies and the development of tools for disaster identification and monitoring.

Since high-impact large-scale events often cover an area beyond traditional regional monitoring operation, remote sensing presents an attractive approach (Schumann et al., 2018). Earth Observation (EO) could provide valuable information across various spatial and temporal scales, especially in data-scarce areas. Significant progress has been achieved in the field of EO in recent years due to a continuous massive effort made by space agencies, scientists, and many others alike. The proliferation of free satellite and geospatial data (Schumann & Domeneghetti, 2016) necessitates action by the flood-expert community. In addition, the heterogeneous datasets already available require not only a profound understanding of the limitations and errors within the data and methods but also the development of more sophisticated algorithms for data processing. Robust frameworks for effective information management and transfer across networks are indeed of fundamental importance. Furthermore, the gap that currently exists between decision-makers and scientists with regard to EO datasets requires the creation of a capacity-building community including scientists as well as stakeholders that can transform EO data into products that provide actionable information for better decision-making (Schumann & Domeneghetti, 2016).

Many space agencies, including NASA and ESA, have increasingly been stepping up to these new challenges in flood monitoring and response by harnessing the full scope of their remote sensing and modelling resources during events, either through ongoing activities at research centers or through projects funding support. The overarching ultimate goal is to connect and demonstrate the integration of EO data and models into societal benefit areas with clear operational capabilities. Currently, significant efforts by experts in multiple domains are of fundamental importance for the creation of decision support tools or systems (DST or DSS) that effectively utilize remote sensing data. For instance, DSSs that collect relevant datasets, including EO and ground data and more recently also social media data (de Bruijn et al., 2019; Jongman et al., 2015), and bring it together in an interoperable service would significantly increase the return on investment for EO missions. A major challenge is to link the data and data service creators, working at a global scale, to the specific local end users. Increasingly, interoperable web services and related technologies are helping, as they allow for instant access to the latest available data and interoperability of various geospatial data systems.

2.2 Literature Review

In terms of flood disaster-related activities, Schumann et al. (2018) summarize the state of the art of existing efforts and systems operating under a public, open-access policy. All of these systems provide unique capabilities but, as noted, nearly all these systems are either voluntary or based on limited project funding, with the exception of the Disaster Charter and the Copernicus program (Schumann et al., 2018). In other words, none of those systems are sustained and managed in the long run without continuous funding and commitment. These EO-based systems range from collecting imagery, through to mapping and prediction of floods. There is however a desperate need to directly link what is being done at multiple organizations and agencies with the end user community to support decision-making prior to and following a flood event.

2.2.1 Heavy Rainfall Monitoring Tools

Recent advances in rainfall measurement allowed scientists to develop early warning systems like ITHACA Extreme Rainfall Detection System (ERDS), a service for the forecasting and monitoring of rainfall events at the global scale available through a WebGIS application (ERDS, 2020). The Centre for Hydrometeorology and Remote Sensing (CHRS) at the University of California, Irvine (UCI), instead, recently developed a Data Portal for quick access of their satellite-based rainfall products (Nguyen et al., 2019). The portal also offers the opportunity to be redirected to iRain, a data visualization platform that provides near real-time information about accumulated rainfall based on the PERSIANN Dynamic Infrared-Rain rate model (PDIR); Nguyen et al., 2020). Moreover, iRain can identify and track extreme precipitation events globally, providing also a short report containing storm statistics. End users can subscribe to intense rainfall alerts after having chosen a rainfall threshold and a location.

2.2.2 Flood Identification Tools

The Global Flood Observatory at the University of Colorado Boulder (CO, USA), known as the Dartmouth Flood Observatory or DFO, tracks, monitors, and archives flood events globally since 1985 and makes data available in various formats including graphics, spreadsheets, and GIS maps (DFO, 2020a). This historic archiving of flood disasters represents crucial information for flood event analytics at global level, such as event dates, duration, geographic area, economic impact, and number of lives lost. In addition, its open-access policy has recently enabled the development of young EO-based flood analytics businesses. The DFO also works in close collaboration with the Global Flood Partnership, which is a participating organization, to build the needed connections at the local level. By comparing, integrating, and providing access to these complementary flood prediction, detection, measurement, and mapping systems, the community provides end users with greatly improved flood information at the global scale.

The Global Flood Monitoring System (GFMS) is an experimental system developed by NASA and University of Maryland. The Tropical Rainfall Measuring Mission (TRMM) Multi-satellite Precipitation Analysis (TMPA) and the Global Precipitation Measurement (GPM) Integrated Multi-Satellite Retrievals for Global Precipitation Measurement (IMERG) data are used as input to a hydrological runoff and routing model characterized by a 1/8th degree lat/lon resolution and a 50°N–50° S spatial coverage (Wu et al., 2014). Even at relatively low resolution, the system allows the end user to have timely information with a quasi-global spatial coverage and quick updates. Time series of water depth and streamflow with a 3-h resolution were obtained at every grid point running the model retrospectively for 15 years. Flood detection is based on a flood threshold derived at grid cell level using surface water storage statistics like the 95th percentile and parameters related to basin characteristics. The system also issues short-term (4–5 days) flood forecasts based on numerical weather prediction (NWP) precipitation data.

Several international initiatives and organizations in which space agencies participate also provide other relevant services and geospatial data. This information "firehose" makes it difficult to coordinate all of the relevant systems during one single event while it becomes impossible with multiple simultaneous events. For instance, response activities to Hurricane Harvey (August 23–September 25, 2017) were still occurring as Hurricane Irma (September 4–October 18, 2017) caused flood damage, and the same was true in Mozambique and neighboring countries during early 2019 in the case of Cyclones Idai and Kenneth. Although each of the response systems provides a "unique" capability, there is to date no global DSS that ingests all the data from existing systems and provides real-time "trusted" information about flood events.

2.2.3 Characteristics of the Main Satellite Emergency Management Mechanisms

The International Working Group on Satellite-based Emergency Mapping (IWG-SEM) (2015) put forward that:

> Satellite-based Emergency Mapping (SEM) is defined as the creation of maps, geo-information products and spatial analyses dedicated to providing situational awareness for emergency management and immediate crisis information for response by means of extraction of reference (pre-event) and crisis (post-event) geographic information/data from satellite or aerial imagery.

Data and products generated by SEM mechanisms are conceived to support civil protection authorities and international organizations in all phases of the emergency cycle (Voigt et al., 2016).

According to the UN-SPIDER Knowledge Portal (UN-SPIDER, 2020), several SEM mechanisms are in place, to support disaster response operations through the provision of maps derived from satellite imagery:

- Copernicus Emergency Management Service (EMS)—Mapping component provides free of charge mapping service in cases of natural disasters, humanitarian crises, and human-made emergency situations throughout the world (Copernicus EMS, 2020). The products of the Mapping component generated in Rapid Mapping mode consist of standard mapping products delivered within hours or days from the activation in support of emergency management activities immediately following a disaster. The Risk & Recovery Mapping mode covers in particular activities dealing with prevention, preparedness, disaster risk reduction, and recovery. The products delivered are based on the provision of satellite imagery from the Sentinel missions and the contributing missions that are made available through the Copernicus Space Component Data Access system (CSCDA). EMS can be triggered only by or through an Authorised User (AU). Authorised Users include National Focal Points in the EU Member States and countries participating in the Copernicus programme, as well as European Commission services and the European External Action Service (EEAS). Copernicus EMS is complemented by an early warning and monitoring component which includes systems for floods, droughts and forest fires.
- The International Charter on "Space and Major Disasters" is composed of space agencies and space system operators who work together to provide satellite imagery for disaster monitoring purposes (The International Charter Space and Major Disasters, 2020). It is also supported by value-added providers, producing maps based on satellite data, and other partners providing data and services for specific regions of the world: Copernicus EMS and UNOSAT are within this group. The Charter can be activated at the request of AUs to support emergencies in their own country, as well as in a country with which they cooperate.
- UNITAR Operational Satellite Applications Programme (UNOSAT) provides satellite

imagery analysis and capacity development to the UN system, UN member states, and its partners (UNOSAT, 2020). UNOSAT benefits from a variety of sources for its satellite imagery: free and open-source, commercial vendors, the Charter (for natural and technological disasters only), in-kind donations. UNOSAT can be activated by UN offices and agencies, government agencies, Red Cross and Red Crescent Movement, international and regional organizations, and humanitarian Non-Governmental Organizations. The service is free of charge for UN sister agencies and humanitarian entities operating in line with UN policies.
- Sentinel Asia is a voluntary and best-efforts-basis initiative led by the Asia-Pacific Regional Space Agency Forum to share disaster information in near real time across the Asia-Pacific region (Sentinel Asia, 2020).
- SERVIR, a joint venture between NASA and the U.S. Agency for International Development in Washington D.C., provides satellite-based Earth monitoring, imaging and mapping data, geospatial information, predictive models, and science applications to help improve environmental decision-making among developing nations in eastern and southern Africa, the Hindu-Kush region of the Himalayas and the lower Mekong River Basin in Southeast Asia (NASA, 2020).
- The Center for Satellite based Crisis Information (ZKI) is an institution of the German Aerospace Center (DLR) providing satellite-based analytics during disasters.

IWG-SEM aims to improve cooperation among SEM mechanisms, to better support emergency management and, in particular, its response phase. This objective is targeted through the implementation of mapping guidelines and the definition of standard formats to quickly exchange information on mapping activations among the different mechanisms (IWG-SEM, 2014). Through the aggregation of GeoRSS feeds generated by every single mechanism, it is possible to obtain an overview of the different past and ongoing activations.

2.2.4 The Use of Optical Remote Sensing in Monitoring Flood Damage on Agriculture

In agricultural areas, optical remote sensing is used to map and quantify damages affecting crops/orchards. This task can be achieved: (a) by comparing two single significant acquisitions preceding and following the event; (b) looking at an entire time series (TS) of synthetic spectral indices (SI) covering a mid- to long-term period including the event. SI are mathematical aggregations of bands from EO data operating like proxies of time-varying features of vegetation (vigor, health, water content, etc.). A widely used SI in vegetation monitoring is the Normalized Difference Vegetation Index (NDVI; Tucker, 1979). Seasonal changes in NDVI TS proved to successfully track phenology (e.g., Pettorelli et al., 2005; Reed et al., 1994) as NDVI well correlates to vegetation productivity (Bai et al., 2008; Fensholt et al., 2006; Running et al., 2004) being mainly responsive to canopy chlorophyll content.

2.3 Material and Methods

2.3.1 Heavy Rainfall Detection

The study uses data drawn from two different analyses that are currently implemented in ERDS. The first one is a near real-time rainfall monitoring based on GPM IMERG Early run data, a global (90°N–90°S) rainfall measurement retrieved from satellite characterized by a 0.1° spatial resolution and a 30 min temporal resolution, available with a 4 h latency (Huffman et al., 2019). The second analysis is based on a weather prediction model called Global Forecast System (GFS). GFS model runs four times a day, providing data characterized by a global spatial coverage and a 0.25° spatial resolution. Heavy rainfall alerts are issued after having performed a comparison between accumulated rainfall and rainfall threshold (Mazzoglio et al., 2019). Threshold values are data and time dependent. Threshold values applied to GPM data are different to GFS ones. The time-dependent threshold values increase as the aggregation interval increases.

2.3.2 Flood Detection

Since 2011, the DFO has collaborated with the NASA Goddard Space Flight Center (GSFC) and other US and international organizations in distributing near real-time flood maps from satellite imagery, primarily from Moderate Resolution Spectroradiometer (MODIS) onboard NASA's Aqua and Terra satellites (Brakenridge & Anderson, 2006; Schumann et al., 2018), and also related EO data. The DFO also hosts a database of virtual stations of discharge records from 1998 until present based on an integration of microwave radiometry measurements and global hydrological modelling. The hydrological model (Water Balance Model, WBM; Fekete et al., 2002) is used for calibrating the microwave signal change (due to changes in surface water area occupying a pixel) to discharge. Specifically, these "River and Reservoir Watch" virtual stations (DFO, 2020b) gauge the state of potential large river flooding daily based on changes in the brightness temperature of the passive microwave signal onboard the AMSR-E, TRMM, AMSR-2, and GPM sensors (Brakenridge et al., 2012).

More recently, the DFO has implemented an open geospatial service protocol. A Web Map Service (WMS) has been set up through which GIS clients can seamlessly ingest the latest up-to-date flood-related products (Fig. 2.1). The developed DSS for global flood information now includes an online GIS visualization interface built on a WMS that leverages existing systems and interoperability tools and standards (DFO, 2020c): the idea is to ensure that systems and tools are not duplicated but leveraged and expanded. The online DSS has been complemented by a free mobile application ("DFO Floods") available on Android and Apple OS, which has the same functionalities and in addition allows the user to link in other free interoperable data services as well as offering offline capabilities by downloading map layers directly onto the mobile device. With these recent developments, decision-makers are able to pull data and products from the DSS at low bandwidth and latency, and also directly request from the DFO tailored information layers as needed for their operations, which was the case for instance for the Harvey event in 2017. Other relevant information is made available alongside remotely sensed flood data, such as output layers from ECMWF forecasts and from global flood hazard models (Dottori et al., 2016).

2.3.3 Satellite Emergency Mapping Mechanisms Response to Cyclones Idai and Kenneth

Tropical Cyclone Idai made landfall as a Category 2 storm during the night of 14 March 2019 near the city of Beira, a city with approximately 500,000 inhabitants in Mozambique. Heavy rainfall, storm surges, and strong winds caused significant damage to critical infrastructure and crops: it has been calculated that more than 100,000 homes were affected. Local authorities have reported 150 fatalities and more than 1500 injured, and the National Institute for Disaster Management (INGC) estimated that 600,000 needed urgent assistance (Copernicus EMS, 2019a).

Copernicus EMS, the Charter, UNITAR-UNOSAT, and DLR-ZKI were all activated. In the framework of the Charter only, 320 satellite acquisitions have been performed, even if not all used because of limitations (e.g., cloud coverage) or almost overlapping characteristics (similar sensors, footprints, and acquisition date/times). Figure 2.2 shows the footprints of satellite acquisitions. Radar images have generally a wider swath and larger overall coverage: they are intended to capture as far as possible the full extent of the flooded areas (delineation products). Earlier products have been generated mainly on this category of images, thanks to the all-weather capabilities of those sensors, capable of penetrating the cloud coverage. Optical images based analysis came in a second stage, with clearer skies, and was concentrated on the areas around

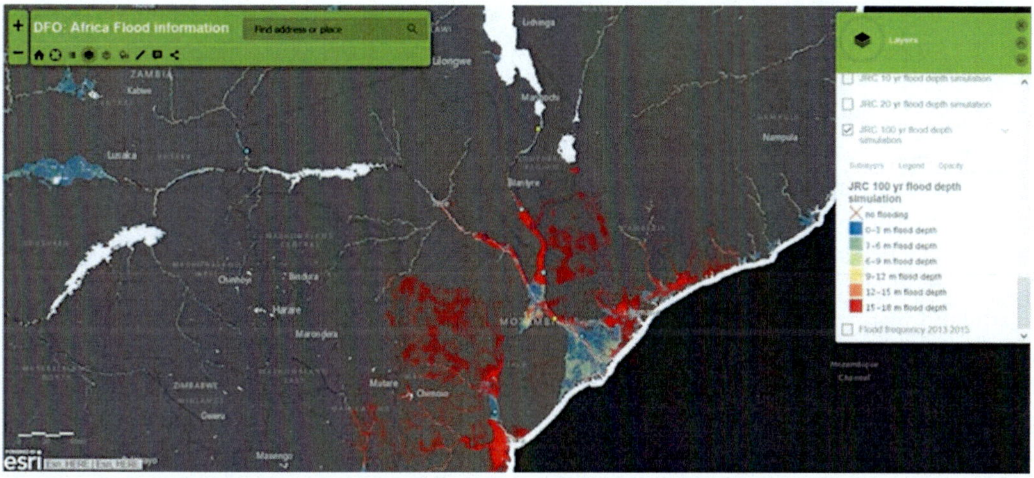

Fig. 2.1 DFO's WMS. Map overlaying an open map layer shows in red flooding mapped during the events in Mozambique and neighboring countries in early 2019 up to March. Regular waters are depicted in light gray. River and Reservoir Watch stations are represented in colors relating to severity of flow. Flood depths from the JRC global flood hazard model of the 100-years return period flows are also shown

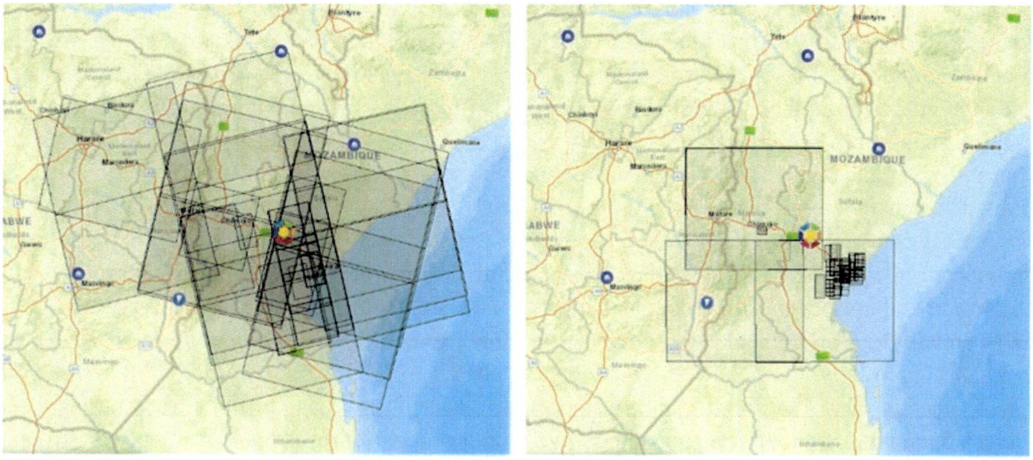

(a) (b)

Fig. 2.2 Satellite footprints of radar (**a**) and optical (**b**) images acquired in the framework of the Cyclone Idai activation (The International Charter Space and Major Disasters, 2019a)

Beira, to identify actual damages to infrastructures (grading products).

After passing through Comoros, Cyclone Kenneth made landfall on 25 April 2019 near the city of Pemba, in Mozambique: the area is sparsely populated and this is the main reason for the lower impact of this second cyclone in about a month hitting the same country. The Copernicus EMS Rapid Mapping module was activated on 26 April to provide Delineation and Grading Maps over several Areas of Interest along the coast of the Cabo Delgado province in Mozambique, and Grading Maps over the northern part of Comoros. This was one of the first activation for which the product portfolio already included the "First Estimate Product" (FEP) which aims to provide a rough estimation of the disaster impacts within 2 h from image reception and can be used to redefine Areas of Interest, that therefore are no more a hard constraint predefined by the user (Copernicus EMS, 2019c).

2.3.4 Analysis of Effects of Flooding Events on Agriculture

Flooding is a very local phenomenon in nature. Consequently, when investigating damages, spatial information should be as much detailed as possible. In particular, in agriculture, the level of detail should be appropriate to the local average size of crop fields. Average crop field size can enormously vary around the world. Moreover, when vegetation is investigated, the temporal resolution of data is a very important feature to be taken into account. Among current EO satellite missions supplying free optical data having an appropriate temporal resolution, the Copernicus Sentinel-2 (S2) appears to be the most appropriate one. S2 records 13 spectral bands in the range 400–2500 nm, with a ground sampling distance (GSD) of 10 m, 20 m, and 60 m (depending on the band), a 290 km swath, and a nominal temporal resolution of 5 days. In particular, Level-2A product from Sentinel Data Hub, ortho-projected and bottom-of-atmosphere calibrated, well fits requirements of multitemporal analysis. Currently, S2 data suffer from two great limitations for vegetation monitoring: (a) the first observable growing season (GS) is the 2016 one and, therefore, very short TS are available; (b) its temporal resolution cannot be enough in those areas around the world where cloud cover is too frequent. Oppositely, Terra/Aqua MODIS data, having a temporal resolution of 1 day (or more) and a GSD up to 250 m, offer a good alternative. MODIS sensors have been operative since 2000 making freely available pre-processed products like the MOD13Q1/MYD13Q1 datasets containing composite maps of vegetation SI like NDVI and EVI (Enhanced Vegetation Index) with a time step of 16 days. Unfortunately, MODIS geometric resolution appears to be inadequate for a field-level investigation, moving the investigation from a local to a regional one. The ideal approach would be therefore a mixed one where MODIS-like data can be jointly used with S2-like ones.

Cloud cover analysis is mandatory to test if this approach could be reasonably successful. Analysis should drive to map occurrences of clouds, informing about local reliability of damage estimates. Comparison of time and position of cloud occurrences with growing calendars of main crops/orchards in the area can ensure about the feasibility of the analysis. Crop calendars define the period along the year when crops develop, reach maturation, and are, finally, harvested. Knowledge about agronomic practices is desirable, as well.

When working with VI (vegetation index) TS, a typical workflow can be the following:

(a) Data selection/download from available archives and VI TS composition (stack)
(b) Data masking, regularization, and smoothing aimed at selecting only vegetated pixels, removing "bad" observations from local VI temporal profiles, recovering profile continuity, and minimizing residual noise
(c) TS modelling by detecting annual phenological cycles (e.g., Fast Fourier Transform analysis) that are singularly fitted by an interpolating model (e.g., asymmetric double logistic function; Hmimina et al., 2013)
(d) Accuracy assessment to test properness of model

Output data are image stacks mapping modelled values of VI with a regular time step.

Phenological metrics (PM) represents synthetic parameters (Eklundh & Jönsson, 2017) that characterize VI temporal profile. PM can be used to test if any significant variation has been induced onto crops phenology after a catastrophic event or as a consequence of climate change. Main PMs are:

- Start of Season (SOS), the Day of the Year (DOY) when GS starts
- End of Season (EOS), the DOY when GS ends
- Length of Season (LOS), the number of days separating EOS from SOS
- MAX_VI, the maximum VI value within GS
- MAX_DOY, the DOY when MAX_VI occurs
- Season Amplitude (SA), the difference between the maximum and minimum VI value within GS
- Rate of Increase, which is evaluated at the start of the season as the difference between

the left 20% and 80% VI levels and the related difference in time

- Rate of decrease, which is evaluated at the end of the season as the ratio of the difference between the right 20% and 80% VI levels and the related difference in time
- Total productivity, the yearly integral of the fitted VI profile
- Seasonal productivity or small integral (SMI), the integral calculated considering the fitted VI profile within SOS and EOS

The availability of an updated map of crops is basic for a damage estimate. Often, available land use maps are not proper, being not updated and, frequently, supplied with a too high degree of class aggregation. Consequently, an up-to-date map of local crops has to be specifically generated for the period preceding the flooding event. Auxiliary information can come from official sources like FAO databases. Crop maps can be obtained by image classification according to two different approaches: one based on spectral signatures of crops in specific and limited times along the year, and the other based on a continuous monitoring of a spectral proxy (like VI) of the vegetative behavior (biomass production) of crops along with their GS. An example of differences in land cover mapping given by these approaches, obtained by a K-Means Iterative Minimum Distance unsupervised classification with a posteriori class meaning interpretation, is shown in Fig. 2.3.

2.4 Results and Discussions

2.4.1 Heavy Rainfall Detection: The Case of Cyclones Kenneth and Idai

The Extreme Rainfall Detection System (ERDS) provided timely alerts before Cyclones Idai and Kenneth landfall using GPM data as input (Table 2.1). Figure 2.4 shows the most affected areas, in terms of number of alerts provided by ERDS, along Idai track in the period between 5 and 24 March 2019, while Fig. 2.5 shows the most affected area between 25 April and 1 May 2019 (during the progression of Kenneth).

2.4.2 Flood Detection and Damage Assessment: Cyclones Kenneth and Idai Case Study

During high-impact flood disasters, the DFO maps flooding also from other satellites, such as the Landsat series and SAR satellite missions, and aggregates these products to several map formats that assist flood response teams through situational awareness across large-scale coverages (Fig. 2.6).

Based on the available satellite images, in response to Cyclone Idai, Copernicus EMS generated 59 different products (9 reference, 36 delineation, and 14 grading maps) covering three different countries: Malawi, Mozambique, and Zimbabwe (Copernicus EMS, 2019b). The Charter provided additional 20 maps and reports covering Mozambique (The International Charter Space and Major Disasters, 2019c) and 10 in Zimbabwe (The International Charter Space and Major Disasters, 2019d), thanks to the collaboration with UNOSAT, SEDEC/MDR, SERTIT, DLR/ZKI, and UoL/NCEO, acting as value-added providers. Complementarities between mechanisms allowed to perform comprehensive impact analyses (ERCC – DG ECHO, 2019b).

As far as Cyclone Kenneth concerns, Copernicus EMS generated 2 FEPs, 12 delineation, and 35 grading map products in Mozambique and Comoros. The Charter has been activated over Mozambique only, providing 198 satellite acquisitions (The International Charter Space and Major Disasters, 2019b), with similar characteristics to the one described and displayed in the case of Cyclone Idai. Two products were generated by Charter value-added providers. One map released by Copernicus EMS was based on imagery acquired in the framework of the Charter thereby acting also as a value-added provider.

Quantitatively estimating the impact of SEM in disaster risk management is a complex task, as many technological trends and operational organizations have to be considered and controlled.

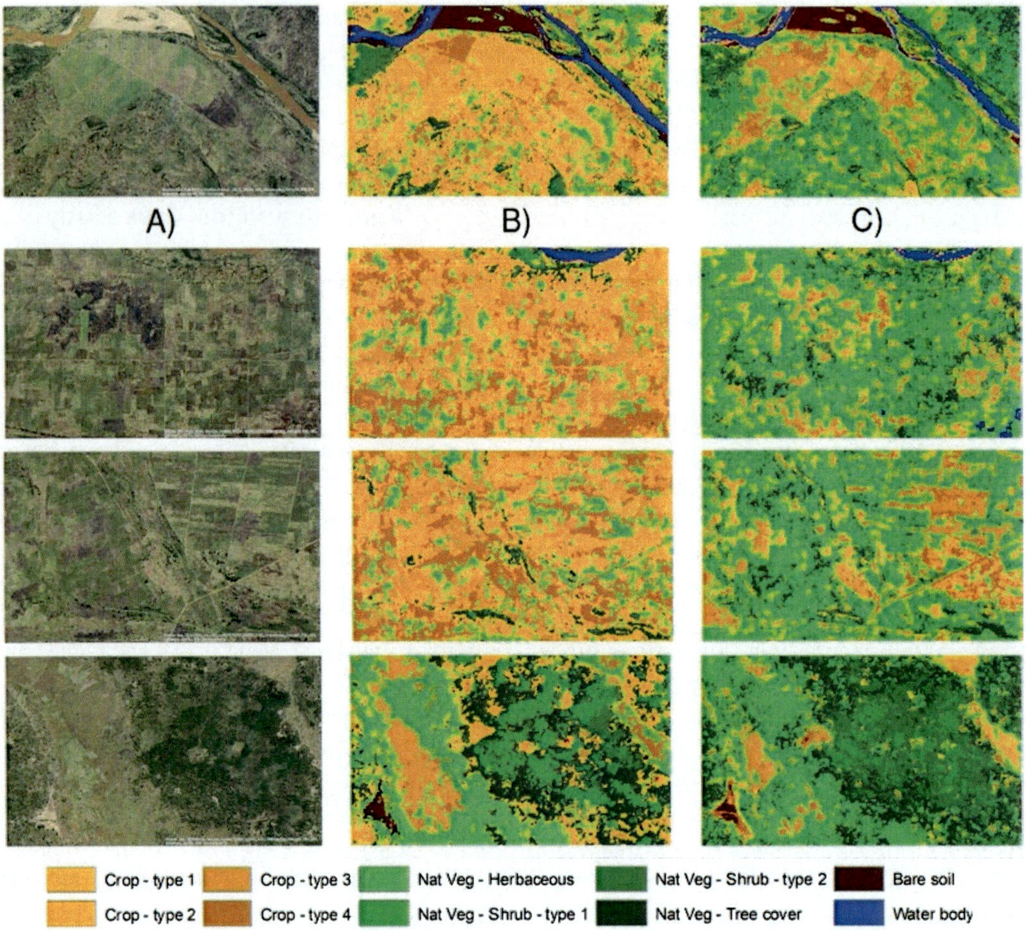

Fig. 2.3 Examples of land cover maps obtained through unsupervised classification based on: (**b**) NDVI monthly profile (August 2017–November 2018) derived from Sentinel-2 imagery; (**c**) VIS+NIR spectral signature extracted from Sentinel-2 multispectral monthly composite (July 2018). A high spatial resolution WorldView-3 image (DigitalGlobe Vivid mosaic; acquisition date 8/2/2018) is also reported for reference purposes (**a**). A K-Means Iterative Minimum Distance algorithm has been adopted, considering the following parameters: number of classes = 12; maximum number of iterations = 100 with change threshold value = 1%; standard deviation to use around the class mean = 0.05. Reported examples show that VI temporal profile-based classification better mapped crops, especially in those areas with high levels of land fragmentation and mixture of cultivated and natural vegetation classes. In fact, some crop types and natural vegetation classes (mainly herbaceous) present a low spectral separability if tested in a single monthly image (July, in this study)

Demand for products and services generated by SEM initiatives and capacity building are generally rising worldwide and this can be considered as a qualitative indicator (Voigt et al., 2016). The added value of such products is widely recognized by "first responders" especially in case of major disasters, those requiring the mobilization of the international community: in most cases, they represent the only source of information on remote and inaccessible areas (Denis et al., 2016).

The full potential of these types of mechanisms will be exploited only through a better integration between services and responders. Related to Cyclone Idai, the Reliefweb repository alone contains 613 mapping and infographic products (as of 01/06/2020): up to the 22 March 2019, the day of the first damage assessment map released by Copernicus EMS,

Table 2.1 Date and time of the first alert provided by ERDS using GPM data as input for the different aggregation intervals in the region where Idai made its second landfall on the coastal area between Beira and Quelimane (Mozambique) on 14 March 2019 at 23:30 UTC and in the region where Kenneth made its landfall on the coastal area near Pemba (Mozambique) on 25 April 2019 at around 16 UTC

Aggregation interval	Cyclone IDAI Date of the first alert	Cyclone Kenneth Date of the first alert
12 h	14/03/2019 10:30 UTC	25/04/2019 05:30 UTC
24 h	14/03/2019 11:00 UTC	25/04/2019 13:30 UTC
48 h	14/03/2019 11:00 UTC	25/04/2019 16:00 UTC
72 h	14/03/2019 14:00 UTC	25/04/2019 22:00 UTC
96 h	14/03/2019 16:00 UTC	26/04/2019 14:30 UTC

the three involved SEM mechanisms produced more than 20 maps having the water extents as main information. On 16 March 2019, WFP Logistic Cluster published the first of a series of maps highlighting access constraints (limitation affecting the usage of roads, ports and airports): those maps, released daily, never explicitly included the water extent layer released from the SEM mechanisms (WFP, 2019). On 18 March 2019, MSF published a map reporting incidence reports for some districts. On the same day, the European Commission DG ECHO released the first map (ERCC – DG ECHO, 2019a) combining meteorological information (cyclone path and rainfall accumulation), flooded areas, and incident reports: following this last product, several organizations started producing new maps combining water extents with local assets. From this example, it becomes evident that a better integration and exchange of information and data between services and responders would have allowed to streamline map production and to optimize resources, limiting the number of products reporting redundant information and increasing the timeliness in delivering ready-to-use products from the responder's perspective.

2.4.3 Damages to Vegetation

A MODIS-based analysis was devoted to compare vegetation behavior in terms of phenology during its growing season, before and after the flooding event with the aim of testing mid/long-term effects on agricultural production. S2 data were used to map existing crops at the moment of flood. The study area is presented in Fig. 2.7.

In this work, cloud analysis involved 20 years of MODIS MOD13Q1 data, a composited product summarizing observations within a 16-day period. The MOD13Q1 quality layer (QL) reports information about the quality of each pixel by assigning a code that defines the eventual type of failure at that position, including cloud presence. It is, therefore, possible to generate a map of cloud occurrences and derive some spatially distributed information like total, mean, minimum, and maximum occurrences along a year or in specific months (Fig. 2.8).

When trying to estimate damages affecting crops/orchard in a flooded area (Fig. 2.9), two types of phenomena have to be taken into account: one concerning short-term effects of flood, basically affecting yearly yield and, consequently, limited in time; and the other concerning long-term effects on crops/orchard that need more GSs (years) to somehow recover.

A short-term damage estimate can be obtained intersecting a map of the flooded area, possibly coming from SEM mechanisms, with the previously mentioned map of crops. The area size of flooded crops multiplied by the average theoretical yield (kg/ha), obtainable from an official database (e.g., FAO), permits to estimate the annual loss of production and, consequently, convert it into an economical "georeferenced" loss. A long-term damage estimate (if present) can be differently investigated by multi-annual analysis based on long VI TS (MODIS scale). Comparison between crops GS of years preceding and following the flooding event permits to get information about this issue. In particular, PM comparison, often operated by anomaly (yearly PM value divided (or minus) the mean PM value along with TS) computation and mapping is one of the most

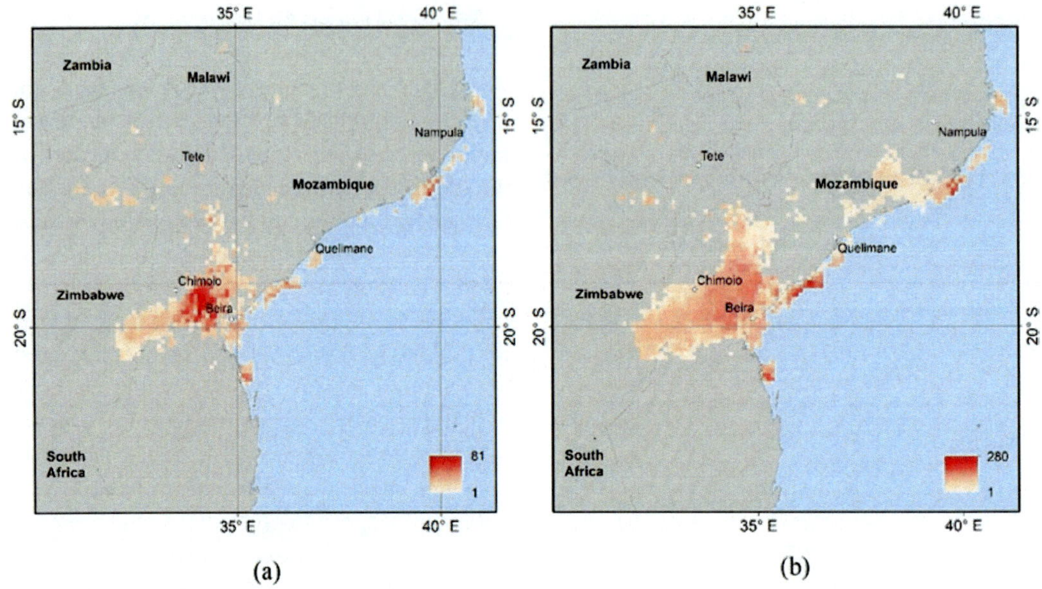

Fig. 2.4 Number of alerts provided by ERDS during the period between 5 and 24 March 2019 using GPM data as input and aggregation intervals equal to 12 h (**a**) and 48 h (**b**). Most of the alerts were issued in the area near Beira, where the combination of heavy rainfall and river overflow created an "inland ocean"

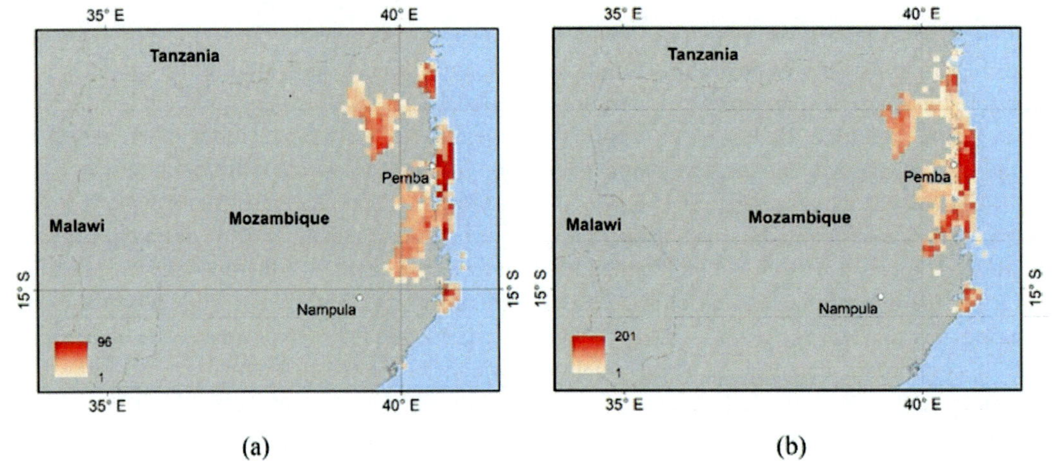

Fig. 2.5 Number of alerts provided by ERDS during the period between 25 April and 01 May 2019 using GPM data as input and aggregation intervals equal to 48 h (**a**) and 96 h (**b**)

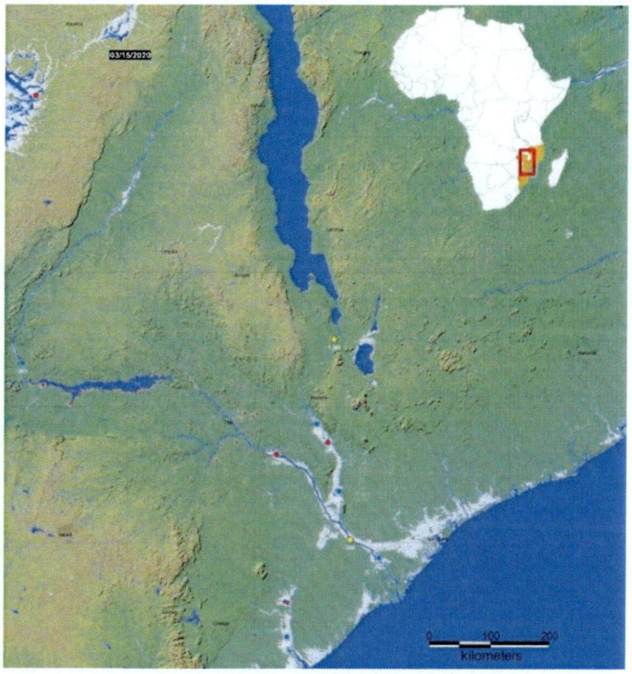

Fig. 2.6 Mozambique: large-scale flood mapping from various satellites, including MODIS, Landsat, and ESA Sentinel-1. Blue areas: mean annual flood. Red areas: flooding in excess of mean annual flood. Light gray areas: maximum observed flooding, 1993–present. The colored dots are interactive and contain River and Reservoir Watch information

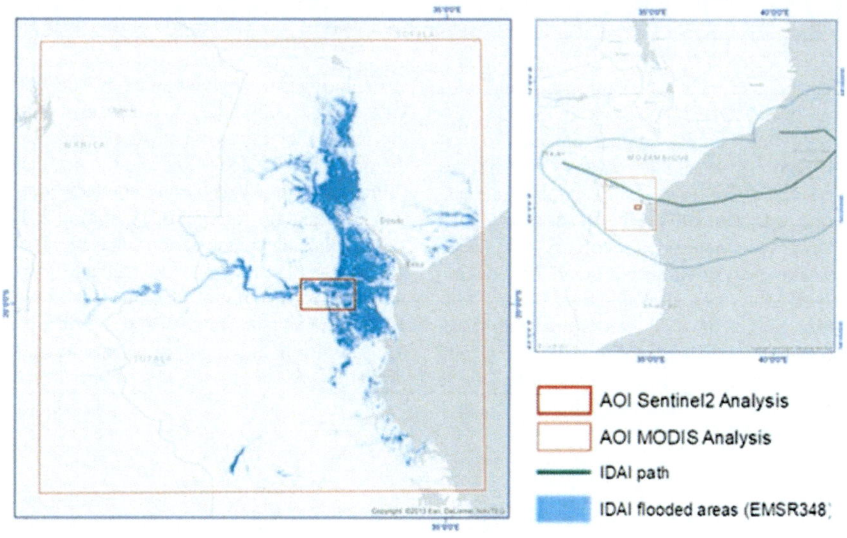

Fig. 2.7 Study area. Regional investigation was achieved by MODIS data (pink rectangle) at a mid-term time scale (2000–2019). Local investigations, aimed at mapping crop types for economical estimation of damages, were carried out in the red squared area by S2 data in the period August 2017–December 2018

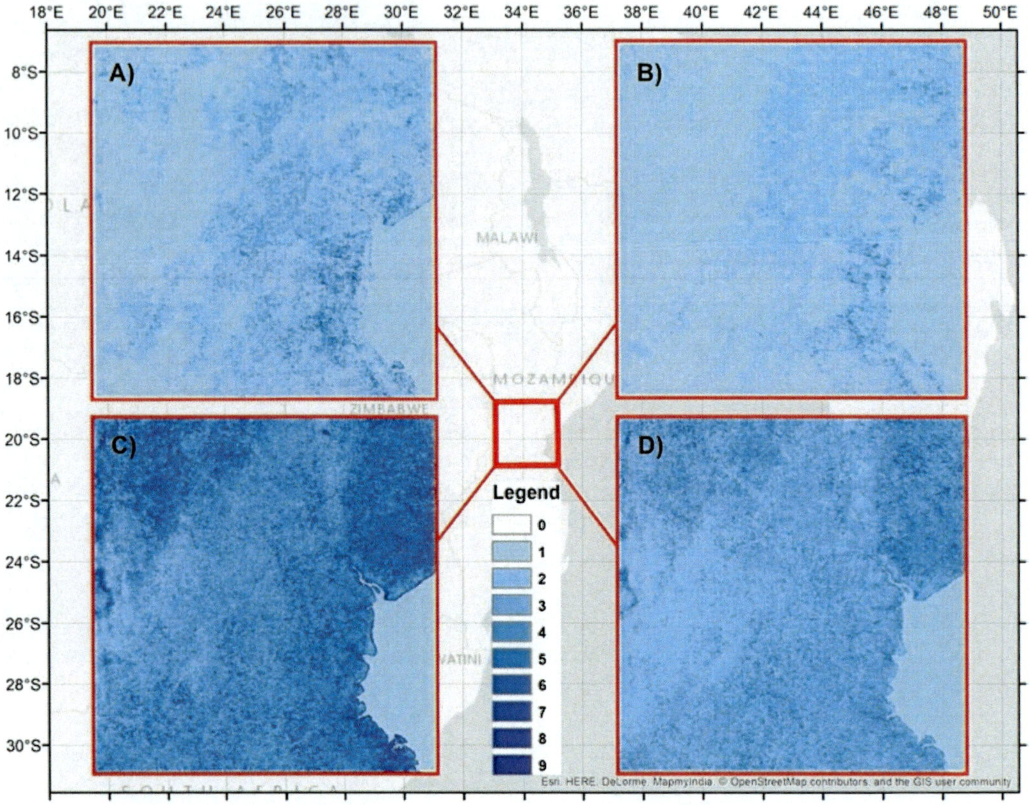

Fig. 2.8 Cloud cover maps showing occurrences of clouds (derived from MOD13Q1 dataset) in the S2 investigated area. (**a**) Map of total number of "bad" occurrences in 2019; (**b**) map of number of "bad" occurrences in February–May 2019; (**c**) map of max number of "bad" occurrences in a year from 2000 to 2019; (**d**) map of max number of "bad" occurrences in February–May from 2000 to 2019. The analysis showed a favorable situation (low cloud cover). Intra annual cloud cover shows, averagely in the area, a maximum value of 6 (out of about 23 yearly maps) occurrences of "bad" observations, with very local peaks of 6–9 occurrences. This value can be considered "not critical" for phenological analysis aimed at exploring the behavior of crops along their growing season, especially if they do not occur sequentially. To test this issue, a focus was done with reference to the period of the year when Cyclones Idai and Kenneth reached the area (March and April 2019). Four months (about 8 MOD13Q1 images) ranging from February to May 2019 where considered and the same period analyzed for all the years when MOD13Q1 data were available for (2000–2019). Averagely, a maximum of 4 "bad" observations (out of 12) was found with some local peaks of 6, confirming that a successful multitemporal analysis was possible in that period

adopted approaches. An example is reported in Fig. 2.10.

2.5 Conclusions

The chapter critically reviewed some of the most relevant EO-based techniques for near real-time identification and monitoring of heavy rainfall events. According to Target 11.5 of the 2030 Agenda for Sustainable Development, it is necessary to significantly reduce both the number of people affected by natural disasters and economic losses (UN General Assembly, 2015). Remote sensing represents a viable solution for addressing this challenge, especially in areas without extensive monitoring networks. Even if the coarse spatial resolution affects the reliability of hydrological studies at the local scale, remote sensing products are able to provide accurate

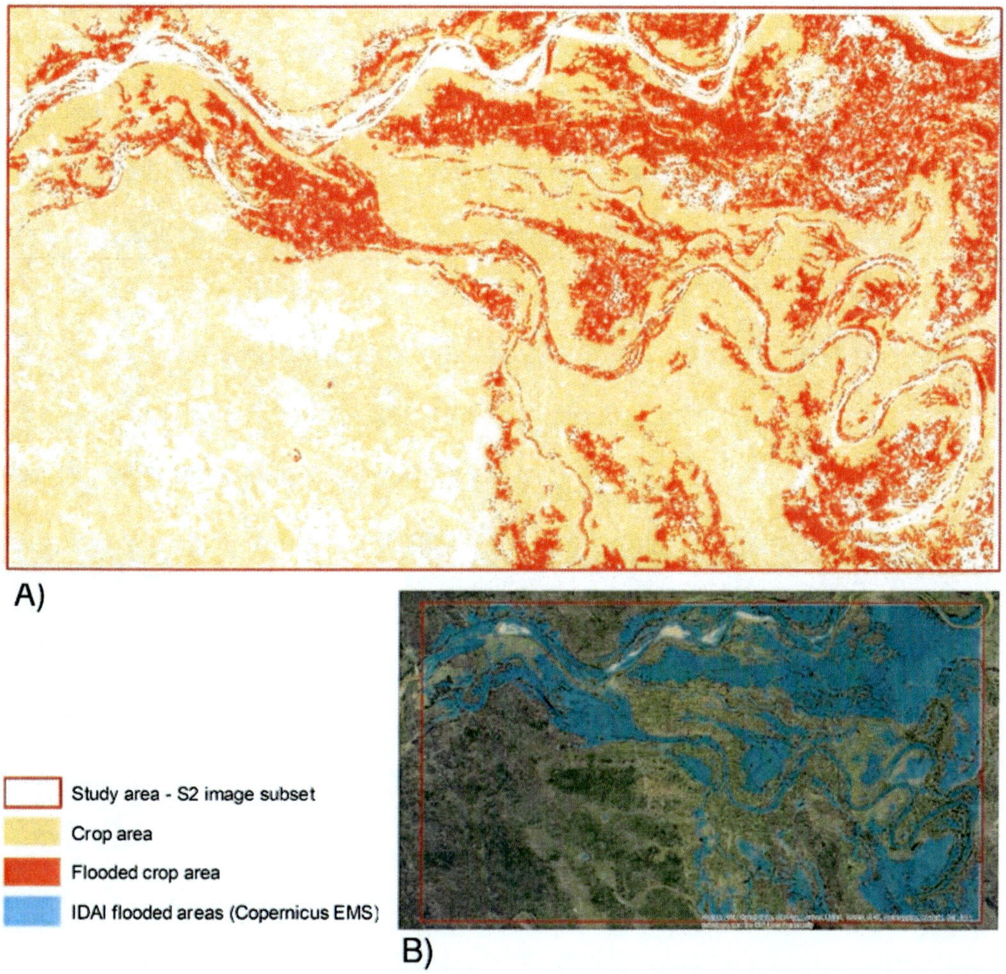

	Crops in study area (S2 image subset) [ha]	Flooded crops in study area (S2 image subset) [ha]	Cereal Yield (2017) [kg /ha]	Seasonal (2017-2018) loss of production [t]
Crop type 1	1330,38	246,81		
Crop type 2	12715,65	3639,12		
Crop type 3	3498,50	1378,17		
Crop type 4	8703,24	4019,93		
TOTAL	**26247,77**	**9284,03**	**872**	**8095,67**

Fig. 2.9 Map of flooded crop areas related to Cyclone Idai in 2019 (**a**). Cultivated areas have been mapped through NDVI temporal profile-based classification (2017–2018 growing season). Flooded area borders were defined in the frame of Copernicus EMSR384 specific operations (**b**). An estimate of related seasonal loss of production has been produced considering the most recent value of cereal (generic) yield (kg/ha) in Mozambique, according to the World Bank collection of development indicators (World Bank, 2020)

measurement for the identification of areas affected by extreme rainfall. The role of global sensing and modelling systems in addressing local disasters, however, requires clarification. Much orbital sensing provides global or near-global geographic coverage. For surface water,

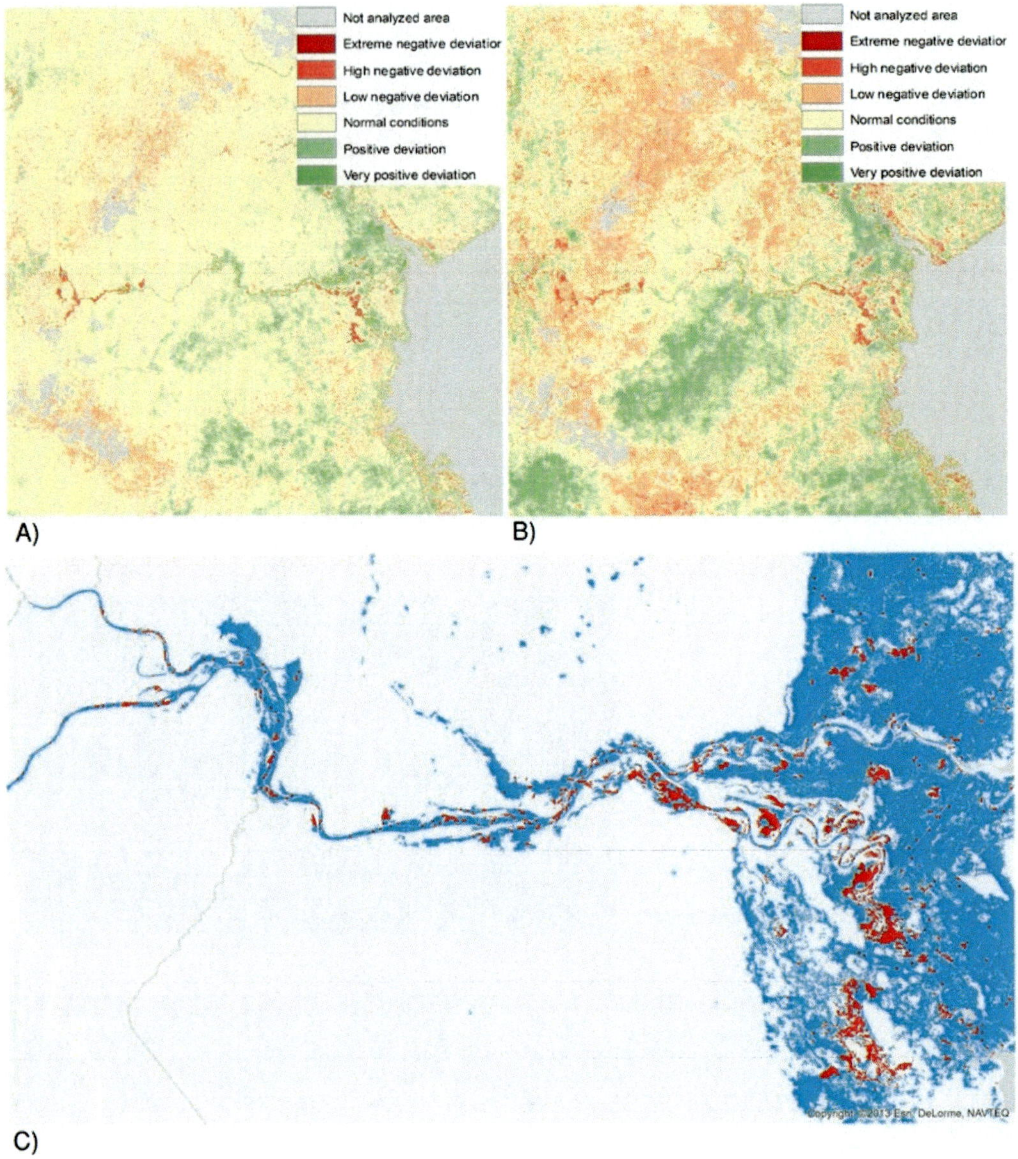

Fig. 2.10 An example of anomaly maps of SMI (**a**) and LOS (**b**) computed with reference to the MOD13Q1 dataset for the 2018–2019 growing season, which was affected by the Cyclone Idai. Maps show the spatial distribution of percent deviations of selected phenological metrics from to the historical average values (estimated from the 2000 to 2018 time-series). These maps, produced in near real-time or immediately after the examined growing season, can support the identification of critical areas where major damages to vegetation productivity are expected. Similar maps can be generated with reference to other significant PM for monitoring purposes and vegetation restoration evaluation. For the purpose of the examined case study, vegetated patches falling in the flooded areas (in blue in **c**) showing extreme negative LOS deviations were mapped (in red in **c**)

these data are increasingly being coupled to global hydrological models. The upcoming NASA/CNES Surface Water Ocean Topography (SWOT) science mission will be used, in part, to help calibrate and improve such global models. Also, the GPM mission, and its planned follow-ons, are to establish the improvements in global precipitation data that will allow increasingly accurate model predictions, including flood predictions for developing nations (Webster, 2013). An overarching challenge is to harness these new global "disruptive technologies" for the local public good. These new capabilities are novel but also overlap with existing governmental functions such as flood alerts and warnings.

Remotely sensed images can deliver key products for early warning, for the planning of life-saving activities and for impact assessment in the phases immediately following a hydrometeorological disaster. The DFO is both providing near real-time products and collecting trusted information from other services involved in the Global Flood Partnership. Recent system developments have taken the DFO into the era of interoperability and are now offering a whole suite of different data products also from other organizations, including many EO flood products but also forecasts and flood hazard simulations, with the objective to help mitigate and respond to flood disasters at the global level. Future developments may include complementing EO and model layers with social media streams to help verify areas affected or pinpoint target regions for satellite image acquisitions. Of course, this functionality will require the highest level of interoperability since all of it needs to integrate seamlessly with end user operation systems and mapping platforms and ideally also be accessible on any device. In this context, strengthening public-private partnerships would allow access to high-end capabilities leveraging advanced data interoperability standards and service protocols developed by the geospatial industry sector. The ability of communities to understand and use this information is also an important aspect.

Matching capabilities and needs is also a key aspect for SEM mechanisms. The IWG-SEM role in the definition of both mapping guidelines and standard formats/products could foster better agreement between maps produced by different organizations involved in the mechanism. Solving restrictive data licensing could further increase data availability in the aftermath of disasters.

The role of satellites is not limited to the emergency phase. Activities in the recovery phase, especially in areas where the agricultural sector is dominant, can take advantage of the possibility of performing reliable analysis with wide spatial coverage and frequent updates from satellites. Short-term damage assessment can take advantage of the information produced by SEM mechanisms without necessarily planning further satellite acquisitions whereas a similar analysis performed with in situ observation would require a more massive resource-intensive effort.

References

Bai, Z. G., Dent, D. L., Olsson, L., & Schaepman, M. E. (2008). *Global assessment of land degradation and improvement 1: Identification by remote sensing* (Report 2008/01). Retrieved October 30, 2020, from https://www.isric.org/sites/default/files/isric_report_2008_01.pdf

Brakenridge, R., & Anderson, E. (2006). Modis-based flood detection, mapping and measurement: The potential for operational hydrological applications. In J. Marsalek, G. Stancalie, & G. Balint (Eds.), *Transboundary floods: Reducing risks through flood management* (pp. 1–12). Dordrecht, Netherlands: Springer.

Brakenridge, G. R., Cohen, S., Kettner, A. J., De Groeve, T., Nghiem, S. V., Syvitski, J. P. M., et al. (2012). Calibration of satellite measurements of river discharge using a global hydrology model. *Journal of Hydrology, 475*, 123–136. https://doi.org/10.1016/j.jhydrol.2012.09.035.

Copernicus EMS. (2019a). *Copernicus EMS monitor major Tropycal Cyclone in Mozambique*. Retrieved October 30, 2020, from https://emergency.copernicus.eu/mapping/ems/copernicus-ems-monitors-major-tropical-cyclone-mozambique

Copernicus EMS. (2019b). *EU provides further support for Mozambique following Cyclone Idai*. Retrieved October 30, 2020, from https://emergency.copernicus.eu/mapping/ems/eu-provides-further-support-mozambique-following-cyclone-idai

Copernicus EMS. (2019c). *Copernicus EMS monitors impact of Cyclone Kenneth in Mozambique and Comoros*. Retrieved October 30, 2020, from https://emergency.copernicus.eu/mapping/ems/copernicus-ems-monitors-impact-cyclone-kenneth-mozambique-and-comoros

Copernicus EMS. (2020). *Service overview*. Retrieved October 30, 2020, from https://emergency.copernicus.eu/mapping/ems/service-overview

de Bruijn, J. A., de Moel, H., Jongman, B., de Ruiter, M. C., Wagemaker, J., & Aerts, J. C. J. H. (2019). A global database of historic and real-time flood events based on social media. *Scientific Data, 6*, 311. https://doi.org/10.1038/s41597-019-0326-9.

Denis, G., de Boissezon, H., Hosford, S., Pasco, X., Montfort, B., & Ranera, F. (2016). The evolution of earth observation satellites in Europe and its impact on the performance of emergency response services. *Acta Astronautica, 127*, 619–633. https://doi.org/10.1016/j.actaastro.2016.06.012.

DFO. (2020a). *Dartmouth Flood Observatory*. Retrieved October 30, 2020, from http://floodobservatory.colorado.edu

DFO. (2020b). *River and reservoir watch*. Retrieved October 30, 2020, from http://floodobservatory.colorado.edu/DischargeAccess.html

DFO. (2020c). *DFO web map service*. Retrieved October 30, 2020, from http://floodobservatory.colorado.edu/WebMapServerDataLinks.html

Dottori, F., Salamon, P., Bianchi, A., Alfieri, L., Hirpa, F. A., & Feyen, L. (2016). Development and evaluation of a framework for global flood hazard mapping. *Advances in Water Resources, 94*, 87–102. https://doi.org/10.1016/j.advwatres.2016.05.002.

Eklundh, L., & Jönsson, P. (2017). *Timesat 3.3 Software Manual*. Sweden: Lund and Malmö University. Retrieved October 17, 2020, from http://web.nateko.lu.se/timesat/docs/TIMESAT33_SoftwareManual.pdf.

ERCC – DG ECHO. (2019a). *Tropical Cyclone Idai impact overview*. Retrieved October 30, 2020, from https://erccportal.jrc.ec.europa.eu/getdailymap/docId/2860

ERCC – DG ECHO. (2019b). *TC Idai: EU response*. Retrieved October 30, 2020, from https://erccportal.jrc.ec.europa.eu/getdailymap/docId/2869

ERDS. (2020). *Extreme Rainfall Detection System*. Retrieved October 30, 2020, from http://erds.ithacaweb.org

Fekete, B. M., Vörösmarty, C. J., & Grabs, W. (2002). High-resolution fields of global runoff combining observed river discharge and simulated water balances. *Global Biogeochemical Cycles, 16*, 10–15. https://doi.org/10.1029/1999GB001254.

Fensholt, R., Sandholt, I., & Stisen, S. (2006). Evaluating MODIS, MERIS, and vegetation indices using in situ measurements in a semiarid environment. *IEEE Transactions on Geoscience and Remote Sensing, 44*(7), 1774–1786. https://doi.org/10.1109/TGRS.2006.875940.

Hmimina, G., Dufrene, E., Pontailler, J. Y., Delpierre, N., Aubinet, M., Caquet, B., et al. (2013). Evaluation of the potential of MODIS satellite data to predict vegetation phenology in different biomes: An investigation using ground-based NDVI measurements. *Remote Sensing of Environment, 132*, 145–158. https://doi.org/10.1016/j.rse.2013.01.010.

Huffman, G. J., Bolvin, D. T., Nelkin, E. J., & Tan, J. (2019). *Integrated multi-satellitE retrievals for GPM (IMERG) technical documentation*. Retrieved October 30, 2020, from https://pmm.nasa.gov/sites/default/files/document_files/IMERG_doc_190909.pdf

IWG-SEM. (2014). *Technical specification of GeoRSS, Version 1*. Retrieved October 30, 2020, from http://www.un-spider.org/sites/default/files/20141126_GeoRSS_technical_specifications.pdf

IWG-SEM. (2015). *Emergency mapping guidelines, Working Paper Version* 1.0. Retrieved October 30, 2020, from http://www.un-spider.org/sites/default/files/IWG_SEM_EmergencyMappingGuidelines_v1_Final.pdf

Jongman, B., Wagemaker, J., Revilla Romero, B., & Coughlan De Perez, E. (2015). Early flood detection for rapid humanitarian response: Harnessing near real-time satellite and Twitter signals. *ISPRS International Journal of Geo-Information, 4*(4), 2246–2266. https://doi.org/10.3390/ijgi4042246.

Mazzoglio, P., Laio, F., Balbo, S., Boccardo, P., & Disabato, F. (2019). Improving an extreme rainfall detection system with GPM IMERG data. *Remote Sensing, 11*(6), 677. https://doi.org/10.3390/rs11060677.

NASA. (2020). *About SERVIR*. Retrieved October 30, 2020, from https://www.nasa.gov/mission_pages/servir/overview.html

Nguyen, P., Shearer, E. J., Tran, H., Ombadi, M., Hayatbini, N., Palacios, T., et al. (2019). The CHRS data portal, an easily accessible public repository for PERSIANN global satellite precipitation data. *Scientific Data, 6*, 180296. https://doi.org/10.1038/sdata.2018.296.

Nguyen, P., Shearer, E. J., Ombadi, M., Gorooh, V. A., Hsu, K., Sorooshian, S., et al. (2020). PERSIANN dynamic infrared-rain rate model (PDIR) for high-resolution, real-time satellite precipitation estimation. *Bulletin of the American Meteorological Society, 101*, E286–E302. https://doi.org/10.1175/BAMS-D-19-0118.1.

Pettorelli, N., Vik, J. O., Mysterud, A., Gaillard, J. M., Tucker, C. J., & Stenseth, N. C. (2005). Using the satellite-derived NDVI to assess ecological responses to environmental change. *Trends in Ecology & Evolution, 20*(9), 503–510. https://doi.org/10.1016/j.tree.2005.05.011.

Reed, B. C., Brown, J. F., VanderZee, D., Loveland, T. R., Merchant, J. W., & Ohlen, D. O. (1994). Measuring phenological variability from satellite imagery. *Journal of Vegetation Science, 5*, 703–714. https://doi.org/10.2307/3235884.

Running, S. W., Nemani, R. R., Heinsch, F. A., Zhao, M., Reeves, M., & Hashimoto, H. (2004). A continuous satellite-derived measure of global terrestrial primary production. *Bioscience, 54*(6), 547–560. https://doi.org/10.1641/0006-3568(2004)054[0547:ACSMOG]2.0.CO;2.

Schumann, G. J. P., & Domeneghetti, A. (2016). Exploiting the proliferation of current and future satellite observations of rivers. *Hydrological Processes, 30*, 2891–2896. https://doi.org/10.1002/hyp.10825.

Schumann, G. J. P., Brakenridge, G. R., Kettner, A. J., Kashif, R., & Niebuhr, E. (2018). Assisting flood disaster response with earth observation data and products: A critical assessment. *Remote Sensing, 10*(8), 1230. https://doi.org/10.3390/rs10081230.

Schumann, G. J. P. (2019). The need for scientific rigour and accountability in flood mapping to better support disaster response. *Hydrological Processes, 33*, 3138–3142. https://doi.org/10.1002/hyp.13547.

Sentinel Asia. (2020). *About Sentinel Asia*. Retrieved October 30, 2020, from https://sentinel-asia.org/aboutsa/AboutSA.html

The International Charter Space and Major Disasters. (2019a). *Charter geobrowser tool, activation 598*. Retrieved October 30, 2020, from https://cgt.disasterscharter.org/en/598

The International Charter Space and Major Disasters. (2019b). *Charter geobrowser tool, activation 606*. Retrieved October 30, 2020, from https://cgt.disasterscharter.org/en/606

The International Charter Space and Major Disasters. (2019c). *Cyclone Idai in Mozambique*. Retrieved October 30, 2020, from https://disasterscharter.org/web/guest/activations/-/article/cyclone-in-mozambique-activation-598

The International Charter Space and Major Disasters. (2019d). *Cyclone Idai in Zimbabwe*. Retrieved October 30, 2020, from https://disasterscharter.org/web/guest/activations/-/article/cyclone-in-zimbabwe-activation-599

The International Charter Space and Major Disasters. (2020). *About the Charter*. Retrieved October 30, 2020, from https://disasterscharter.org/web/guest/about-the-charter

Tucker, C. J. (1979). Red and photographic infrared linear combinations for monitoring vegetation. *Remote Sensing of Environment, 8*(2), 127–150. https://doi.org/10.1016/0034-4257(79)90013-0.

UN General Assembly. (2015). *Transforming our world: The 2030 Agenda for Sustainable Development, A/RES/70/1*. Retrieved October 30, 2020, from https://www.refworld.org/docid/57b6e3e44.html

UN-SPIDER. (2020). *UN-SPIDER Knowledge Portal*. Retrieved October 30, 2020, from http://www.un--spider.org/space-application/emergency-mechanisms

UNOSAT. (2020). *UNOSAT Rapid mapping service*. Retrieved October 30, 2020, from https://www.unitar.org/maps/unosat-rapid-mapping-service

Voigt, S., Giulio Tonolo, F., Lyons, J., Kučera, J., Jones, B., Schneiderhan, T., et al. (2016). Global trends in satellite-based emergency mapping. *Science, 353*(6296), 247–252. https://doi.org/10.1126/science.aad8728.

Webster, P. J. (2013). Improve weather forecasts for the developing world. *Nature, 493*, 17–19. https://doi.org/10.1038/493017a.

WFP. (2019). *Mozambique access constraints as of 16 March 2019*. Retrieved October 30, 2020, from https://reliefweb.int/sites/reliefweb.int/files/resources/moz_op_accessconstraints_a3p_20190316.pdf

World Bank. (2020). *Cereal yield (kg per hectare)—Mozambique*. Retrieved October 30, 2020, from https://data.worldbank.org/indicator/AG.YLD.CREL.KG?locations=MZ

Wu, H., Adler, R. F., Tian, Y., Huffman, G. J., Li, H., & Wang, J. (2014). Real-time global flood estimation using satellite-based precipitation and a coupled land surface and routing model. *Water Resources Research, 50*, 2693–2717. https://doi.org/10.1002/2013WR014710.

3 ICT Readiness for Flood Risk Reduction and Management: Lessons from Eastern Cape Province's Port St Johns Municipality, South Africa

Godwell Nhamo, Lazarus Chapungu, and Simbarashe Nyika

Abstract

The Eastern Cape and KwaZulu-Natal in South Africa experienced severe floods in April 2019 and the negative impacts were significant. This chapter seeks to determine information communication and technology (ICT) readiness for flood risk reduction and management at the local municipality of Port St Johns in the Eastern Cape Province. The chapter further investigates the resilience of ICT infrastructure to flooding. ICT in this article refers to technology used for regular, everyday tasks like sending emails; making video calls; searching on the internet; using a tablet or mobile phone and even the older technology, like TVs and radios. The data was collected mainly through a household questionnaire survey. This was complemented by means of document analysis and field observations. The key findings were that the mobile phone was the most prevalent ICT device followed by TV and radio. Several barriers to the effective use of ICT for flood risk reduction and management were observed and these included the lack of capacity to buy data, the availability of a reliable network and low levels of literacy of the elderly. Since South Africa and the world are moving swiftly in scaling up the use of mobile phones, for flood risk reduction and management, this implies that network coverage be improved at the municipality of Port St Johns. Community training tailor-made for flood and other disaster risk reductions and management should also be introduced.

Keywords

ICT readiness · Flood risk reduction · Port St Johns · South Africa · Mobile phone · SMS

3.1 Introduction

Climate-induced hazards such as floods, droughts and tropical cyclones are jointly accountable for a plethora of disaster fatalities and economic losses globally (McCallum et al., 2016; UNISDR, 2015; IPCC, 2012). Floods have affected more people than any other natural disaster. For example, about 2.3 billion people were affected directly by floods in the period 1995–2015

G. Nhamo (✉) · L. Chapungu
Institute for Corporate Citizenship,
University of South Africa, Pretoria, South Africa
e-mail: nhamog@unisa.ac.za

S. Nyika
Walter Sisulu University, Mthatha, South Africa
e-mail: snyika@wsu.ac.za

G. Nhamo, L. Chapungu (eds.), *The Increasing Risk of Floods and Tornadoes in Southern Africa*,
Sustainable Development Goals Series, https://doi.org/10.1007/978-3-030-74192-1_3

(McCallum et al., 2016). In recent years about 73.1 million individuals became flood victims. Floods were responsible for 47% of disaster-related losses since 2015 (See, 2019), and 34.2 million people were affected by flooding in 2018 (Louvain, 2019).

Exposure and sensitivity to flooding have increased over the years owing to the increased frequency and severity of flood events. The increased vulnerability to flooding can be attributed to the growing trend of urbanisation, which alters land use and exposes more people to natural hazards (GFDRR, 2015). Communities residing within river basins and closer to large water bodies have been greatly affected (McCallum et al., 2016; Hallegatte, 2011; Mileti, 1999). The poor and vulnerable communities have been the most affected.

Information is fundamental prior to, during and after a flood event (Hassan, 2015). Thus, ICT is a key lifeline that has enormous potential to preclude possible fatalities. Worldwide, there has been significant progress with regard to the development of ICT applications used for flood reporting. The applications have proved to be convenient in saving lives and reducing disaster-related economic and social costs. In Kenya, for example, the Ushahidi[1] platform was developed for crowdsourcing and it proved useful for flood reporting in Australia (Koswatte et al., 2018; Okolloh, 2009) and producing a 2013 floods crisis map for the Czechia (Pánek et al., 2017). In 2015, floods that occurred in Chennai were reported through the use of the Flooded Streets application (See, 2019; Naik, 2016). Another ICT application, the Flood Hazard Maps, was used to map flood-prone areas in the Philippines, (Victorino et al., 2016). Mobile applications for flood monitoring and reporting have been developed in Argentina, France and New Zealand (Le Coz et al., 2016). These are among the several examples of global efforts to use ICT-based systems for flood risk reduction and management.

Yap (2011) argues that responding effectively to flood events requires all parties to efficiently obtain authentic and precise data. Thus, efficient and effective flow of information to relevant stakeholders is indispensable in coordinating response activities, such as evacuations. This points to the fact that effective communication is critical (Wattegama, 2007) within the entire disaster cycle that includes early warning, search and rescue, relief and response and recovery and reformation. To this end, several ICT resources have been used in the management of flood disasters. The internet is currently one of the most essential sources of information. The use of mobile phones, commonly referred to as "smart phones", as well as social media technologies have increased significantly over time, becoming a major vehicle for transmitting information worldwide (Alexander, 2014).

Web 2.0 technologies are also an important element in disaster risk reduction (DRR) and management communication. These technologies allow immediate feedback and real-time group chats from ground zero as the disaster develops. Some built-in capabilities of Web 2.0 applications include videos, images and enabling appropriate relief strategies to be implemented in quasi-real-time. In this way, stakeholders and disaster managers can estimate the magnitude of the disaster and victims and attract the attention of many relief efforts from relevant organisations (Huang et al., 2010).

Nowadays, ICT is widely used for flood risk reduction. Bradford et al. (2012) put forward that management options that focus on nonstructural issues such as better land use planning, resettling, event prediction and warning are increasingly being advanced by the use of ICT. ICT facilitates the functionality of citizen observatories such as social media and sensor technologies, which have been regarded as a means of generating pertinent information and advancing demand-driven policy responses. Identifying interventions that strengthen disaster preparedness and mitigate losses is a growing policy challenge (Allaire, 2016). Hence, ICT is highly likely to improve disaster communication, especially in the case of flooding events, which results in saving lives and reducing disaster-related losses. The author goes further indicating that social media platforms such as Facebook and Twitter are a form of ICT that has a colossal presence worldwide with more

[1] Ushahidi is a crowd sourcing platform used for reporting any disturbance within communities.

than one billion users. This has seen agencies involved in flood disaster management taking advantage of the use of these platforms to devise ways of using them during flood emergencies. Disaster communication efforts would require an understanding of the impact of ICT-founded social media platforms since this has implications on the incorporation of online applications. When confronted with a flood catastrophe, individuals worldwide have shown increased propensity for using ICT-related technologies and resources to access latest information about the prevailing disaster and appropriate actions to take. ICT-founded platforms are important due to their ability to be easily searched, perform real-time updates and self-publishing capabilities (Allaire, 2016).

The background to the chapter is that the Port St Johns local municipality located in the Eastern Cape Province of South Africa experienced heavy rains, leading to severe floods on 22 April 2019 (Port St Johns Disaster Risk Management Satellite Centre, 2019). The heavy rains also affected parts of KwaZulu-Natal (Ministry of Cooperative Governance and Traditional Affairs, 2019) where more than 50 people died (Business Live, 2019). Based on the call for support identified by the municipality to which ICT functionality would be needed, there was a need for temporary shelter, school uniforms and stationeries, sponges, blankets and food parcels (Port St Johns Disaster Risk Management Satellite Centre, 2019). According to the records of the Port St Johns Disaster Risk Management Satellite Centre, 882 households (victims) were affected, of which 656 had their belongings partially damaged, 226 were left homeless and 121 were tenants (renting at Green Farm and around town) after the disaster that hit 11 wards. Other areas in the Eastern Cape that were affected which are not part of this work is Lusikisiki, where two school children drowned in the floods, and Mbizana, where a 39-year-old woman died. From the then Minister of Cooperative Governance and Traditional Affairs' description of the flooded area in Port St Johns, "The settlement is located in a valley that receives water from the high areas that surround it. From the assessment, Green Farm will always provide challenges due to the topography of the area and its location" (Ministry of Cooperative Governance and Traditional Affairs, 2019 online).

Several questions for scientific inquiry can be raised: Can access to ICT resources increase resilience towards flood impact? What is the level of ICT readiness in southern African cities regarding flood risk reduction and management? Given the paucity of knowledge on this, any work that investigates ICT readiness for flood risk reduction and management in a flood-prone region merits attention. This may assist in activating the processes of interrogating policy prescriptions that are key in enabling the use of ICT to ward off flood-related consequences. It is given these gaps and information needs that this chapter explores ICT readiness for flood risk reduction and management, drawing lessons from Eastern Cape Province's Port St Johns, South Africa.

In light of the background information, this chapter investigates the extent of ICT readiness for disaster risk reduction (DRR) and management in Port St Johns local municipality of the Eastern Cape Province, South Africa. The research also investigates the damage to ICT infrastructure and its resilience to the April 2019 floods the area experienced. The chapter sets the objective to determine the extent of ICT readiness for flood risk reduction and management in Port St Johns. The next section details the relevant literature. This is followed by a section on materials and methods used in generating and analysing the data before presenting and discussing the findings. A conclusion is also drawn from the work and some policy recommendations are made.

3.2 Literature Review

Davies (2020) notes that floods sweep away bridges, houses and trees and destroy infrastructure in their paths. They too can cause deaths, displacements and economic losses. ICT infrastructure is not spared either. It is envisaged that climate change is exacerbating flood occurrence

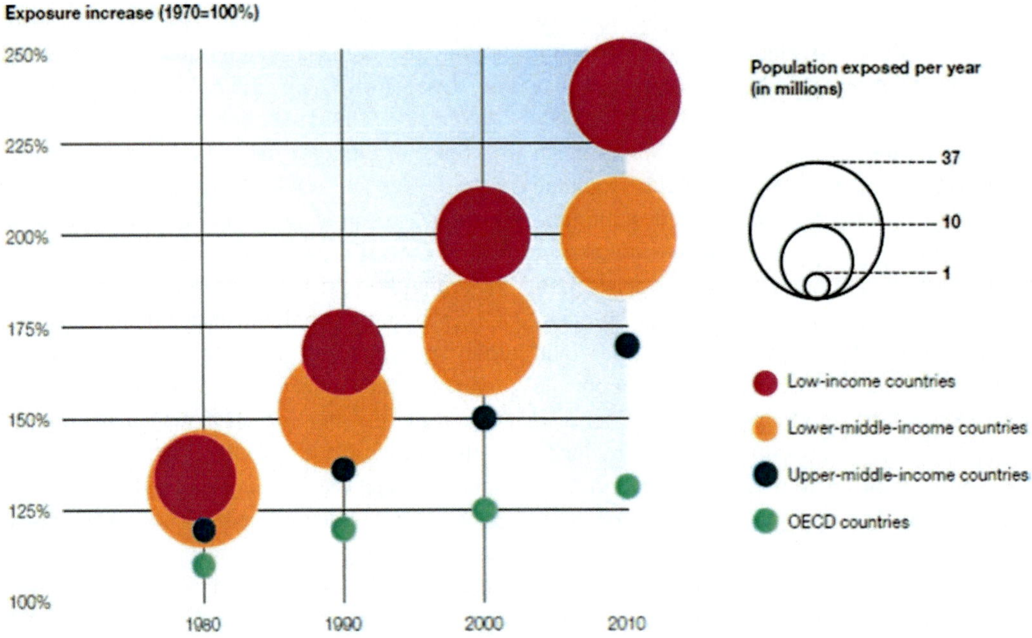

Fig. 3.1 Global Flood exposure trend. (Source: Adapted from UNISDR (2011: Online))

resulting in worldwide increase in the number of people affected (Neelson, 2017). Trends in global flood exposure are shown in Fig. 3.1.

The frequency and severity of floods have increased in southern Africa, even outside the most recent wave of floods driven by cyclones Kenneth and Idai in 2019, resulting in the displacement of thousands of individuals and massive destruction of property. The number of people exposed and vulnerable to flood hazards has increased over time. A series of scientific models predict that more frequent and severe tropical cyclones associated with flood events will occur in the future due to climate change. The need to fortify communities and build their resilience cannot be overemphasised (Chapungu, 2020).

3.2.1 Definition and Classification of Floods

Floods can be classified into three categories: coastal, fluvial and pluvial (Below et al., 2009). Yang and Liu (2020) put forward that pluvial flooding takes place when the amount of rainfall received exceeds infiltration rate, or when the capacity of storm drains is exceeded by flood water. Falconer et al. (2009) added that pluvial flooding stem from overland flow spawned by precipitation before the runoff is channelled into any watercourse. The frequency of pluvial flooding has increased over the years, resulting in an increase in the impacts and threats to life and property (Yang and Liu 2020). Climate change is predicted to be one of the forces behind the increased frequency of pluvial floods (Hurford et al., 2012; Kaźmierczak & Cavan, 2011). Fluvial flooding is the flooding that occurs when the amplitude of a river system to contain the load is exceeded by the amount of rainfall received (Yang and Liu 2020). Yang and Liu (2020: 3) further aver that:

> It occurs either as "overbank flooding", which happens when water rises to overflow the edges of a river; or as "flash flooding" characterised by an intense, high-velocity torrent of water that occurs in a river within six hours or less.

Flash floods culminate from either high-intensity rainfall or abrupt bursting of a dam or levee. Archer and Fowler (2018) assert that flash floods are forceful and have the dexterity to push huge debris, which may exacerbate their destructive nature. Their abruptness means short warning and response time compared to other types of floods. Mountainous terrains are prone to flash floods, and flooding in these regions is highly destructive due to steep topographic conditions which intensify flash-flood force and impetus to damage low-lying areas where human settlements, infrastructure and invaluable property are normally concentrated (Liu et al., 2018). Port St Johns is one such area.

Coastal flooding is another type of flooding. It mainly occurs in low-lying areas driven by wind waves and elevated water levels (Wolf, 2009). In most cases, it is generated by massive waves, towering tides, storm surges, and mean sea level anomalies. Precipitation and river flow contribute to coastal flooding in deltas and estuaries (Wolf, 2009). Climate change is predicted to influence the intensity and frequency of tropical cyclones and sea level rise; consequently, this will amplify the magnitude and rate of coastal flooding (Heberger et al., 2011; Dawson et al., 2009). Densely populated coastal municipalities are at high risk to flooding, yet their amplitude to mitigate or reduce the associated impacts is significantly low (Woodruff et al., 2013). ICT could play a pivotal role in enhancing preparedness to deal with the consequences associated with flooding. However, for most coastal cities in southern Africa, the level of ICT readiness for flood risk reduction remains veiled in obscurity. Port St Johns is no exception.

3.2.2 The Need for Information on Flood Risk Reduction and Management

The need for access to information by all stakeholders in a flood-prone area before, during and after a flood event cannot be overemphasised (ADPC, 2011). However, this information must be accurate, reliable and timely; otherwise individuals and institutions will make fundamental emergency resolutions based on unverified information which may cause further harm. In a flood situation, information must be shared with the general public and other stakeholders. ICTs facilitate the ease of sharing the information to reduce risks and manage the flood disaster. ADPC (2011: 28) aver that:

> Flood risk reduction and management measures have different information needs for different audiences. Baseline data about a country or region affected by a flood is a necessity to enable prevention, mitigation, preparedness planning and recovery planning. Disaster response, rehabilitation and reconstruction activities require real-time information about the impact of a flood event and the resources available to combat it. For stakeholders to respond swiftly and effectively, the information needs to be readily collected, processed, analysed and shared.

ICT also plays a key role in ensuring availability of information which is critical during the entire disaster cycle. Intensive risk mapping and assessment could assist in gathering such baseline information (ADPC, 2011). Figure 3.2 shows the information management cycle within an ICT system.

The entire information management cycle is facilitated by an ICT system. Data are measurements or observations made from events and/or activities taking place at grassroots level, for example, the number of internally displaced persons (IDPs). The data is analysed in a computerised system so that data is processed into information which will be used as knowledge for risk reduction and management. The knowledge of the flood risk and management options will result in practical initiatives from which further data can be collected and the cycle goes on. The information management process is not one dimensional but occurs in a cycle (ADPC, 2011). The different information needs in flood hazard management is presented in Table 3.1.

3.2.3 ICT in Flood Disaster Response

When individuals are caught up in a flood emergency, they have acute need for information. Usually, households as well as individuals lack accurate information with regard to flood risk (Allaire, 2016). Flood disasters that occur

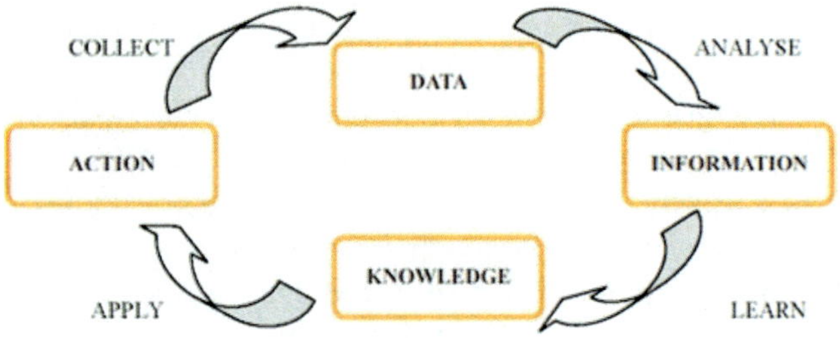

Fig. 3.2 Information management cycle. (Source: Adapted from ADPC (2011: 28))

abruptly have the propensity to separate individuals from their families. The flood victims also run short of shelter, adequate food and basic necessities. They get more confused and scared by events unfolding around them (ADPC, 2011).To this end, ICT-based programming tailored to the needs of such people can provide essential information lifeline. Following the December 2004 Indian ocean tsunami, the affected individuals expressed apprehension that they lacked information about aid and related processes. This translates to the fact that they did not know the options available to them. The information required following a disaster takes an incremental approach starting from basics such as: What transpired? Where are other people? Over time, other critical information needs develop: Where can we find food and water? How can we access hospitals? Individuals and households begin to crave for more critical details. Interventions would require to manage expectations (ADPC, 2011). ICT can be very useful in expectation management during a flood emergency. Allaire (2016) puts forward that information deprivation during flood disaster response causes stress and exacerbates trauma. ICT plays an important role through transmitting timely, accurate objective information that saves lives and bolster recovery efforts. ADPC (2011) argues that the power of ICT recline in its effectual analysis, application and management of information.

3.2.4 ICT in Flood Early Warning Systems

UNEP-DEWA (2006: online) defines an early warning system as "The provision of timely and effective information that allows individuals exposed to hazard to take appropriate action to avoid or reduce their risk and prepare for effective response". Haritharaj et al. (2020) put forward that there are mainly four types of elements involved in early warning systems: risk knowledge, monitoring and warning, dissemination and communication and response capability. ICT plays a central role in all the elements. Failure of any of these elements will cause the failure of entire system. Subin and Sujatha (2020) note that the use of ICT in flood early warning systems is reflected in the surge in numbers of individuals having access to mobile phones commonly known as "smart phones". In 2019, mobile phone subscriptions rose to 8.3 billion. The rate of use is increasing rapidly in developing countries. Asia Pacific alone had 2.8 billion people subscribed to mobile services. The number of subscribers continues to grow annually (GSMA, 2020). An increase in mobile connectivity has activated growth of various platforms used for information sharing in different forms including texts, pictures and voices. Consequently, mobile phone-based applications are on the increase which help with access to flood disaster information. Mobile phones play a crucial role at all stages of flood disaster risk reduction and management cycle. This includes dissemination of information on early warnings during the pre-disaster period, information exchange during flood disaster and information sharing in the immediate aftermath to facilitate recovery.

Although a lot still needs to be done in the developing world concerning the use of ICT in flood early warning systems, developed regions, specifically Europe, have made significant progress in this regard. ICT is quite instrumental in early flood warning systems through using geographic information systems (GIS) and remote sensing (Onafeso & Samuel, 2012). Stephenson and Anderson (1997) noted that effective flood

Table 3.1 Different information needs in flood hazard management

Information needs	Possible actions based on available information	ICT methods and platforms
Flood Mitigation		
• Development plans and decisions at the national, local and community level • Social, demographic and economic characteristics • Land use plans, environment management plans • Utility services network information • Hazard and vulnerability maps • Risk zones • Geologic and hydrometeorological information • DRM plans	• Identify spatial and temporal variation in hazard severity, occurrence and likelihood and/or variations in vulnerability • Identify service and infrastructure assets and gaps • Identify and communicate high risk "hotspots" where disaster impacts are most likely to be severe • Identify appropriate structural and/or nonstructural mitigation measures and prioritise resources • Evaluate appropriateness of land use and development plans • Target public outreach campaigns and choose appropriate messages, sources and channels • Recommend appropriate codes and ordinances • Promote risk education among decision-makers, highlighting how development decisions can impact risks	• Geographic information Systems • Remote sensing • Social media platforms • Public media houses (Radio and TV Channels) • Internet websites • Emails • Early warning systems
Preparedness		
• Country hazard profiles • Locations of shelters and critical infrastructure • Hazard and vulnerability maps • Risk zones • Populations at risk • Access to telecommunication and electricity services • Equipment, emergency personnel and volunteers for disaster response	• Identify spatial and temporal variation in hazard severity, occurrence and likelihood and/or variations in vulnerability. Identify appropriate places for resource stockpiling, staging areas, evacuation routes and emergency operations centres • Identify service and infrastructure assets and gaps • Improve warning strategies by identifying appropriate channels, sources and messages before an event • Improve evacuation planning by identifying potential zones, shelters, routes and the location of populations with special evacuation needs • Develop and visualise a hazard and impact scenario during an exercise • Conduct public education campaigns, including incorporating disaster risk awareness into school curricula • Conduct emergency exercises and drills	• Early warning systems • Databases • Geographic information Systems • Remote sensing • Social media platforms • Public media houses (Radio and TV Channels) • Internet websites • Emails
Response		
• Hazard and vulnerability maps • Geospatial information on the flood event: "Where is it? What is in the area? How do I get there?" • Situation update: affected populations, persons needing rescue, roads, shelter, etc. • Information regarding the latest developments in the relief efforts	• Target alerts using appropriate channels, sources and messages • Anticipate likely impacts across areas of interest • Anticipate likely short-term needs across areas of interest Identify and communicate appropriate shelter and mass care locations • Familiarise disaster response teams with areas of interest • Provide a baseline to clearly characterise short-term disaster impacts • Provide a baseline to monitor progress of response activities • Conduct loss and damage assessment • Assist the public to connect with family, friends and colleagues in stricken areas	• Geographic information systems • Remote sensing • Social media platforms • Public media houses (Radio and TV Channels) • Internet websites • Emails
Recovery and Reconstruction		

(continued)

Table 3.1 (continued)

Information needs	Possible actions based on available information	ICT methods and platforms
• Damage and needs assessment • Same information needs as for mitigation	• Identify and communicate locations for recovery assistance centres • Provide a baseline by means of which new hazards and/or patterns in vulnerability can be identified. Evaluate appropriateness of redevelopment plans • Identify appropriate mitigation measures • Identify appropriate changes in preparedness and response activities • Provide a baseline to clearly characterise long-term disaster impacts. • Provide a baseline by means of which r progress of recovery activities can be monitored	• Geographic information systems • Remote sensing • Social media platforms • Public media houses (Radio and TV Channels) • Internet • Emails

Source: Adapted from ADPC (2011: 29–30)

risk reduction and management uses ICTs, particularly remote sensing, satellite and cellular mobile and geographic information systems (GIS). The real-time nature of the dissemination mechanism of ICTs facilitate quick decision-making processes to create resilience and bring down the magnitude of risk and uncertainty regarding natural hazards. In addition, the Sendai Framework on Disaster Risk Reduction (2015–2030) proposed the intensive application of in situ and space-based earth observations, space technologies, remote sensing, geographic information systems, hazard modelling and prediction, weather and climate modelling and forecasting in early warning systems (UNDRR, 2015).

Observations of flood events are done through weather radar satellites and automatic hydrometeorological stations. The two types of satellites used are polar orbital and geostationary satellites. The current ICT-based flood early warning systems include Global Disaster Alert and Coordination System (GDACS), European Flood Awareness System (EFAS), Community-Based Flood Early Warning System, Numerical Weather Prediction (NWP), Flash Flood Guidance System (FFGS) and Urban Early Warning System (Haritharaj et al., 2020). GDACS is a web-based disaster information system which gives warning alerts for floods, using different colour codes according to the intensity of occurrence. White alert indicates a minor event; green alert indicates a moderate event; orange alert indicates potential local disasters and the red alert indicates potentially severe disasters. EFAS provides increased preparedness for riverine floods across Europe. The system displays information as colour-coded overview maps and finally sends out warning emails to the corresponding EFAS partners. Flood early warning information is given to partners up to 10 days in advance. A community-based flood early warning system is an ICT-enabled system. It comprises a flood sensor attached to the transmitter to detect rising water levels. When water reaches a critical level, a wireless signal system transmits information to the receiver. Flood warning messages are disseminated through mobile phones to appropriate agencies and vulnerable communities downstream.

An urban early warning system is another ICT-based early flood warning system, which is technically complex. Haritharaj et al. (2020: 31) describe the system as a system in which

> The Sensor Monitoring module receives data streams from the sensors installed in the dike. Raw sensor data is filtered by the AI Anomaly Detector that identifies abnormalities in dike behaviour or sensor malfunctions. The Reliability Analysis module calculates the probability of dike failure in case of abnormally high water levels or an upcoming storm and extreme rainfalls. If the failure probability is high, then the Breach Simulator predicts the dynamics of a possible dike failure, calculates water discharge through the breach and estimates the total time of the flood. After that, the Flood simulator models the inundation dynamics.

A numerical weather prediction (NWP) is another system which is ICT-based and used for the prediction of weather, using atmospheric models and computational techniques. Here atmospheric motions are expressed in differential equations. The integrated form of these equations will form an atmospheric model for weather prediction. It is usually done by the National Centre for Medium Range Weather Forecast (NCMRWF). A widely used NWP model is the Weather Research and Forecasting (WRF) model which is having improved forecast at shorter duration. FFGS was invented by the India Meteorological Department (IMD) for predicting the possibility of flash floods up to 6 h in advance (AICRPAM, 2018). This system gives alerts to the disaster relief forces as well as residents. The system involves computation of local on daily basis and along with the forecasted rainfall to predict the flooded condition. For this purpose, India is divided into 27,000–28,000 micro watersheds of about 50–70 km^2 each with a soil hydrology model which will run on a daily basis to assess the runoff potential.

3.2.5 ICT Readiness in South Africa: With Comparison to Her BRICS Peers

It is critical to consider some perspectives linked to ICT readiness, especially access to electricity, broadband and mobile cellular subscriptions and the percentage of individuals using the internet in South Africa and also compare such to its peers. The statistics concerning access to electricity over the years in South Africa as compared to its peers to the BRICS (Brazil, Russia, India, China and South Africa) block is shown in Fig. 3.3. From the data, South Africa compares favourably to its peers. Since 2001, the country has increased access to electricity for its population from about 70.2% to about 91.2% in 2018. Brazil, China and Russia stood at 100% as from 2018. South Africa is also way ahead of the sub-Saharan Africa (SSA) averages of 26% in 2000 which rose to 47.7% in 2018.

Data sets on fixed-broadband subscriptions per 100 inhabitants, mobile telephone subscriptions per 100 inhabitants and the percentage of individuals using the internet and fixed telephone subscriptions per 100 inhabitants were analysed. The findings were that South Africa has very low penetration concerning fixed broadband subscriptions. The subscriptions were merely 3 per 100 inhabitants and only slightly higher than India that had 1.3 subscriptions per 100 inhabitants (Fig. 3.4). South Africa's scenario leaves a huge gap compared to China with 26.9 subscriptions per 100 inhabitants, Russia with 21.4 subscriptions per 100 inhabitants and Brazil with 13.7 subscriptions per 100 inhabitants.

However, when it comes to mobile phones, South Africa surpasses all its BRICS peers. The country had 18.5 subscriptions per 100 inhabitants in 2000, which increased to 159.9 subscriptions per 100 inhabitants in 2018 (Fig. 3.5). Effectively, this implies that there are households and/or individuals with more than one subscription to a mobile phone line. This remains a critical pillar if the mobile phone is to be used for flood risk reduction and management, including early warning systems. South Africa is closely followed by Russia that had 2.2 subscriptions in 2000, increasing drastically to 157.4 subscriptions per 100 inhabitants in 2018 followed by China with 115.5 subscriptions per 100 inhabitants, Brazil with 98.8 subscriptions per 100 habitants and India with 86.9 subscriptions per 100 inhabitants in 2018. Generally, there is a clear, rapid and solid trend in growth in mobile phone subscriptions per 100 inhabitants, making it a good platform to harness for flood and other disaster risk reduction and management.

The rate of coverage for individuals using the internet was also investigated. The results reveal that South Africa is not doing badly, since it is ranked in the third place according to 2017 data. The country had about 5.3% of individuals using the internet in 2000, a figure that rose to more than half, at 56.2% in 2017 (Fig. 3.6). Russia had 76% of individuals using the internet in 2017, while Brazil had 67.5%. China had 54.3% of individuals using Internet and India came in a far distance with 34.5% in 2017. For completeness, and since there are still households and/or individuals with fixed telephone lines, the last analysis performed looked at this aspect.

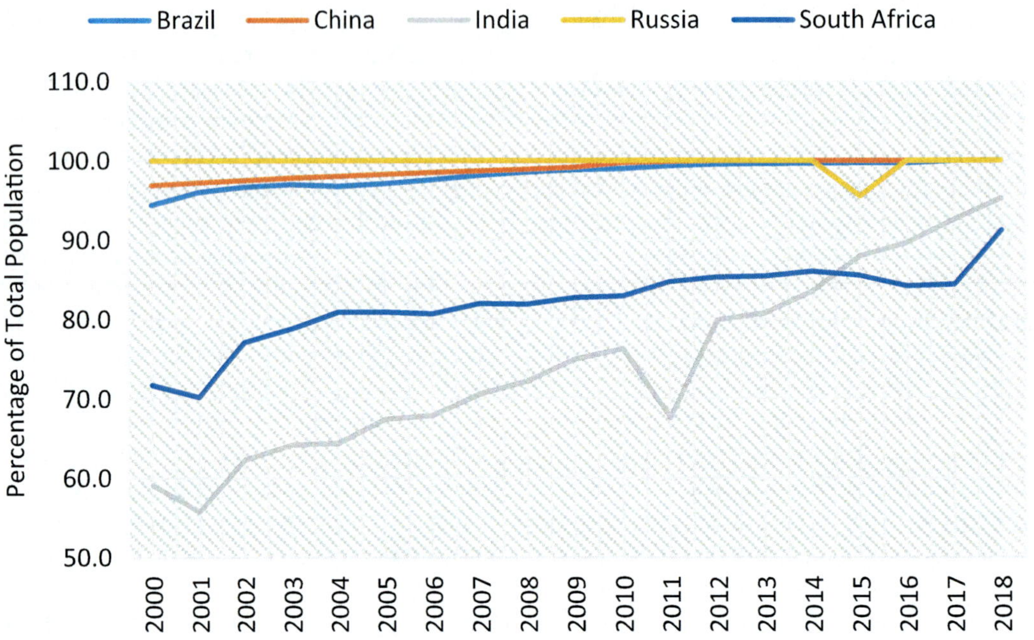

Fig. 3.3 Access to electricity (2000–2018). (Source: Authors, Data from WDI (2020))

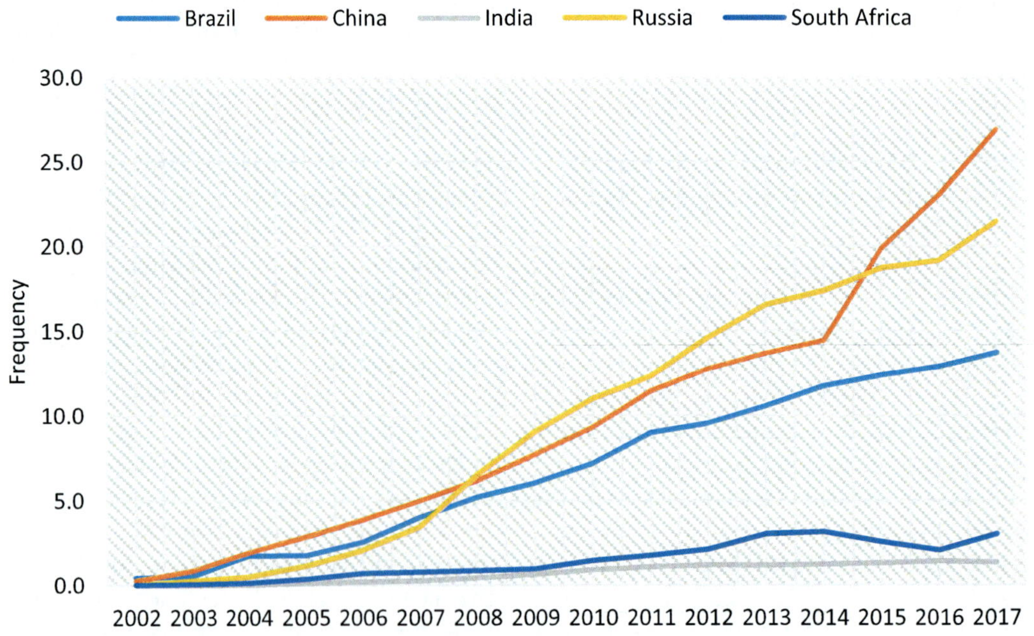

Fig. 3.4 Fixed-broadband subscriptions per 100 inhabitants. (Source: Authors, Data from ITU-D (2020))

The general trend was that fixed telephone line subscriptions were now going down across BRICS (Fig. 3.7), following some peaks along the way. South Africa remained with the lowest subscriptions throughout from 2000 to 2018. In 2000, there were about 11 subscriptions per 100 inhabitants and this continued declining steadily to 5.8 subscriptions in 2018. A drastic reduction in subscriptions is also observed in China and Russia. Peaking at 31.8 subscriptions per 100

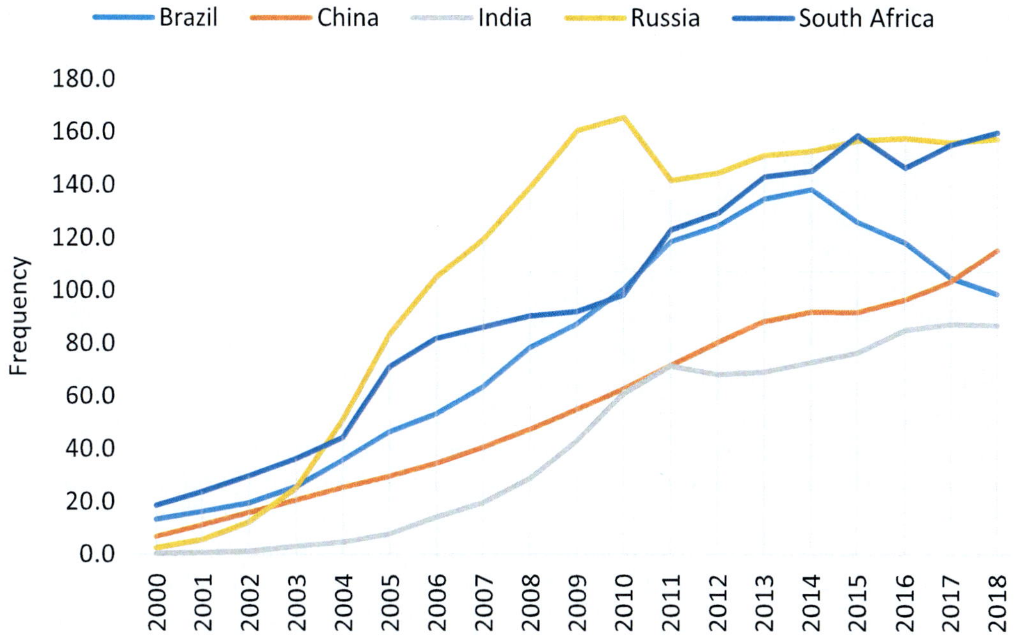

Fig. 3.5 Mobile phone subscriptions per 100 inhabitants. (Source: Authors, Data from ITU-D (2020))

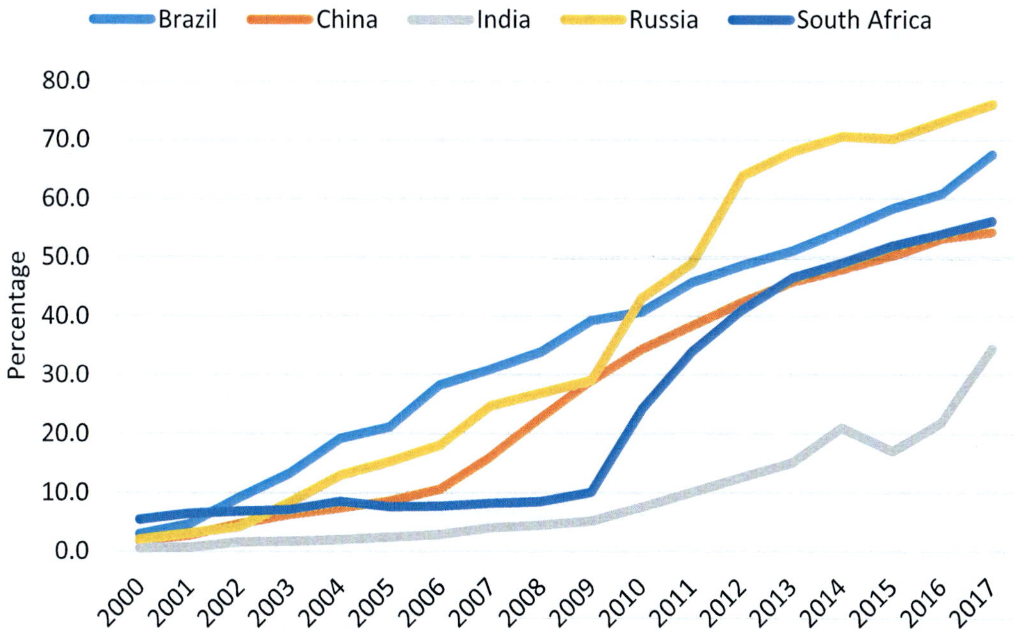

Fig. 3.6 Percentage of individuals using the Internet. (Source: Authors, Data from ITU-D (2020))

inhabitants in 2008 from a low of 21.9 in 2000, Russia's subscriptions dropped to 20.7 subscriptions in 2018. A similar pattern is reflected for China that had 11.2 subscriptions in 2000, rising to a peak of 27.5 in 2006, before declining to 13.5 subscriptions per 100 inhabitants in 2018. Brazil's graph is flattening, although also on a decline, rising from a low of 17.7 subscriptions per 100 inhabitants in 2000 to a peak of 22.4 in 2013 before starting to decline to 18.3 subscriptions per 100 inhabitants in 2018. India's trend remains at a very low subscription base marked by 3.1 subscriptions per 100 inhabitants in 2000 which grew to the peak of 4.4 in 2005 before it continuously declined to 1.6 subscriptions per 100 inhabitants in 2018. Given the confirmed trend of falling subscriptions for fixed telephone lines, this will not be a viable option to consider for future flood risk reduction and management in South Africa and possibly in many other countries. This scenario is solidified by the popularity of Industry 4.0 (Fourth Industrial Revolution) that has grasped the world (Nhamo et al., 2020).

The next section focuses on the materials and methods used in generating and analysing data.

3.3 Materials and Methods

The research focused on Port St Johns local municipality in the Eastern Cape Province of South Africa. This way, the research narrows down to one particular case study (Allaire, 2016). This study adopted the mixed methods research design as the strategy of inquiry, which uses quantitative and qualitative approaches. The qualitative approach involved gathering data from official municipal documents, while the quantitative involved gathering facts and figures through household questionnaire surveys and field observations. A review of previous studies focusing on relevant thematic areas was also performed to contextualise the study.

The data were mainly generated from a household questionnaire survey with a realised sample of 197 household respondents after 28 recorded dropouts. This gave a response rate of 87.56%. From the realised sample, 78.68% of households were from the urban set-up, with the remaining 21.32% from the rural set-up. Households from the urban set-up could also be grouped under urban (formal) which took up 13% of the total and urban (informal) which, in turn, took up 87%. The use of surveys in studying ICT and DRR in general and ICT and floods in particular is a common phenomenon. Muktar et al. (2018) used a questionnaire survey in the evaluation of ICT access, use and preferences for livelihood resilience that included floods in the Malaysian fisher folk sector. Allaire (2016) did the same in researching the 2011 Bangkok floods, sampling 469 households. Other methods of data generation included on the ground observations and reflections during interviewing sessions.

Seventy percent of the respondents were household heads, with 15% being children of

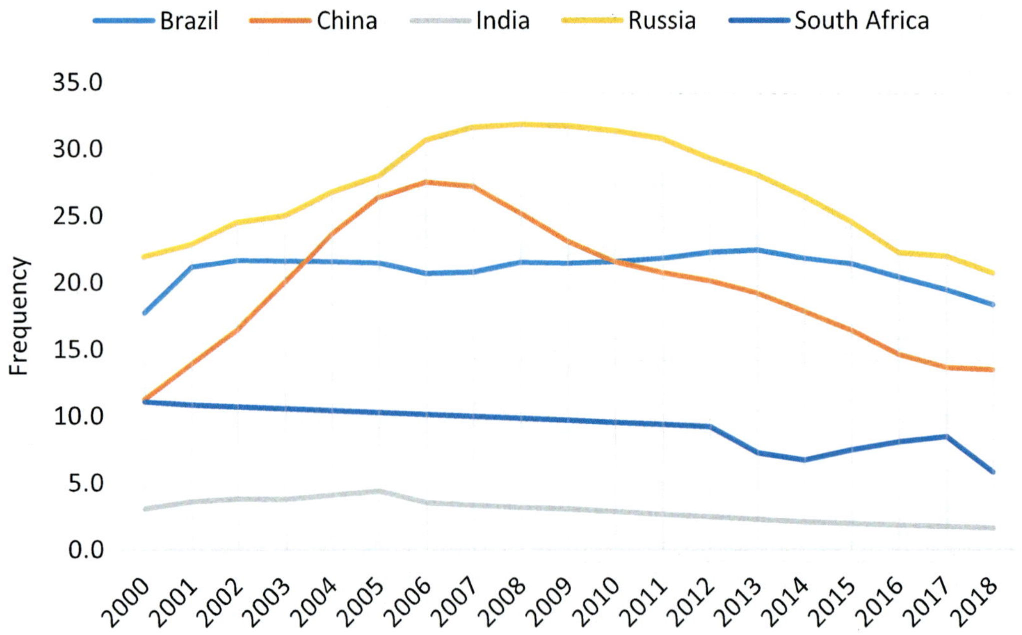

Fig. 3.7 Fixed telephone subscriptions per 100 inhabitants. (Source: Authors, Data from ITU-D (2020))

18 years and above. The remaining share of respondents were relatives or workers (6% apiece) or in the other category (4%). As for gender, 69% were female and the remainder male. One of the respondents wished not to disclose gender, which was one of the options for the question. The age groups of respondents are shown in Fig. 3.8.

A total number of 18 people were involved during the fieldwork that took place from 2 to 13 December 2019 as fieldworkers. These people included 13 researchers and/or authors involved in the project and five hired fieldworkers that also assisted in face-to-face administration of the household questionnaire survey. The administering of the household questionnaire survey focused more on the worst hit areas that included the Green Farm informal settlements (Siqathule, 2019; Ministry of Cooperative Governance and Traditional Affairs, 2019), with households sampled on a cluster bases drawing from the list provided by the municipality as indicated in the introduction. Document analysis and field observations were also employed. The location of the study areas is shown in Fig. 3.9.

The questionnaire survey was administered on an offline QuestionPro platform. The fieldworkers worked offline in the field during the day and uploaded the responses onto the QuestionPro platform in the evenings. Before rolling out the household survey, a half-day training workshop on researching disaster areas was undertaken on the 2 December 2019. This was facilitated by the senior academics and a psychologist in the research team. These members had also been involved in the July 2019 3-day workshop for authors that took place in Ekurhuleni, South Africa, which also comprehensively looked at undertaking research in disaster areas. Data were also obtained from the World Bank's World Development Indicators (WDI) and ITU-D and this covered South Africa's peers from Brazil, India, China and Russia—the BRICS trading and partnership block.

Prior to rolling out the household questionnaire survey, a pilot study was done. Several matters came up that needed redressing. The questionnaire was noted to be lengthy, taking +/−60 min to complete. The time would be rather long, resulting in the respondents losing interest. This was addressed accordingly by eliminating repetition in some questions and deleting some questions. Grammatical errors were also picked up and corrected. Other necessary adjustments were made to fine-tune the questions. There were also other challenges relating to access of some wards as the area is predominantly rural with gravel roads with settlements spaced.

Broadly, ICT readiness in South Africa was considered interfacing it with some SDGs indicators from the United Nations Inter-agency and Expert Group on SDG Indicators (IAEG-SDGs) Tier Classification of 15 October 2018 (IAEG-SDGs, 2018). Up to 7 ICT-related SDGs indicators drawn from five SDGs, namely SDG 4 (Quality Education), SDG 5 (Gender), SDG 7 (Sustainable Energy), SDG 9 (Industry and Infrastructure) and SDG 17 (Partnerships) (United Nations, 2015), were pulled out and these include those in Box 3.1. Furthermore, data were generated from the Telecommunication Development Sector (ITU-D) publicly available online resources.

Box 3.1: ICT-Related Indicators

- 4.4.1 Proportion of youth and adults with information and communication technology skills, by type of skills.
- 4.a.1 Proportion of schools with access to the internet for pedagogical purposes and/or computers for pedagogical purposes.
- 5.b.1 Proportion of individuals who own a mobile telephone per 100 people.
- 7.1.1 Proportion of population with access to electricity.
- 9.c.1 Proportion of population covered by a mobile network, by technology.
- 17.6.2 Fixed internet broadband subscriptions per 100 inhabitants, by speed.
- 17.8.1 Proportion of individuals using the Internet (from any location), all individuals (%).

Source: Authors, based on IAEG-SDGs (2018: 1–33).

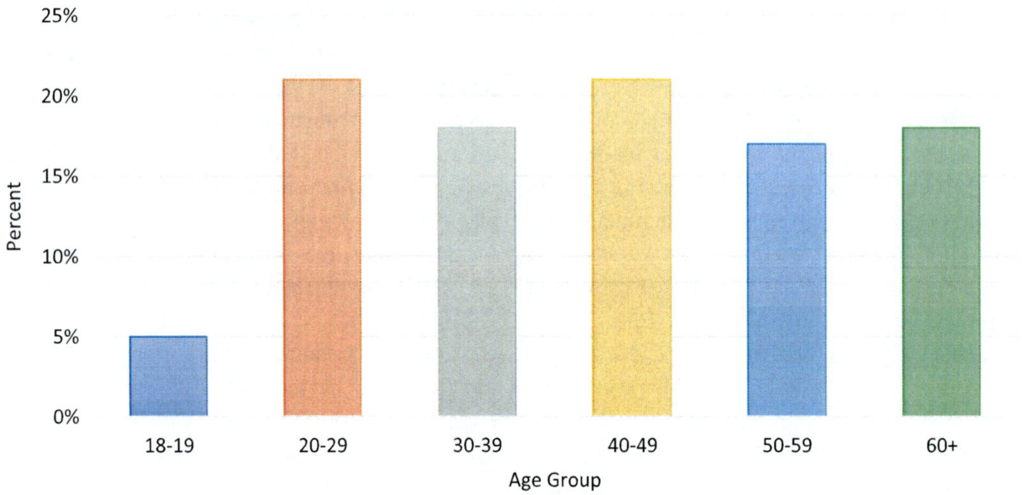

Fig. 3.8 Age groups of respondents (n = 194). (Source: Fieldwork 2019)

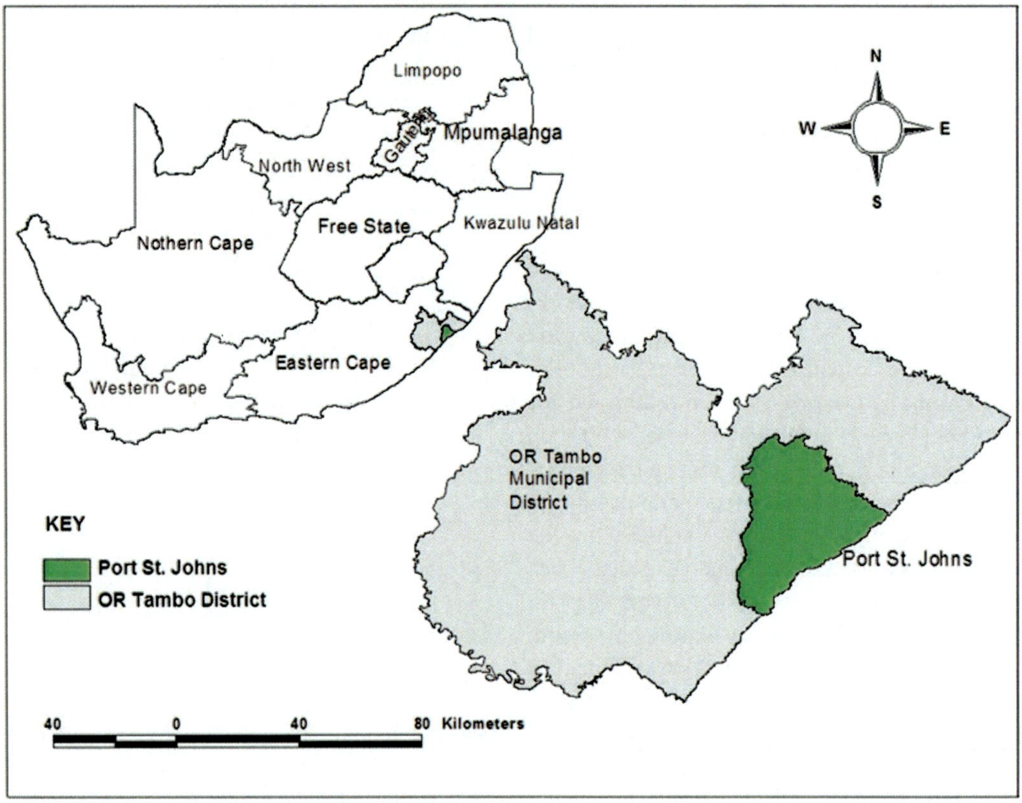

Fig. 3.9 Study location. (Source: Authors (2020))

As for data analysis, data from the household questionnaire survey were partially analysed, using QuestionPro, while other data sets were exported into MS Excel for plotting graphs. To enrich the analysis, relevant literature was infused, providing further clarity and solidifying the narratives. The next section presents and discusses the key findings with the hope to inform policy on ICT and flood risk reduction and management in South Africa.

3.4 Presentation of Key Findings and Discussions

This section is divided into four main sub-sections: (1) access to electricity, (2) prevalence of ICT devices for flood risk reduction and management; (3) barriers to effective use of ICT for flood risk reduction and management; and (4) communicating evacuation information during the floods. Each of the three sub-sections will now be discussed in detail in the following paragraphs.

3.4.1 Access to Electricity and Device Charging

The phenomenal growth in access to electricity in South Africa, as reflected in the Port St Johns local municipality, was confirmed during the fieldwork in 2019 with 95% of households (n = 195) indicating they were connected to the national grid. From the field observations in Port St Johns, it emerged that there were several homes with illegal electricity connections, especially in the Sinuka area. In the context of this study, illegal connections still allow residents to operate their ICT devices and help them communicate. Mujuzi (2020) confirms the widespread challenge of electricity theft and/or illegal connections in South Africa that have resulted in billions of US dollars having been lost in revenue to the national energy utility, Eskom. Apart from theft, residents in many previously disadvantaged communities, particularly those from Soweto in Gauteng, openly refuse to pay their bills. The author goes further to identify an additional gap in the law as electricity theft is not a statutory offence. However, prosecutors across the country have been advocating to have electricity thieves punished. Hence, there have been conflicting High Court decisions.

A question was slotted in the survey to determine where households charged their devices. Up to 97% of the respondents (n = 197) indicated that they charged their mobile and other devices, using electricity from the grid. This piece of information is necessary because there are times when floods and other disasters lead to prolonged blackouts (Pant et al., 2018). In fact, 68% of household respondents (n = 188) in Port St Johns indicated that they experienced blackouts following the floods. However, there was no significant damage to ICT infrastructure worth reporting in this chapter.

3.4.2 Prevalence of ICT Devices for Flood Risk Reduction and Management in Port St Johns

The household respondents were asked to indicate the nature of ICT devices they had in their possession. Among such were the following: fixed telephone lines, mobile phones, fax machines, personal computers, laptops, tablets and/or iPads, radios and/or TV. Up to 97% of the respondents (n = 197) revealed that they did not own a landline and the same applied to fax machines and personal computers. Ninety-five of the respondents further revealed that they did not have a tablet and/or iPad, yet 57% and 45% owned a television and radio, respectively. However, for the sake of interest 94% of the households surveyed indicated that they had a mobile phone. This percentage tallies well with the penetration rate that was very high for South Africa as a country presented earlier. The national penetration for 2018 from the ITU-D was 159.9 subscriptions per 100 inhabitants.

This finding remains central to future ICT-related flood risk reduction and management within Port St Johns and possibly in South Africa. As high as 68% of households had two or more mobile phones (Fig. 3.10). Mobile phone subscriptions are also in line with the earlier findings from Statistics South Africa (StatsSA) reported by the Eastern Cape Socio-Economic Consultative Council (ECSECC). The ECSECC (2017) pre-

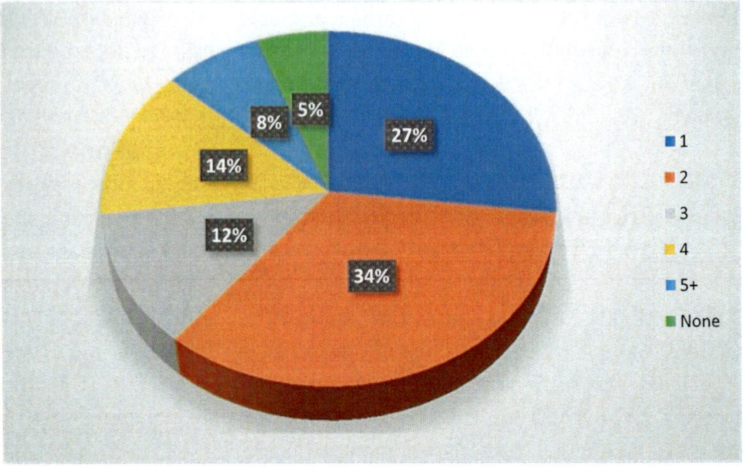

Fig. 3.10 Availability of mobile phones per household (n = 197). (Source: Fieldwork 2019)

sented a 90.8% access to cellular phones from Port St Johns local municipality in 2016, an increase from 76% in 2011. In addition, access to fixed landline phones declined to 1.1% in 2017 compared to 2.3% in 2011.

Additionally, the researchers asked the household respondents to indicate if they had smartphones (a phone that has WhatsApp, Facebook, Twitter, etc.). Up to 69% of the respondents confirmed that they had smartphones. Furthermore, although 64% of the respondents revealed that they only use one mobile network service provider, 23% used two different service providers, while three used three different service providers in their household. The common service providers in South Africa are MTN, Vodacom and Cell C.

3.4.3 Barriers to Effective Use of ICT for Flood Risk Reduction and Management

Several main barriers to the effective use of ICT for flood risk reduction and management in the identified Port St Johns communities were identified and included the lack of capacity to buy data, availability of a reliable network and low levels of literacy in the elderly. The bulk of the households (84%) earned less than R4400[2] per month (about US$270 as of November 2020). The rest of the statistics is reflected in Fig. 3.11. As for employment (n = 194), more than half of the respondents were unemployed, 15% were employed either full time or part time, 14% indicated that they were self-employed, while 2% were unable to work and 1% retired. Given this scenario at grassroots level, consultations revealed that many survived on government social grants. The challenge of connection costs in predominantly rural areas of South Africa is not new.

A project focusing on affordable connectivity for voice and data in the Eastern community of Mankosi between Port St Johns and Coffee Bay by a University of Western Cape researcher revealed similar findings (McLeod, 2015). With an average personal income of R388 per month, data would be out of reach to many. However, in line with the need to harness ICT for flood risk reduction and management, some positives were noted. Up to 87% of the population in the area had access to mobile services at least once a week with individual community members spending 17 min on average on calls. Only 8% were found to connect to the internet. Overall, an estimated 22% of the income was spent on communication, particularly calls (Ibid.).

With the household survey, it was observed that some households did not have mobile data for accessing WhatsApp messages or the internet. Most of these issues pointed to the inability

[2]The exchange rate stood at US$1 to R16.40 as of 20 October 2020.

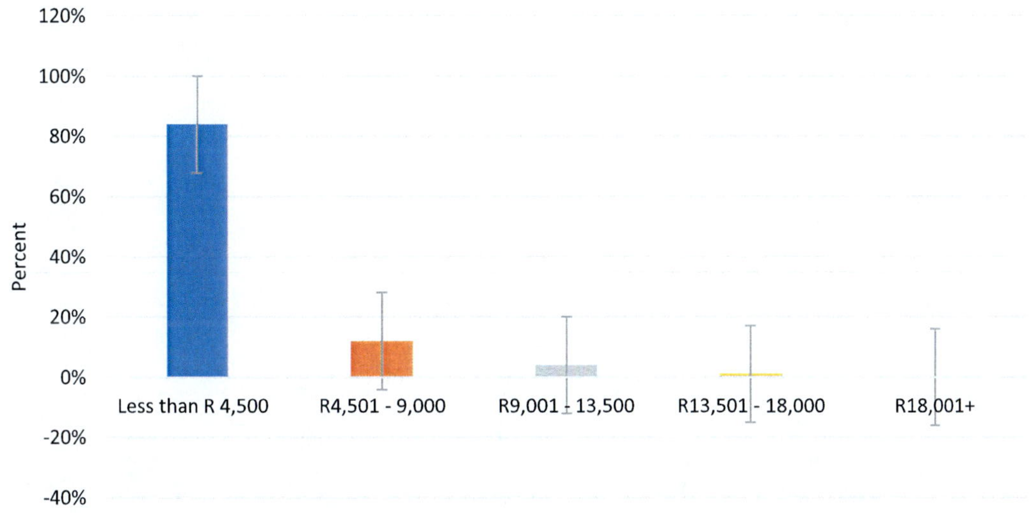

Fig. 3.11 Estimated household income (*n* = 196). (Source: Fieldwork 2019)

of these households to afford data to be connected. Some respondents mentioned that their children sometimes sent them money. However, this was not enough to buy airtime and/or data all the time. This reduced mobile phones to message receiving devices. There was also the issue of literacy. Most elderly folks who responded to the survey indicated that they could not read the messages, as they were illiterate. As such their children, grandchildren and/or relatives of school-going ages read the messages for them. Hence, this remains a barrier since during disasters such as floods, messages need to be relayed directly to beneficiaries.

The researchers asked households to reflect on the everyday availability of signals across several ICT platforms. The findings for those critical to flood risk reduction and management in Port St Johns are presented in Fig. 3.12. By far, the platform regularly available is the mobile phone, followed by the TV and radio; connectivity remains a challenge. For example, in one rural settlement where the household questionnaire survey was administered, there was a huge problem with the cellular network coverage, which meant that communication, using different ICTs, were problematic. Generally, network coverage in Port St Johns local municipality was bad. When the research team arrived, they tried calling the councillors for appointments and it was a huge struggle to get in touch with most of them. Making calls was difficult, even to team members, especially during the day.

The challenge of connectivity in Port St Johns was typical of predominantly rural and mountainous areas in the Eastern Cape and other locations in the country. To this end and in 2018, Vodacom, one of the main service providers, had to channel US$21 million towards ramping up network coverage and improving connectivity in deep rural Eastern Cape (African Leadership Magazine, 2018). New 3G sites were to be deployed, resulting in internet connectivity and access to many other linked services such as Twitter, Facebook, WhatsApp and so forth. Port St Johns was one of the beneficiaries to the roll-out. The investment was part of Vodacom's Rural Coverage Network Expansion Programme. The advancement of a digital Eastern Cape Province and South African economy was further emphasised by Vodacom in a press release on 3 October 2019 that updated progress on the Eastern Cape investments (Vodacom, 2019). The 4G mobile, fibre-optic broadband and the Internet of Things (IoT) were all part of enhancing the digital economy. A total of 268 microwave and 88 Fibre-optic links were to be upgraded to enable 4G data capabilities and deployments to base stations.

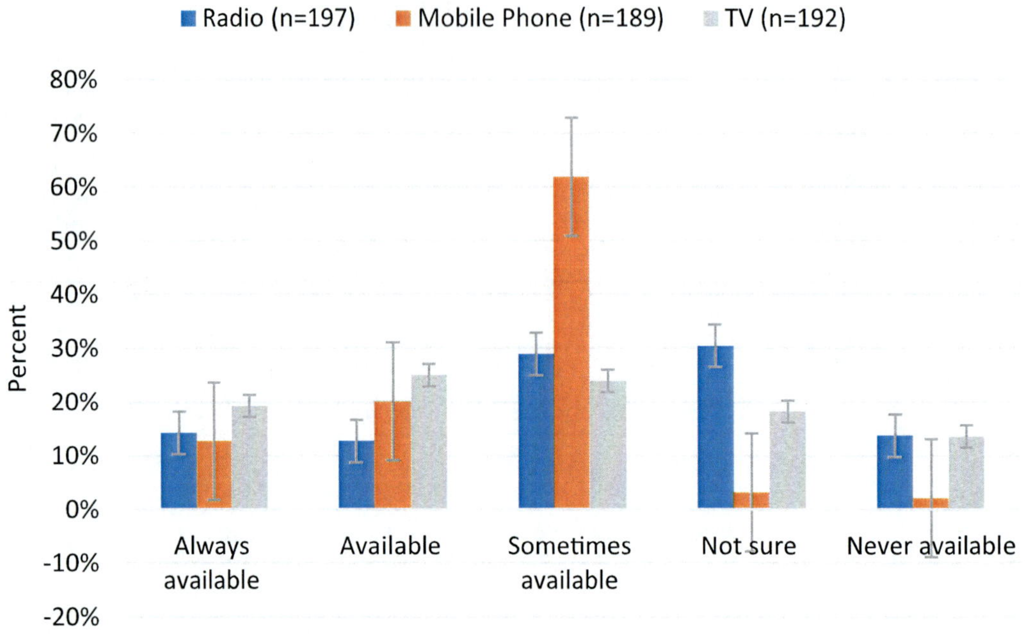

Fig. 3.12 Everyday network availability (signals). (Source: Fieldwork 2019)

Lastly, 30 new sites were to be deployed in 12 rural areas including Port St Johns (Vodacom, 2019).

Although not popular in Port St Johns, the use of social media and other ICTs in DRR and crisis management has been growing (Alexander, 2014; Smith et al., 2017), mainly due to the remarkable rise of ICT (Kaur & Sood, 2019; Kaur & Sood, 2020). Alexander established several ways in which social media is used, including monitoring the disasters, extending emergency relief, crowdsourcing and promoting research. Other areas where ICT is helpful during the flood and other disaster management include recording, exchanging and processing information (Sakurai & Murayama, 2019). However, red flags are also raised (Alexander, 2014) since the use of social media is at times associated with potential negative aspects, like disseminating rumours and undermining the authority. Drawing from the implementation of the European Flood Directive 2007/60/EC demanding public participation that ensures citizens' involvement in the flood management cycle, When et al. (2015) view ICT as an enabler for such, which can be applied across all the stages of the disaster cycle from preparedness (early warning) and search and rescue to response and recovery (Haritharaj et al., 2020). Parajuli and Baral (2018) adds remote sensing technologies to the equation, also advocating to apply across the disaster cycle stages identified herein. The Short Message Service (SMS) is identified as particularly useful.

3.4.4 Communicating Evacuation Information During the Floods

The households (n = 197) were also requested to indicate their preferred language of use for early warning during disasters, such as the Port St Johns floods of April 2019. An estimated 83% indicated that they would prefer a local language to be used. Linked to communication, a question was included requesting responses on which sources of flood risk early warning respondents would trust? Ten options were included. The top five sources (as ranked) included TV announcements (25% of respondents), word of mouth from the relatives (17%), SMS (15%), word of mouth from the local head or chief (13%) and radio announcements (12%). The remaining share went to word of mouth from a local school; government extension worker or councillor; social media (Twitter, Facebook, WhatsApp, etc.); other; warning from Meteorology or other responsible government agency and word of mouth from a local church, NGO/CBO. It is interesting to note that the government agency is almost the least trusted platform.

The authors also included a question that asked respondents to indicate which platform delivered the flood news first to the household and/or individuals who responded. Twenty-one such potential platforms were included to which the respondents ($n = 197$) had to select only one. While six platforms including email, Twitter, local school, non-governmental organisation (NGO) and/or community-based organisation (CBO), business entity and the newspaper did not receive any votes. The top five platforms were: did not receive any message (28%), TV (19%), SMS (17%), radio (9%) and councillor (6%). Further details are shown in Fig. 3.13.

The use of online information in the household and individual citizens' early warning and advisable actions in DRR have been growing globally (Allaire, 2016). The author established that the use of online media helped households to minimise flood losses by an average of 37% during the 2011 Bangkok floods. This translated to an estimated US$3708 per household. Social media was also identified as the preferred and first choice for communicating the situations and damages (Meechang et al., 2020). Sharp and Carter (2020) also tease out how social media is being used to organise volunteers during floods in the United Kingdom. The findings were that the use of social media saved time and reached a broader audience than conventional media. Conventional media in the authors' perspectives include email, telephone, print media and so forth. Social media was particularly effective in the recruitment and management processes of the volunteers. Social media further assisted in mobilising relief and recovery equipment.

As the ICT platforms identified in this chapter are for use during flood risk reduction and management, it was inevitable that the authors seek clarity concerning the states at which victims would be convinced to evacuate and be evacuated. The responses are shown in Fig. 3.14. It emerges that up to 53% of respondents were of the view that they would only evacuate when they see clear signs of being in danger of injury or possible loss of life. Secondly, the respondents were also likely to evacuate when there was an immediate word from their ward councillor (22%). However, there was also a significant cluster of respondents who indicated they would not go anywhere (17%). This information remains necessary for ICT readiness in flood risk reduction and management by officials in Port St Johns and for other government officials in the country. To this end, ICT platforms should be utilised for raising awareness of the dangers of not evacuating during floods and other disasters.

The ICT platforms also remain useful in the documentation and Big Data generation, storage and retrieval, manipulation and dissemination of information perfected by the efficiencies of cloud computing and industry 4.0. For example, databases could be generated in terms of the households affected and the extent of the damage. Hadoop systems could be set up to keep complete videos of the flood occurrences for future settlement planners to estimate or predict possible flood patterns (Mazumder & Dhar, 2018). ICT platforms are also handy in relief and emergency aid distribution. The government and other key stakeholders need information that can be retrieved timely including identifications, gender, age groups, locations and so forth. In the instance of Port St Johns data would have been needed concerning where the internally displaced victims were accommodated. From the household survey, it emerged that the majority were in public halls (38%), followed by houses (13%), churches (8%) and other establishments. Data concerning the length of stay in these accommodation establishments and when the last victims left are all critical for future planning. Lastly, the fact that a greater majority (77%) of the respondents ($n = 194$) did not receive any communication on the flood, in addition to a message they indicated was no clear (11%), is worrying. Ultimately, only 11% indicated the message communicated was clear.

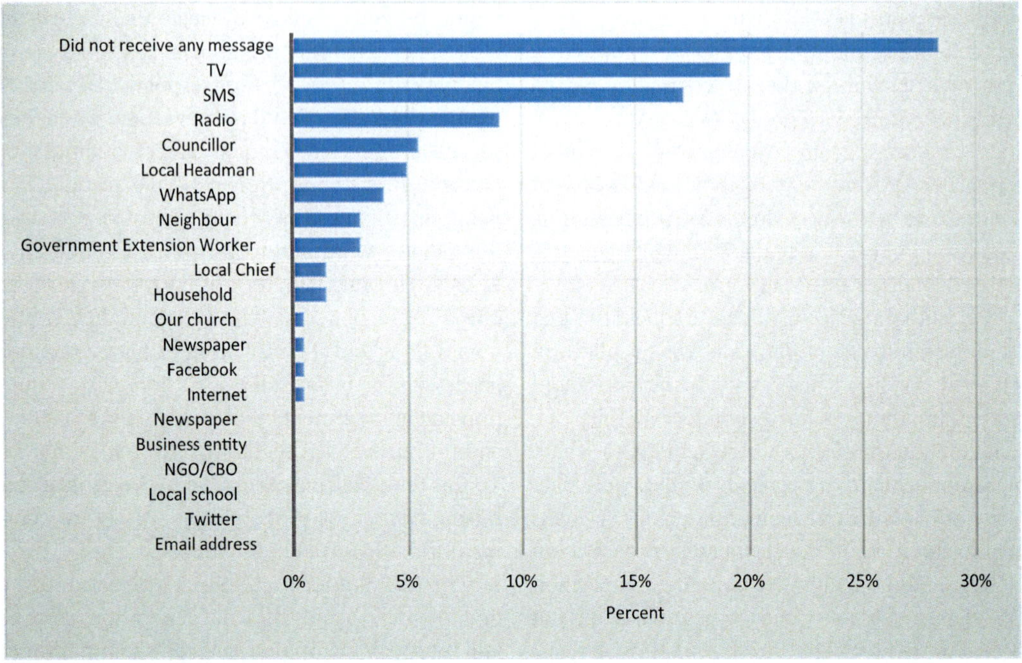

Fig. 3.13 Which platform did you first receive the messages on the flood? ($n = 197$). (Source: Fieldwork 2019)

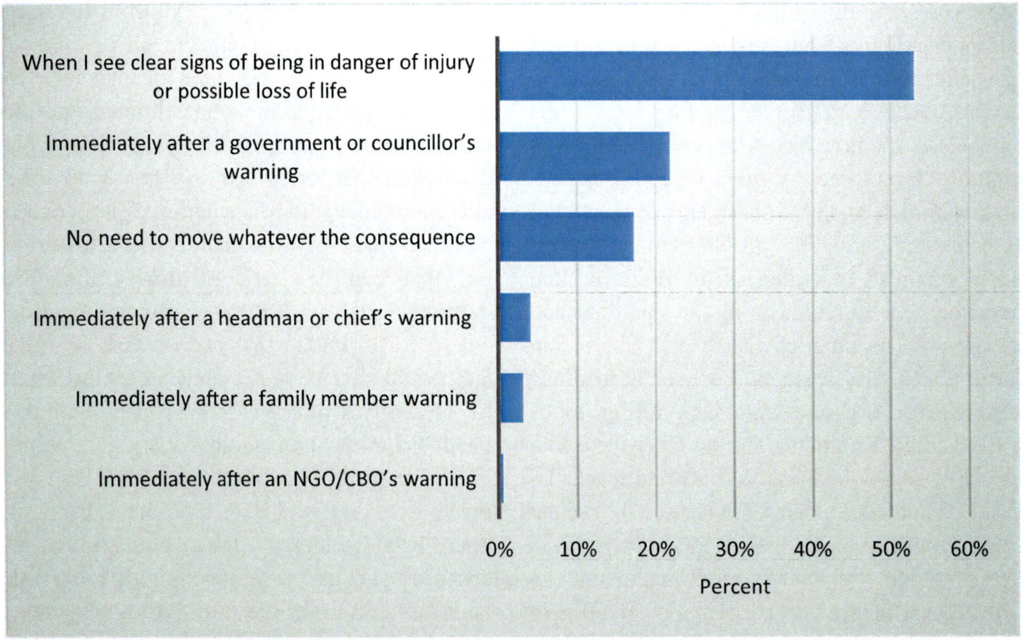

Fig. 3.14 When will you be convinced to evacuate from the flood after a warning? ($n = 196$). (Source: Fieldwork 2019)

3.5 Conclusion

ICT and DRR matters, specifically flood risk reduction and management remain topical. The reason is flooding has been on the increase not only in South Africa, but worldwide. Given the rapid rise and uptake of ICT under the Fourth Industrial revolution, more and more opportunities are emerging to scale up ICT in flood risk reduction and management. The chapter investigated the extent of ICT readiness for DRR and management, specifically floods in Port St Johns local municipality of the Eastern Cape Province, South Africa. The main findings were that the mobile phone was the most prevalent ICT device followed by the TV and radio. The fixed landline telephone looks set to be obsolete as revealed by the BRICS trends from 2000 to 2018 and the Port St Johns results. However, several main barriers to effective use of ICT for flood risk reduction and management included the lack of capacity to buy data, the availability of a reliable network and low levels of literacy in the elderly. Since South Africa and the world are moving swiftly in scaling up the use of the mobile phones, harnessing such for flood risk reduction and management in Port St Johns demands that network coverage be improved, as well as addressing the capacity to buy data and enhance literacy levels. Community training tailor-made for flood and other disaster risk reductions and management should also be introduced.

References

ADPC. (2011). *ICT for disaster risk management*. Republic of Korea: The Academy of ICT Essentials for Government Leaders Module Series, United Nations Asian and Pacific Training Centre for Information and Communication Technology for Development (UN-APCICT/ESCAP).

AICRPAM. (2018). *Annual Report 2017-2018* (67p). Thrissur: Kerala Agricultural University.

African Leadership Magazine. (2018). *Vodacom channels $21m to rural internet connectivity*. Retrieved October 22, 2020, from https://www.africanleadershipmagazine.co.uk/vodacom-channels-21m-to-rural-internet-connectivity/

Allaire, M. C. (2016). Disaster loss and social media: Can online information increase flood resilience? *Water Resources Research, 52*, 7408–7423. https://doi.org/10.1002/2016WR019243.

Alexander, D. E. (2014). Social media in disaster risk reduction and crisis management. *Science Engineering and Ethics, 20*, 717–733. https://doi.org/10.1007/s11948-013-9502-z.

Archer, D. R., & Fowler, H. J. (2018). Characterising flash flood response to intense rainfall and impacts using historical information and gauged data in Britain. *Journal of Flood Risk Management, 2018*(11), S121–S133.

Below, R., Wirtz, A., & Guha-Sapor, D. (2009). *Disaster category classification and peril terminology for operational purposes*. Brussels, Belgium: Centre for Research on the Epidemiology of Disasters.

Bradford, R. A., O'Sullivan, J. J., van der Crats, I. M., Krywkow, J., Rotko, P., Aaltonen, J., Bonaiuto, M., De Dominics, S., Waylen, K. (2012). Risk perception: Issues for flood management in Europe. Natural Hazards Earth System Sciences, 12(7): 2299–2309.

Business Live. (2019). *Cyril Ramaphosa flies to flood-ravaged KZN as over 50 reported dead*. Retrieved October 22, 2020, from https://www.businesslive.co.za/bd/national/2019-04-24-cyril-ramaphosa-flies-to-flood-ravaged-kzn-as-over-50-reported-dead/

Chapungu, L. (2020, May). *Mitigating the impact of cyclone disasters: Lessons from Cyclone Idai*. South African Institute of International Affairs, Policy briefing, Climate Change and Migration.

Davies, R. (2020). *Zimbabwe—Deadly Storms and Floods in Manicaland, Floodlist*. Retrieved November 12, 2020, from http://floodlist.com/africa/zimbabwe-floods-manicaland-november-2020

Dawson, R. J., Dickson, M. E., Nicholls, R. J., Hall, J. W., Walkden, M. J., Stansby, P. K., Mokrech, M., Richards, J., Zhou, J., Milligan, J., et al. (2009). Integrated analysis of risks of coastal flooding and cliff erosion under scenarios of long term change. *Climate Change, 95*, 249–288.

ECSECC (Eastern Cape Socio Economic Consultative Council). (2017). *Port St Johns local municipality socio-economic review and outlook 2017*. East London: ECSECC.

Falconer, R. H., Cobby, D., Smyth, P., Astle, G., Dent, J., & Golding, B. (2009). Pluvial flooding: New approaches in flood warning, mapping and risk management. J. *Flood Risk Management, 2*, 198–208.

GFDRR (Global Facility for Disaster Reduction and Recovery). (2015). *Bringing resilience to scale* (GFDRR Annual Report'14). Washington, DC: GFDRR. Retrieved October 28, 2020, from https://www.gfdrr.org/sites/gfdrr/files/publication/GFDRR%20ANNUAL%20REPORT%202014.pdf

GSMA. (2020). *The mobile economy Asia Pacific 2020*. London, UK: GSM Association.

Hallegatte, S. (2011). *How economic growth and rational decisions can make disaster losses grow faster than wealth. Policy Research Working Paper*. Washington, DC: World Bank.

Haritharaj, S., Visakh, N. U., & Antony, A. C. (2020). Early warning systems: An alert to Flood. *Vigyan Varta, 1*(1), 30–32.

Hassan, A. A. M. (2015). *Role of ICT in natural disaster management of Bangladesh*. A Dissertation for the Degree of Master in Disaster Management, Postgraduate Programs in Disaster Management (PPDM) BRAC University, Dhaka, Bangladesh.

Heberger, M., Cooley, H., Herrera, P., Gleick, P. H., & Moore, E. (2011). Potential impacts of increased coastal flooding in California due to sea-level rise. *Climate Change, 2011*(109), 229–249.

Huang, C., Chan, E., & Hyder, A. A. (2010). Web 2.0 and Internet Social Networking: A New tool for Disaster Management?—Lessons from Taiwan. *BMC Medical Informatics and Decision Making, 10*(57). https://doi.org/10.1186/1472-6947-10-57.

Hurford, A. P., Priest, S. J., Parker, D. J., & Lumbroso, D. M. (2012). The effectiveness of extreme rainfall alerts in predicting surface water flooding in England and Wales. *International Journal of Climatology, 32*, 1768–1774.

IAEG-SDGs (United Nations Inter-agency and Expert Group on SDG Indicators). (2018). *Tier classification for global SDG indicators 15 October 2018*. New York: IAEG-SDGs.

IPCC (Intergovernmental Panel on Climate Change). (2012). Managing the risks of extreme events and disasters to advance climate change adaptation. In C. B. Field, V. Barros, T. F. Stocker, D. Qin, D. J. Dokken, K. L. Ebi, M. D. Mastrandrea, et al. (Eds.), *A special report of Working Groups I and II of the Intergovernmental Panel on Climate Change*. Cambridge and New York: Cambridge University Press.

ITU-D (The Telecommunication Development Sector). (2020). *Statistics*. Retrieved August 18, 2020, from https://www.itu.int/en/ITU-D/Statistics/Pages/stat/default.aspx

Kaur, A., & Sood, S. K. (2019). Analytical mapping of research on disaster management, types and role of ICT during 2011–2018. *Environmental Hazards, 18*(3), 266–285. https://doi.org/10.1080/17477891.2019.1567457.

Kaur, A., & Sood, S. K. (2020). Hydro-meteorological hazards and role of ICT during 2010-2019: A scientometric analysis. *Earth Science Informatics, 13*, 1201–1223. https://doi.org/10.1007/s12145-020-00495-0.

Kaźmierczak, A., & Cavan, G. (2011). Surface water flooding risk to urban communities: Analysis of vulnerability, hazard and exposure. *Landscape and Urban Planning, 2011*(103), 185–197.

Koswatte, S., McDougall, K., & Liu, X. (2018). VGI and crowdsourced data credibility analysis using spam email detection techniques. *International Journal of Digital Earth, 11*, 520–532. https://doi.org/10.1080/17538947.2017.1341558.

Le Coz, J., Patalano, A., Collins, D., Guillén, N. F., García, C. M., Smart, G. M., et al. (2016). Crowdsourced data for flood hydrology: Feedback from recent citizen science projects in Argentina, France and New Zealand. *Journal of Hydrology, 541*, 766–777. https://doi.org/10.1016/j.jhydrol.2016.07.036.

Liu, C., Guo, L., Ye, L., Zhang, S., Zhao, Y., & Song, T. (2018). A review of advances in China's flash flood early-warning system. *Natural Hazards, 92*, 619–634.

Louvain, C. (2019). Natural disasters 2018; Centre for Research on the Epidemiology of Disasters CRED: Brussels, Belgium; USAID: Geneva, Switzerland. p. 8.

Mazumder, S., & Dhar, S. (2018). Hadoop ecosystem as enterprise big data platform: Perspectives and practices. *International Journal of Information Technology and Management, 17*(4), 334–348. https://doi.org/10.1504/IJITM.2018.095060.

McCallum, I., Liu, W., See, L., Mechler, R., Keating, A., Hochrainer-Stigler, S., Mochizuki, J., Fritz, S., Dugar, S., Arestegui, M., Szoenyi, M., Bayas, J. C. L., Burek, P., French, A., & Moorthy, I. (2016). Technologies to support community flood disaster risk reduction. *International Journal of Disaster Risk Science, 7*, 198–204. https://doi.org/10.1007/s13753-016-0086-5.

McLeod, D. (2015). *The challenge of connecting SA's rural poor*. Retrieved October 22, 2020, from https://techcentral.co.za/the-challenge-of-connecting-sas-rural-poor/58306/

Meechang, K., Leelawat, N., Tang, J., Kodaka, A., & Chintanapakdee, C. (2020). The acceptance of using information technology for disaster risk management: A systematic review. *Engineering Journal, 24*(4). https://doi.org/10.4186/ej.2020.24.4.111.

Mileti, D. (1999). *Disasters by design: A reassessment of natural hazards in the United States*. Washington, DC: Joseph Henry Press.

Ministry of Cooperative Governance and Traditional Affairs. (2019). *Minister Zweli Mkhize visits areas affected by floods in Eastern Cape*. Retrieved October 22, 2020, from https://www.gov.za/speeches/minister-zweli-mkhize-see-first-hand-damages-caused-recent-rains-eastern-cape-27-apr-2019

Mujuzi, J. D. (2020). Electricity theft in South Africa: Examining the need to clarify the offence and pursue private prosecution? *Obiter, 41*(1), 78–87.

Muktar, B. G., Man, N., Saleh, J. M., & Daneji, M. I. (2018). Evaluation of ICTs access, use and preferences for livelihood resilience: Results from a survey of Malaysian fisher folks. *The Journal of Agricultural Education and Extension, 24*(4), 377–388. https://doi.org/10.1080/1389224X.2018.1479279.

Naik, N. (2016). *Flooded streets—A crowdsourced sensing system for disaster response: A case study*. In: Proceedings of the ISSE 2016—2016 International Symposium on Systems Engineering. Edinburgh: IEEE. https://doi.org/10.1109/SysEng.2016.7753186.

Neelson, A. (2017). Flood disasters more than double across Europe in 35 years. *The Guardian*. Retrieved from https://www.theguardian.com/environment/2017/jan/19/flood-disasters-morethan-double-across-europe-in-35-years

Nhamo, G., Nhemachena, C., & Nhamo, S. (2020). Using ICT indicators to measure readiness of countries to implement Industry 4.0 and the SDGs. *Environmental*

Economics and Policy Studies, 22, 315–337. https://doi.org/10.1007/s10018-019-00259-1.

Okolloh, O. (2009). Ushahidi, or 'testimony': Web 2.0 tools for crowdsourcing crisis information. In H. Ashley (Ed.), *Change at hand: Web 2.0 for development Participatory learning and action* (pp. 65–70). London: International Institute for Environment and Development.

Onafeso, O. D., & Samuel, K. J. (2012, November). *The role of ICT in the establishment of flood early warning system in Nigeria.* Paper presented at the Climate Change Conference, Centre for Climate Change and Environmental Research, Osun State University, Okuku Campus, Okuku, Nigeria.

Panek, J. Marek, L. Paszto, V. and Valuch, J. (2017). The Crisis Map of the Czech Republic: The nationwide deployment of an Ushahidi application for disasetrs. *Disasters, 41*, 649–671. https://doi.org/10.1111.disa.12221.

Pant, R., Thacker, S., Hall, J. W., Alderson, D., & Barr, S. (2018). Critical infrastructure impact assessment due to flood exposure. *Journal of Flood Risk Management, 11*, 22–33. https://doi.org/10.1111/jfr3.12288.

Parajuli, J., & Baral, S. (2018). Remote sensing schemes mingled with information and communication technologies (ICTs) for flood disaster management. *Journal on Pattern Recognition, 4*(4), 1–12.

Port St Johns Disaster Risk Management Satellite Centre. (2019). *Floods that affected Port St Johns local municipality.* Port St Johns: O.R. Tambo District Municipality.

Sakurai, M., & Murayama, Y. (2019). Information technologies and disaster management: Benefits and issues. *Progress in Disaster Science, 2*, 100012. https://doi.org/10.1016/j.pdisas.2019.100012.

See, L. (2019). A review of citizen science and crowdsourcing in applications of pluvial flooding. *Frontiers in Earth Science, 7*, 44. https://doi.org/10.3389/feart.2019.00044.

Sharp, E. N., & Carter, H. (2020). Examination of how social media can inform the management of volunteers during a flood disaster. *Journal of Flood Risk Management, 16*(4), e12665. https://doi.org/10.1111/jfr3.12665.

Siqathule, S. (2019). *Declare disaster zone, pleads flood victim in Port St Johns informal settlement.* Retrieved October 22, 2020, from https://www.groundup.org.za/article/declare-disaster-zone-pleads-flood-victim-port-st-johns-informal-settlement/

Smith, L., Liang, Q., James, P., & Lin, W. (2017). Assessing the utility of social media as a data source for flood risk management using a real-time modelling framework. *Journal of Flood Risk Management, 10*, 370–380. https://doi.org/10.1111/jfr3.12154.

Stephenson, R., & Anderson, P. S. (1997). Disasters and the information technology revolution. *Disasters, 21*(4), 305–334.

Subin, P., & Sujatha, S. (2020). Witnessing a Disaster: Public Use of Digital Technologies in the 2015 South Indian Floods. *Digital Journalism, 8*(1), 15–31. https://doi.org/10.1080/21670811.2019.1636693.

United Nations. (2015). *Transforming our world: The 2030 agenda for sustainable development.* New York: United Nations Secretariat.

United Nations Environment Programme Division of Early Warning and Assessment (UNEP-DEWA). (2006). *One planet many people: Atlas of our changing environment.* Kenya: Nairobi.

UNDRR (United Nations Office for Disaster Risk Reduction). (2015). *Sendai framework on disaster risk reduction (2015-2030).* New York: UNDRR Secretariat.

UNISDR. (2011). Global Assessment Report on Disaster Risk Reduction 2011. Retrieved from http://www.inisdr.org (Accessed 26 November 2020).

UNISDR (United Nations International Strategy for Disaster Reduction). (2015). *Making development sustainable: The future of disaster risk management. Global assessment report on disaster risk reduction.* Geneva: UNISDR.

Victorino, J. N. C., Estuar, M. R. J. E., & Lagmay, A. M. F. A. (2016). Validating the voice of the crowd during disasters. In K. Xu, D. Reitter, D. Lee, & N. Osgood (Eds.), *Lecture Notes in Computer Science (including subseries Lecture Notes in Artificial Intelligence and Lecture Notes in Bioinformatics) 9708 LNCS* (pp. 301–310). Cham: Springer. https://doi.org/10.1007/978-3-319-39931-7_29.

Vodacom. (2019). *Vodacom Eastern Cape region delivers a substantial return to society and accelerates network infrastructure in deep rural areas.* Retrieved October 22, 2020, from https://www.vodacom.com/news-article.php?articleID=7425

Wattegama, C. (2007). *ICT for disaster management.* Asian and Pacific Training Centre for Information and Communication Technology for Development, pp. 22–26. Retrieved December 06, 2014, from http://www.unapcict.org

WDI (World Development Indicators). (2020). World Development Indicators. Retrieved from https://datatopics.worldbank.or/world-development-indicators/ (Accessed 28 August 2020).

When, U., Rusca, M., Evers, J., & Lanfranchi, V. (2015). Participation in flood risk management and the potential of citizen observatories: A governance analysis. *Environmental Science & Policy, 48*, 225–236. https://doi.org/10.1016/j.envsci.2014.12.017.

Wolf, J. (2009). Coastal flooding: Impacts of coupled wave–surge–tide models. *Natural Hazards, 2009*(49), 241–260.

Woodruff, J. D., Irish, J. L., & Camargo, S. J. (2013). Coastal flooding by tropical cyclones and sea-level rise. *Nature, 2013*(504), 44.

Yang T.H and Liu, W. C. (2020) A General Overview of the Risk-Reduction Strategies for Floods and Droughts, Sustainability 12, 2687; https://doi.org/10.3390/su12072687

Yap, N. T. (2011). *Disaster management, developing country communities and climate change: The role of ICTs.* Institute for Development Policy and Management, Canada, pp. 6–7. Retrieved December 06, 2014, from http://www.niccd.org

Check for updates

4 A Reflection of Hydrological and Other Emerging Perspectives of Lake Kariba's Operation Noah

Godwell Nhamo, Kaitano Dube, and David Chikodzi

Abstract

There has been considerable academic documentation on the manner in which the world mobilised resources to rescue animals from the floodwaters after the closure of Kariba Dam, shared between Zambia (then Northern Rhodesia) and Zimbabwe (then Southern Rhodesia) in 1958. Similarly, there are several publications that critique the manner in which both colonial governments either side of the Zambezi River prioritised animals over the 50,000 plus Tonga speaking people that occupied the Zambezi valley prior to their displacement to make way for the dam. In fact, to this day, there are voices still calling for justice from forced displacements and the subsequent resettlement into less fertile and productive areas of the Zambezi valley. This paper's value-add is to investigate the hydrological perspectives of Operation Noah through historical climate data, thus taking another look at why the floodwaters rose so rapidly within a short period catching many by surprise. Drawn from the Biblical Noah's flood times, Operation Noah was a global rescue plan to save the animals marooned in over 500 islands that formed as the Kariba Dam waters rose. Results show that the unusual hydrological events on the Zambezi River during the operation influenced the abrupt upsurge of water levels in the lake, affecting the rescue plans and increasing the costs. The filling of the dam coincided with the all-time 110 years water flow regime of the Zambezi River and its tributaries. The study concludes that the floodwaters rose so rapidly that it took everybody by surprise and unprepared resulting in the loss of life. We recommend that hydrological and other perspectives be taken into consideration during planning of reservoir construction.

Keywords

Kariba Dam hydrology · Operation Noah · Flood disaster · Animals

4.1 Introduction

There has been considerable academic documentation on the manner in which the world mobilised resources to rescue animals from the floodwaters after the closure of the Kariba Dam shared between Zambia (then Northern Rhodesia)

G. Nhamo (✉) · D. Chikodzi
Institute for Corporate Citizenship, University of South Africa, Pretoria, South Africa
e-mail: nhamog@unisa.ac.za; chikod@unisa.ac.za

K. Dube
Vaal University of Technology, Vanderbijlpark, South Africa
e-mail: kaitanod@vut.ac.za

G. Nhamo, L. Chapungu (eds.), *The Increasing Risk of Floods and Tornadoes in Southern Africa*, Sustainable Development Goals Series, https://doi.org/10.1007/978-3-030-74192-1_4

and Zimbabwe (then Southern Rhodesia) in 1958. The Kariba Dam, a global engineering masterpiece of the twentieth century (Darbourn, 2015), had its two ports through which the whole volume of the mighty Zambezi River used to flow closed on 3rd December 1958 (Smithers, 1959). The dam is shared between Zambia and Zimbabwe and jointly owned and managed through the Zambezi River Authority (ZRA). Although Smithers acknowledged that the closure of the two ports heralded a major step in the finalisation of the project, he also raised a red flag regarding its potential greatest harm to the environment, especially the animals that were being marooned by the floods (Smithers, 1959). As it became apparent that the rising floodwaters were to create about 500 islands, both the Zambian and Zimbabwean authorities had to do something about it following international pressure, especially from the United Kingdom (Critchley, 1959). Hughes (2006: 823) brings a critical eye to the whole project and concludes that "With Kariba, whites imported their hydrological heritage, and they found the lake to be beautiful". The author follows up the 2006 paper with a book publication on the same subject (Hughes, 2010).

Given the historical perspectives of the Kariba Dam, the challenge regarding how the flooding took place and how the wild animals rescue adventure code-named Operation Noah occurred (Nyandoro, 2014; Alibekov et al., 2019), the authors had to first search for eyewitness accounts published in the late 1950 and early 1960. Such eyewitness accounts resulted in a good grasp of what transpired. In these eyewitness accounts are included three academic papers published by the journal *Oryx*'s 1959 and 1960 volumes. The published, detailed accounts included that of Reay H.N. Smithers, then Director of the National Museums of Southern Rhodesia; Lieutenant Colonel R.A. Critchley, then President of the Game Preservation and Hunting Association of Northern Rhodesia; and an account by A.M. Habthooen and J.A. Lock. Therefore, a section is dedicated in the literature review to their accounts.

Since Lake Kariba was the first and the largest manmade tropical dam in the world, experiences during and after its construction are important to document and analyse so that good practices can be adopted in similar projects, and weaknesses noted to minimise their repeat. It is anticipated that, given the current climate change debate, many countries are going to invest in large dams to deal with the growing global water insecurity. Of equal concern is the continued evacuations that are also being experienced in some islands in Kariba due to climate change and variability that results in huge hydrologic fluctuations of water levels in Lake Kariba (Dube & Nhamo, 2020). The literature on Operation Noah mainly documents the animals that were saved, methods of capture used and the budgetary constraints during this period. The hydrological insights of the operation have rarely been analysed—yet, they are also at the core of an in-depth understanding of the problems faced during this period. This paper therefore investigates and documents the hydrological perspectives of Operation Noah, thus taking another look at why the floodwaters rose so rapidly within a short period catching many by surprise following the filling of Lake Kariba from 1958 to 1964.

4.2 Literature Review

This section comes in three sub-sections that focus on the Kariba Dam designs, understanding Kariba Dam's Operation Noah (henceforth Operation Noah), and social justice issues surrounding Operation Noah. However, before getting into the matters highlighted, it is prudent that the link between the Biblical Noah and the flood and Operation Noah and the flood be established. From Genesis Chapter 6 verse 9 to Genesis Chapter 9 verse 17 the Bible records that the earth was corrupt and full of violence in God's eyes (The Holy Bible, 2011). God indicated to Noah that He would destroy all the people. Noah was instructed to make an ark with all the specifications provided. God further highlighted that He would bring a great flood destroying all life under the heavens. Noah was to enter the ark and all his family, including bringing along two of all living creatures, male and female. The rain would fall for 40 days and 40 nights. This happened, with Noah, his family and all the animals in the ark in their two-by-twos being saved. To this day, the world talks of this Biblical great flood as the Noah's flood. Hence, the rescuers of the animals that were trapped in many islands when the Kariba Dam waters were rising code-named the mission Operation Noah.

4.2.1 Kariba Dam Designs and Construction

The Kariba Dam measures about 280–300 km long, has a capacity of about 180 km^3 of water and has a total surface area measuring 5500 km^2 at full supply (World Commission on Dams (WCD), 2000; Habthooen et al., 1960). The dam was constructed with the primary purpose of hydropower generation from both the North and South banks, with a total of 10,000 workers having been part of this huge project (Hughes, 2006). At the design stage, the combined generation capacity of both Kariba North and South power stations was 1320 MW (WCD, 2000). Currently Kariba North has an installed capacity of 1080 MW and Kariba South has an installed capacity of 1050 MW (ZRA 1996) which seems to be the maximum capacity for the two power stations. The construction project stretched for about 5 years from 1955 to 1959. The finer details regarding statistics is presented in Table 4.1.

Table 4.1 Statistics surrounding the Kariba Dam

Parameter	Details
Type of dam	Double curvature concrete arch dam
Height	617 m
Spillway gates	6 gates, 8.8 m wide and 9 m high
Discharge capacity of spillway	9500 m^3/s
Length of reservoir	280 km
Minimum retention level	488.50 m
Minimum operating level	475.50 m
Total storage	180.6 km^3
Islands at full capacity	293
Mean depth	29.18 m
Maximum depth	97 m
Water retention period	3.3 years
Live storage	64.8 km^3
Maximum surface area	5577 m^2
Basin catchment area (Angola, Botswana, Namibia, Zambia and Zimbabwe)	687,059 km^2
Depth of stilling pool	78 m
Volume of stilling pool	410 × 106 m^3
Kariba South Bank Power Cavern	1050 Mw max capacity
Kariba North Bank Power Cavern	4 × 153.5 MW = 615 MW max capacity
Total generation capacity	1320 MW

Source: Modified after WCD (2000: vi) and Magadza (2005: 206)

Built in two phases that included the dam wall and Kariba South Power cavern (Phase I) and the Kariba North Power cavern, the final total cost for Phase I stood at $1.32 billon[1] (WCD, 2000), 97% of the projected costs. The World Bank financed the project, which was completed in 1960 and included additional costs to raise the dam wall by 6 m, and the need to add 100 MW generation capacity. After several delays, Phase II was commissioned in 1776 with the final cost standing at $480 million. During the feasibility and planning processes, flows at Kariba were projected based on those of Big Tree hydrologic gauging station at Victoria Falls (WCD, 2000).

Several unexpected challenges emerged during and after the construction of the Kariba Dam. For example, there were geological issues and earthquake activities. Between 1963 and 1983, several earthquakes occurred, of which 20 were in excess of magnitude 5 on the Richter scale (WCD, 2000); the highest concentration of these were at the dam wall (Magadza, 2005). This is worrisome for dam safety. From Magadza's account, the Zambezi valley is prone to seismic activity, given that it is part of the southern end of the East Africa Rift Valley. Six of these earthquakes happened in 1963 as the lake filled up. Another problem is that there was no environmental impact assessment (EIA) done before dam construction and the 1000 km^2 of vegetation

[1] Converted to constant 1998 US dollars.

cleared for gillnet fishery meant rotting vegetation emitted sulphur dioxide that led to copper corrosions in the powerhouse and downstream (WCD, 2000).

4.2.2 Understanding Operation Noah

Critchley (1959) provides an eyewitness account of what was happening in the final stages of the construction of Kariba Dam prior to filling. From his account, he witnessed 100,000 acres of dry land with bushes being cleared at a cost of more than £1 million. This was to provide fishing grounds and harbour sites. Heavy earth moving machinery were used to clear bush at the rate of 50 acres an hour. The Tilapia were being stocked in the lake at about 10 tons annually from the Zambia Game and Fisheries Department ponds at Chilanga, near Lusaka. The stocking was projected to be completed in 1963 with an annual harvest of over 9000 tons expected once maturity was realised. As the dam wall was closed, immediate disaster struck with hundreds of tiger fish found dead floating. An investigation by a Rex Adams of the Federal Fisheries concluded that their stomachs were too full of large crickets (Smithers, 1959) driven out of their holes from the banks by the rising waters (Scudder, 2005).

On another front, wild animals marooned on disappearing islands were now starving and drowning. From the Zimbabwean side, four game rangers led by Rupert Fothergill established the Herculean Task of Rescue, while Zambia thought there was no need for such an operation on their side (Critchley, 1959). In February 1959, the news of the pending disaster of drowning animals was picked up by the Southern Rhodesian media, leading to Mr. Aitken Cade, Leader of the Opposition in Zimbabwean Parliament and President of the Wild Life Protection Society giving notice of a private motion that noted that the government's interventions had been inadequate. The motion did not proceed, as the government increased the aid significantly. Once it became clear that no significant help from Zambia was likely, at the beginning of March, the Northern Rhodesian Game Preservation and Hunting Association enlisted for assistance from the Fauna Preservation Society in London. This had an immediate impact that witnessed a press conference being held in London on 19th March 1959, appealing for £10,000 to procure boats, engines, camping and medical equipment.

As indicated earlier, Operation Noah was a mission to rescue wild animals trapped on many islands of the Kariba Dam from the rising waters (Alibekov et al., 2019). As of 3rd December 1959, the inaugural rescue team was in place (Habthooen et al., 1960). Prior to the Kariba Dam filling up, two game reserves had been established in Zimbabwe, namely Matusadonha and Chizarira (Critchley, 1959). However, what emerged later was that it was not a joint operation between the two countries. The details of teams and equipment from the two countries are shown in Fig. 4.1. Zimbabwe (Rhodesia) had already released £17,500 for the upcoming season's rescue work. Eventually, as this was taking place, the Zambian government fulfilled its pledge to increase support and voted £30,000 to buy more boats and increased European and African staff (Ibid.). Scudder (2005) brings up a figure of £100,000 as total expenditure from the Zimbabwean government as of 1960. As the money started coming in from the UK, an all-metal parent ship worth £6200 was ordered from Salisbury (now Harare) and delivery was due in October 1959 in good time for next season's rescue operations (Critchley, 1959).

By September 1959, the floodwaters had risen to 449.58 m and the Kariba Dam's maximum set at 484.63 m (Critchley, 1959) and this was projected to be reached 5 years later in 1964. If the rains were heavy, this could be reached earlier and this indeed happened as such in 1963 (Scudder, 2005). Operation Noah was projected to last until 1962. By September 1959, 150 of the 500 islands had been submerged already (Critchley, 1959). As part of the preparation for the next rescue season, operation camps were being built at 80 km intervals on the Zimbabwean side. The rescue operations involved the use of animal scare tactics trying to drive them to either "safer" islands or to dry ground outside the lake

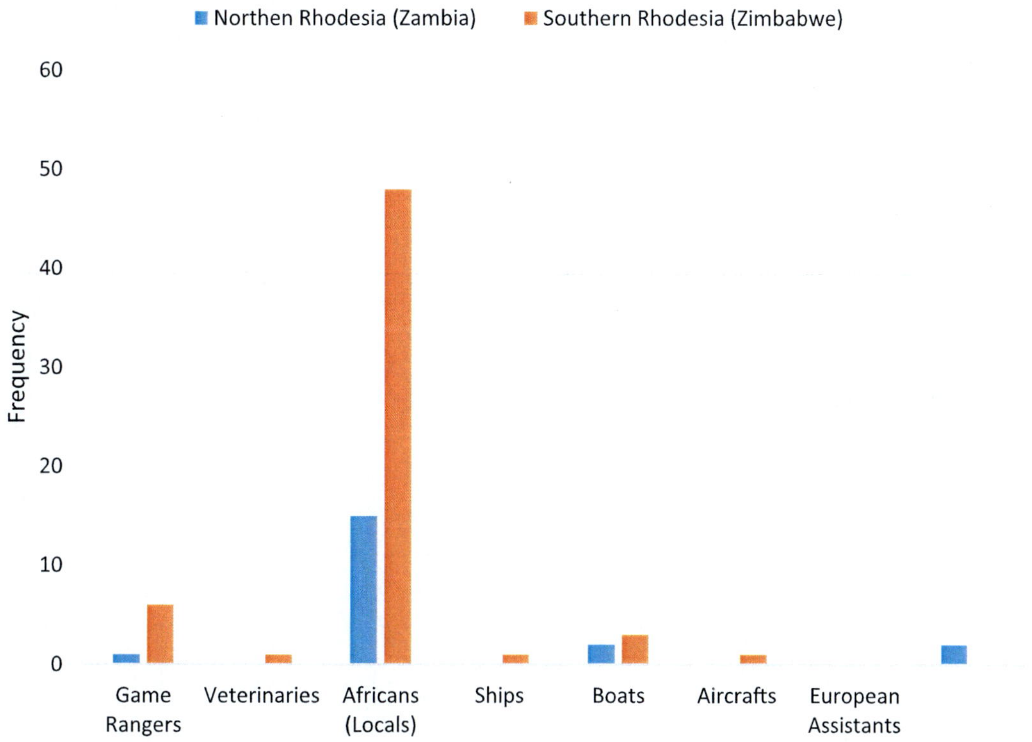

Fig. 4.1 Rescue teams composition and equipment. (Source: Authors, Data from Critchley (1959: 102))

boundaries. It also involved the use of large nets to trap the animals and bring them to safety in the new game reserves. The rescue missions were dangerous as the Zambezi valley was infested with crocodiles (McGregor, 2005) and an unprecedented concentration of snakes on the islands (Smithers, 1959). As of 1980, an estimated 40,000 wild crocodiles were still in existence in and around Lake Kariba (McGregor, 2005).

By 31 October 1959, a total of £9500 from the UK had been forwarded to Colonel Critchley's rescue fund (Critchley, 1959). By February 1959, Operation Noah had rescued more than 300 animals (Smithers, 1959), the largest and common species being impalas. The details regarding the saved animals as of 31 July 1959 are shown in Fig. 4.2. The figures presented are also supported by Habthooen et al. (1960), who indicated that by 30 June 1959, more than 2000 animals had been saved. These figures were drawn up for 35 species, ranging from squirrels to elephants. The count started from as low as one bush baby, 2 jackals, 3 hyenas, 5 civet cats to as high as 585 warthogs and 1866 impalas.

It emerged that some animals did not run away from the rising waters, as it is believed they were used to the Zambezi valley waters' seasonal floods. In addition, most birds including guinea fowl, francolin and other smaller species remained in their habitats and eventually died of hunger and exhaustion (Critchley, 1959). Saving the bigger game was not easy. In any case, many from this category are dangerous, among such, the lions, elephants, buffalos, giraffes, leopards, rhinos, etc. In fact, the rhinos proved particularly difficult (Habthooen et al., 1960). Smithers (1959) believed rescuing these was simply out of question. Additional work by the WCD (2000) revealed that between 4000 and 6000 large animals and several small ones ended up being saved. Forty black rhinoceros were relocated to Hwange National Park. Evidence of Operation Noah is dotted across the Zambezi Valley, Kariba Town and other surroundings. Figures 4.3 and 4.4

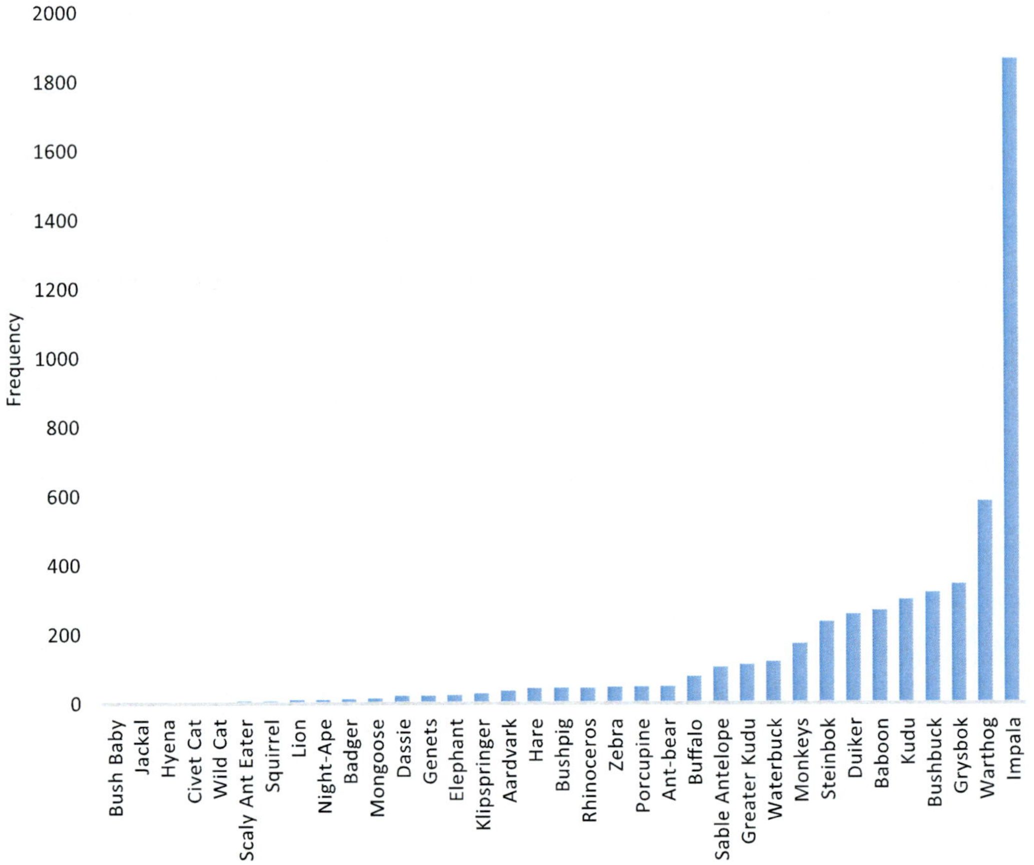

Fig. 4.2 The animals finally saved from both countries. (Source: Authors, Data from Operation Noah memorial plaque; Critchley (1959: 107))

show two memorial sites for this great historical event.

The Santa Babra Memorial Chapel at Kariba Heights was designed and constructed in the shape of a coffer dam and memorialises the lives of 86 workmen who lost their lives in accidents during the construction of the Kariba Dam and power station. Their names are inscribed on a marble inscription in the chapel. The Operation Noah memorial plaque, a short distance from the Santa Babra Chapel at Kariba Heights, was commissioned by the then Southern Rhodesia Prime Minister, Ian Smith. Ultimately, nobody knows the exact figures of how many animals died during the Kariba Dam disaster. However, the number of crocodiles and aquatic birds increased because of an extension to their habitat. Although hippos and buffaloes were reported to have initially declined, subsequently their numbers are said to have peaked (WCD, 2000). The next section briefly looks at the social (in)justice issues from Operation Noah.

4.2.3 Social (In)Justice Matters Surrounding Operation Noah

The entry point to this section is that more than 100 men died during the construction phase of the Kariba Dam (WCD, 2000). These were men caught in machines, electrocuted, crushed from rock falls, killed on the roads or fell to death from collapsing scaffolding. In fact, working conditions were so bad that on 25 February 1959, an estimated 5650 workers went on strike (Tischler, 2013). In addition, about 50,000 Tonga speakers

Fig. 4.3 The Santa Babra Memorial Chapel in Kariba Heights. (Source: Authors)

had to be resettled (Critchley, 1959), who had been residents in the Gwembe valley of the Middle Zambezi (Magadza, 2005). The extent of the injustices done during the displacement of these people is highlighted in a publication by the Zambezi River Authority (ZRA) titled Kariba Dam's Operation Noah Re-launched (ZRA, 1996).

Kariba Dam was selected as one of the ten case studies worldwide to contribute to the debate on the development effectiveness of large dams by the WCD in 1999 (WCD, 2000). The WCD records that the 1951 pre-project document reported 29,000 people to be resettled on a £4 million budget, far much less than the amounts released to rescue the wild animals (Magadza, 2005). The comparative budgets for rescuing the animals and relocating people was £968 per animal and £50 per person. As the decision to have this project managed separately by each government in their jurisdiction was made, the project was removed from the main project. Ultimately, 29,000 more people needed to be relocated, bringing the total to 57,000 at the same budget (WCD, 2000; Scudder, 2005). However, Magadza (2005) put the figure at about 80,000 people. To this end, depending on the source, the figures for the Tonga people who had to be resettled range between 50,000 and 80,000. Whatever the figures put forward, one thing is certain: the Tonga people changed overnight from being the "River People" to being refugees in their own territory (Harnish et al., 2019). The Tonga people had always been at peace in their Gwembe valley, where they practiced sustainable agriculture and subsistence hunting and gathering. They applied their knowledge on seasonal rains and floods to grow crops throughout the year (Magadza, 2005). They cultivated the rich alluvial soils of the floodplains. Yet the dam meant they had to be relocated to semi-arid areas and tsetse fly-infested new areas, bringing them into instant food deficit.

Fig. 4.4 Operation Noah memorial in Kariba Heights. (Source: Authors)

It can therefore be argued that the forced relocation of the Tonga people from the Gwembe valley to make way for a dam project meant to provide electricity (a luxury that the people could not even dream of) was a synchronised sabotage to their livelihoods—hence the feeling of resentment.

As Magadza (2005) reports, the Tonga people still exhibit a resentful attitude towards wildlife management. From a study by McGregor (2005) that took place between February and September 2001, it emerged that there was continued conflict between artisanal fishermen and crocodiles in the Binga area of Zimbabwe. The fisherman poison crocodiles in nests and/or trap the creatures in gillnets where they drown or get killed with spears. Magadza (2005) notes that the lack of awareness of Lake Kariba project, limited and/

or lack of infrastructure and social disruptions were among some conflict points during both the construction and relocation of the Tonga people. The Tonga people were only made aware of the project less than a year prior to flooding. As these people were used to the floods, they could not comprehend the concept of a permanent flood. As such, there is great suspicion that many perished in floodwaters as they resisted moving.

The resistance from the Tonga people also resulted in the use of excessive force by the colonial regime, including discharging of firearms by police (Magadza, 2005). The design amendments to raise the wall by an additional 5 m meant more communities were endangered and these were not in the original plan of relocation (WCD, 2000). The dam separated a once unitary ethnic Tonga group between the two countries, that to date have to travel for hundreds of kilometres if they wish to see relatives either sides of the colonial barrier—Lake Kariba. The Kariba border post remains the main border along the 300 km stretch, which may mean people travelling for more than 600 km and for days to visit relative less than 100 or so kilometres on the other side of the lake. In one instance, a community under one chieftaincy found itself split in two, with marginalised resettlements more than 200 km apart (Harnish et al., 2019).

The resettlement packages were not good and this caused resentment among the people. For example, in June 1958, a group of people in the Chiefdom of Chipepo stoned a district commissioner (WCD, 2000). A similar incident happened 3 months later in the same area, with anti-resettlement protests that attacked police. The police retaliated by opening fire, killing eight people and injuring 32 others. As the resettlement took place in 1957 and 1958, people to be resettled "were treated like animals or things rounded up and packed into lorries" in preparation to move to their new destinations (Ibid: xii). Secondary challenges were caused by the lack of water, lack of basic infrastructure like electricity and being prohibited and/or restricted to access the resources of the lake, including fishing (Mashingaidze, 2013). Having noted these challenges, a number of projects were launched as Operation Noah was relaunched (ZRA, 1996). Among such projects were those focusing on sustainable health, promotion of tourism, education, fishing and access to irrigation. Sadly, most of the initiatives were donor-funded with no direct revenue from Lake Kariba operations (Magadza, 2005), and as late as July 2018, many communities in Binga had not been connected to the national power grid. Harnish et al. (2019) place on record power lines passing over a Mulonga senior headman's homestead. In their view, the authors argue that the scenario is a symbol of the promised benefits of Kariba Dam that are yet to materialise for former Gwembe Valley Tonga development refugees.

Drawing from the discussions in the three subsections focusing on the Kariba Dam designs, understanding Kariba Dam's Operation Noah and social justice issues surrounding Operation Noah, a gap on documenting more on the hydrological perspectives of Operation Noah remains. The next section presents the materials and methods used in the paper.

4.3 Materials and Methods

For the purposes of addressing the set research objective, an analysis of the hydrological records before, during and after Operation Noah was done by applying the flood frequency analysis and the flood annual exceedance probability as derived from the Big Tree station upstream at Victoria Falls. This data was compared with net inflow data obtained from the Zambezi River Authority, which shows the amount of water volumes that ultimately gave rise to the water levels that were witnessed. These statistical measures are widely performed for predicting and quantifying the discharge of floods at a particular location of a river on desired time intervals. The analysis requires discharge records to be at least 10 years and above, but 30 years of data with no gaps can be seen as the best time series (Holmes, 2014). The Big Tree station is located just upstream of the Victoria Falls on the Zambezi River, which provides 75% of the flow into the lake. It was selected because it also provided flow data in the design of Lake Kariba. It is also the only station that had a long-term time-series data before and after the construction of Lake Kariba. The location for Operation Noah (the Kariba Dam area), shared between modern day Zambia and Zimbabwe, is shown in Fig. 4.5.

The recurrence interval is a statistical assessment of the average time that passes between floods of a certain magnitude (Bhat et al., 2018). Flood frequency analysis is termed as the dimensionless mechanism most widely utilised to relate the magnitude of extreme flood events to their frequency of occurrence through the use of probability distributions, based on past recorded peak discharge data at specific gauging stations along a river (Chow et al., 1988). Flood recurrence intervals are computed using historical stream peak flows. To this end, the annual peak flows for the selected Big Tree station were obtained from the Zambezi River Authority. These were then ranked according to magnitude, from the largest to the smallest. The flood recurrence interval was determined using the Weibull equation:

$$R = m + 1 / n \qquad (4.1)$$

where R is the flood recurrence interval, m is the number of years on record and n is the rank of the record in the array.

A graph was then derived by plotting peak discharge for each year of the record versus recurrence interval. The annual exceedance probability of a flood is the likelihood of a flood of a particular magnitude being equaled or exceeded in a year. This helps to provide an estimate of the intensity of a flood event, as high magnitude flood events have a low probability within a hydrological season of being equaled or surpassed. It can also be viewed as the likelihood expressed as a percentage that a flood of that magnitude or greater will occur in a calendar year

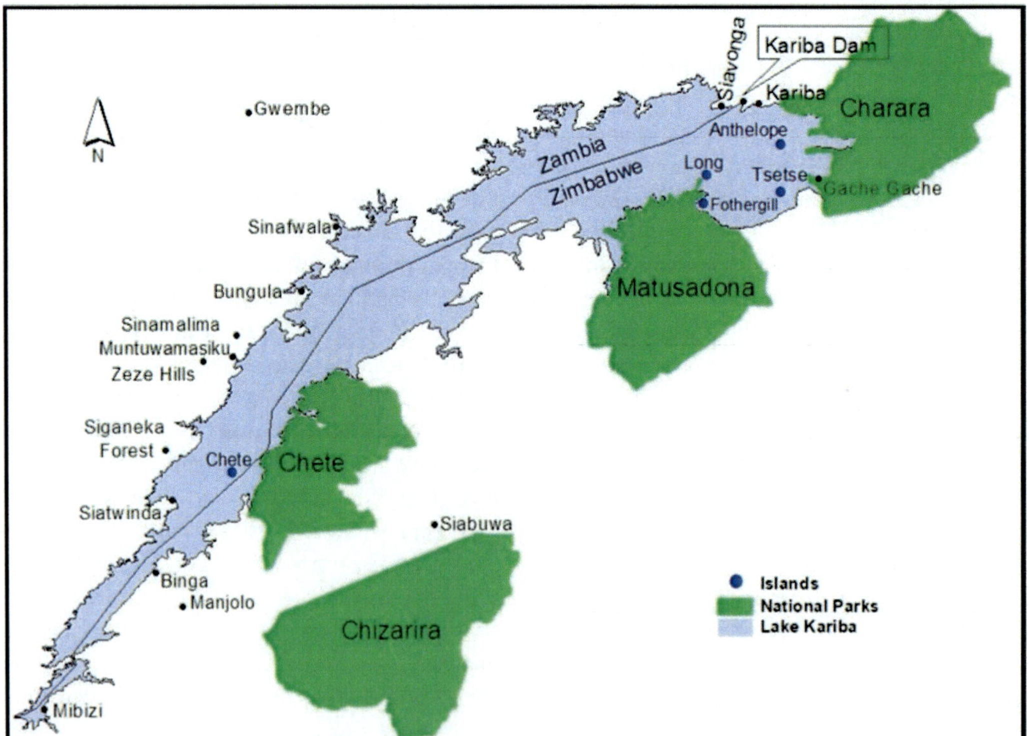

Fig. 4.5 Operation Noah study area. (Source: Authors)

(Goring et al., 1997). The exceedance probability was calculated using the formula:

$$P = 1/T \tag{4.2}$$

where P is the annual exceedance probability (in percentage) and T is the return period of the flood. The next section presents data and discusses the key findings.

4.4 Presentation of Data and Discussion of Findings

Magadza (2005) presents three sub-areas that are traditionally associated with the Zambezi basin. These are the Upper Zambezi (upstream of the Victoria Falls), the Middle Zambezi (from the Victoria Falls to Cahora Bassa Dam) and the Lower Zambezi (downstream of the Cahora Bassa Dam into the Indian Ocean). What is of interest is the fact that the Upper Zambezi has rich flood plains and wetlands, particularly the Barotse Flood Plain of Zambia, the Chobe swamps from Botswana and the Caprivi wetlands of Namibia. Drawing from Vas (2000, cited in Magadza, 2005), the largest flood in the Zambezi River basin took place in March 1958. Again in the 1970s, the reservoir was almost full when "intense, prolonged precipitation occurred in large areas of the basin and ultimately caused the largest flood to enter Lake Kariba" (Magadza, 2005: 211). This had immediate downstream impacts at Cahora Bassa, which was caught by surprise given that the war of liberation in Zimbabwe was at its peak and it took almost 2 weeks to have the message relayed. Communication had to go through Portugal. As the Kariba spillways discharge reached Cahora Bassa, the authorities immediately opened four flood gates there.

Based on the analysis done by the WCD (2000), it is clear that the engineers and planning teams did not anticipate what happened in 1957 and 1958. Although the final Kariba design was for the safe passage of a one in 10,000-year flood, the original design was for a 3-month flood with a volume of 68 km^3; this was projected based on the 1924–1955 flow data series. However, during construction there was a flood in 1957 with a peak of 8200 m^3/s. This forced a revision of the design flood value from 68 to 74 km^3. In 1958, another flood hit and a 3-month volume of 61 km^3 occurred, resulting in the spillway capacity increased to 92 km^3.

Figure 4.6 shows the annual peak flow hydrograph of the Zambezi River from 1924 to 2020 at the Big Tree hydrological gauging station. This is a key hydrological station within the upper Zambezi catchment because it has the longest time-series data, was used in formulating the designs of Lake Kariba and has a less than 2-week lag time with Lake Kariba. As highlighted in Fig. 4.6, before the construction of the reservoir began, the highest peak flows of 5998 cm^3/s and 5734 cm^3/s were both recorded in the seasons of 1947/48, 1951/52 and 1952/53. These figures were based on available hydrological data on the highest flood peaks, from which scenarios on the rate of dam flooding were derived. This therefore implies that the odds of getting flood magnitude outside this range were very low and negligible.

Despite the odds of getting a flood above 5998 cm^3/s being low, the time period of Operation Noah (1958–1964), when the dam was filling up, recorded the highest flood magnitude in the history of the Zambezi River at the Big Tree station. This was unprecedented even under the most pessimistic stochastic hydrological projections, given that the 1957/58 season's 9435.56 cm^3/s flow was close to twice the previously known highest peaks. Furthermore, in the 1962/63 season, the station recorded a flood peak of 6801.83 cm^3/s, which ranks fourth in the all-time floods of the Zambezi River. During the same period, a peak inflow of water was recorded in Lake Kariba (Fig. 4.7). The highest inflows were recorded in the months of February, March and April, in ascending order. This could be indicative of severe rainfall activity that could have been witnessed in that particular period which were unprecedented. Almost a similar episode plays out in the years around 1978. Other peaks that can be recorded were seen in 1968, 1988, 2000, 2007 and in 2016 through to 2017, although these were a bit lower.

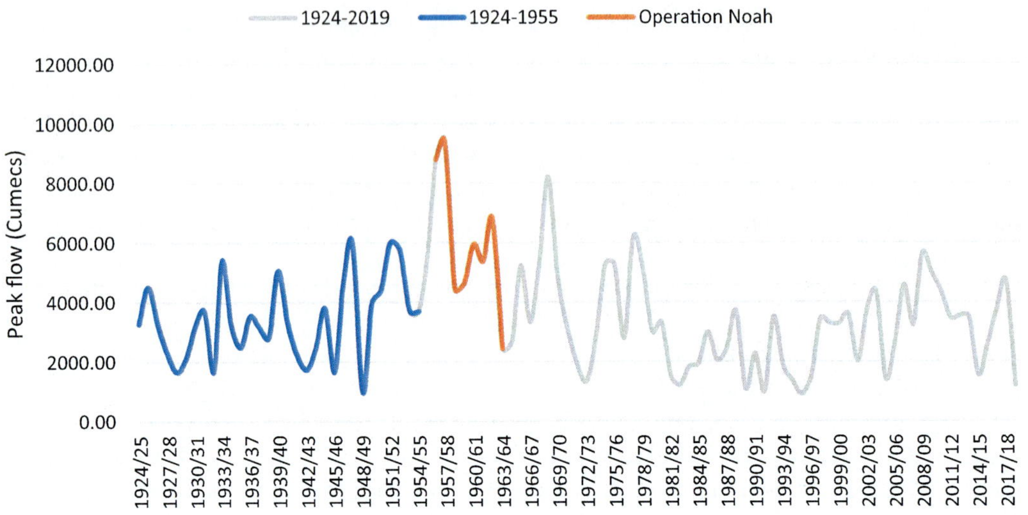

Fig. 4.6 Zambezi River annual peak flow hydrograph at the Big Tree station. (Source: Authors)

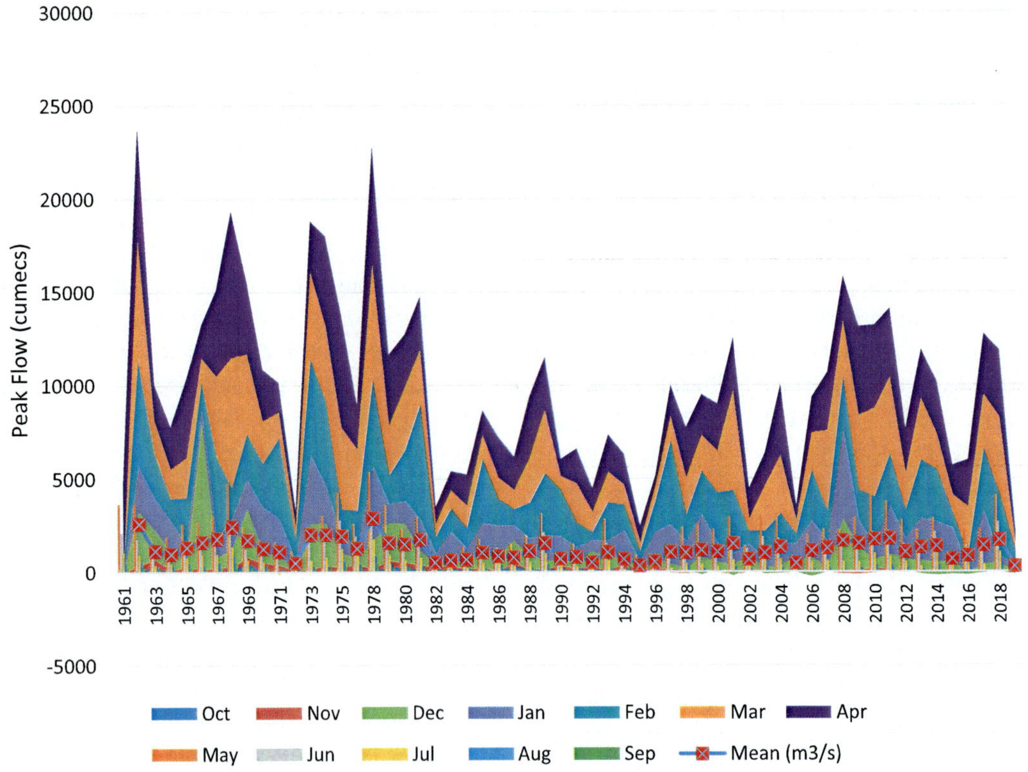

Fig. 4.7 Peak inflow of water recorded in Lake Kariba. (Source: Authors)

Looking at the whole time series of peak flows from 1924 to 2019, it can be argued that the period of Operation Noah had the highest flood magnitude on the upper and middle sections of the Zambezi basin. Considering that the governments of Zambia and Zimbabwe were grappling with the resettlement of over 50,000 people from the basin and also financing the completion of the dam and related infrastructures, they also could have been taken by surprise. The rate at which the dam levels rose was too swift and allocating resources, even for animal rescue was extremely difficult. The rise of the lake due to the unusual floods also took the animals by surprise, as indicated in the literature, given that they were only used to the generic annual flood regimes of the Zambezi River. Figure 4.8 shows a recreation of the surface area that was covered by water from 1960 to 1962 on the Zimbabwean side of Lake Kariba. It shows that as each year passed during Operation Noah, there was an increase in the surface area that the lake was covering, where new animal habitats were being inundated by water and previously available islands disappearing. The figure also depicts the controversy of large hydro-electrical dam construction. As at 1960, the huge surface area underwater constitutes the dead storage of the lake. This implies water that cannot be used for the production of power yet inundating large tracts of land and displacing thousands of animal and human communities. Only the water levels depicted in blue (1962) show the live storage of the dam. Lake Kariba has 114 km^3 of dead storage, equivalent to almost 2% of the Southern Africa's entire region's annual runoff (Ohlsson, 1995) and more than five times the annual freshwater withdrawals for the region (21.23 km^3), yet very little of this water can be used without jeopardising power generation.

Figure 4.9 shows the steep rise in the levels of Lake Kariba during Operation Noah. By 1961, the lake had surpassed the minimum level of 475.5 m required for the generation of electricity, and by 1962, it was almost reaching its maximum level of 488.5 m. The steep rise in the level of the lake also coincided with steep increases in the discharge and the volume of water entering the lake, as shown from previous figures.

Figure 4.10 shows the return periods of floods of different magnitudes on the Zambezi River as recorded at the Big Tree station. The three top floods during Operation Noah were those from 1957/58 of magnitude 9435.56 cm^3/s, the 1962/63 flood of magnitude 6801.83 cm^3/s and the 1960/61 of magnitude 5998.95 cm^3/s and these had return periods of 106, 26.5 and 11.8 years, in that order. These flows were also outside the 1 in 10-year flood magnitude, which constitute close to 90% of the Zambezi River flow regimes. It is also noted that on the Zambezi River any flood magnitude above the 1 in 10-year flood causes significant flooding in the lower catchment. For some rivers, like the Jhelum in the Kashmir basin, flows having recurrence interval of 5 years or more are enough to exceed the carrying capacity of the river and cause significant flooding (Bhat et al., 2018). The actual flow levels during

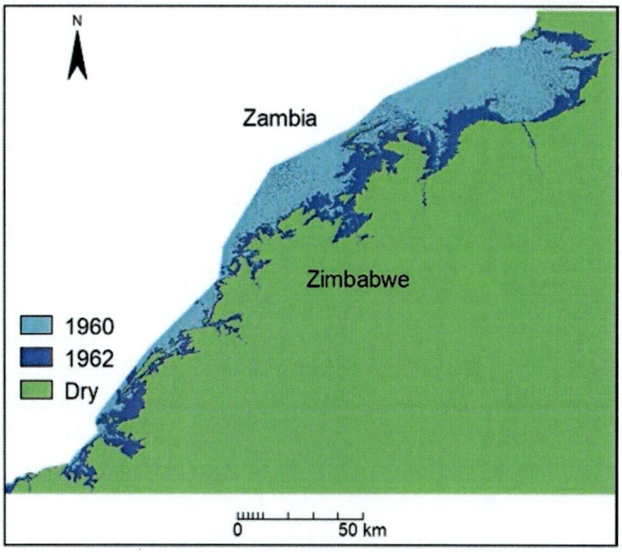

Fig. 4.8 Areas inundated by water from 1960 to 1962 during Operation Noah on the Zimbabwean side. (Source: Authors)

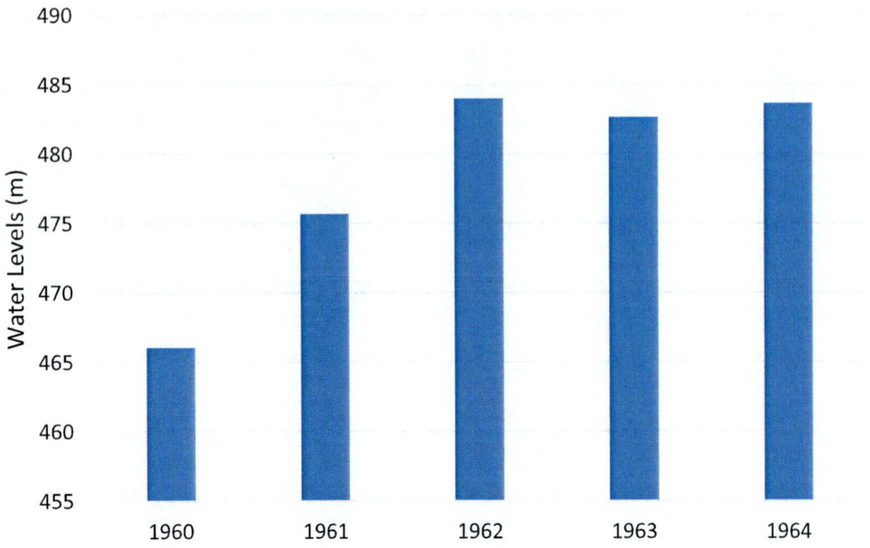

Fig. 4.9 Rise of water levels in Lake Kariba during Operation Noah. (Source: Authors)

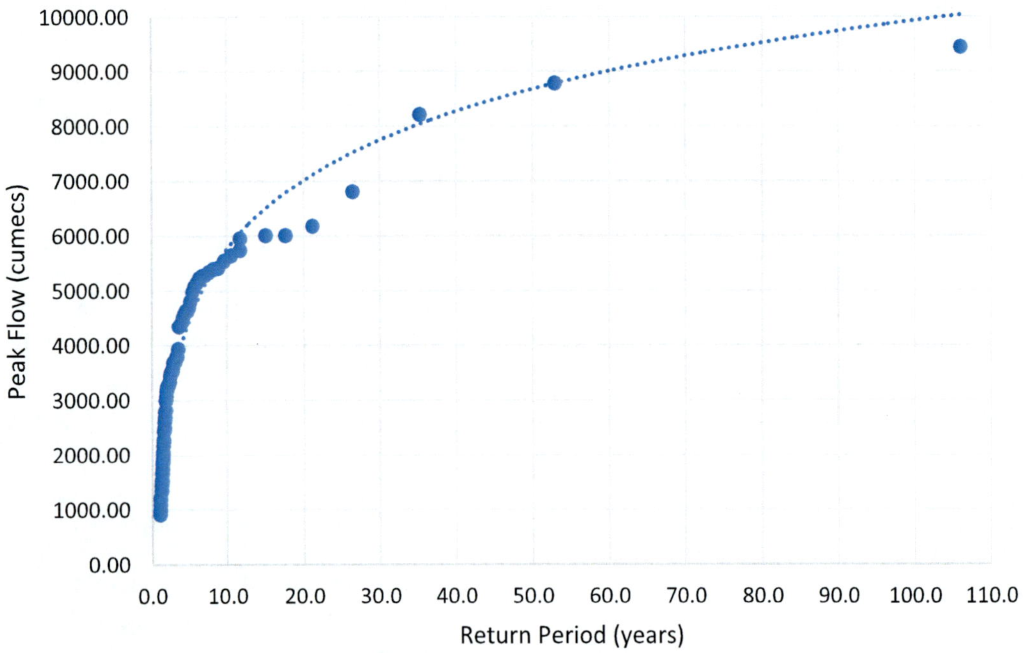

Fig. 4.10 Flood return periods for the Zambezi River 1924–2019. (Source: Authors)

Operation Noah contrast sharply to the anticipated highest magnitude floods from the records, which were the 1947/48 and the 1951/52 flood magnitude of 5998.95 cm^3/s with a return period of 17.7 years. This further reinforces the argument that the peak flows that were observed during Operation Noah years were unusual and uncharacteristic of the river, hence catching both the animals and the responsible authorities by surprise with little response time.

Figure 4.11 shows the annual exceedance probability of peak floods in the Zambezi River

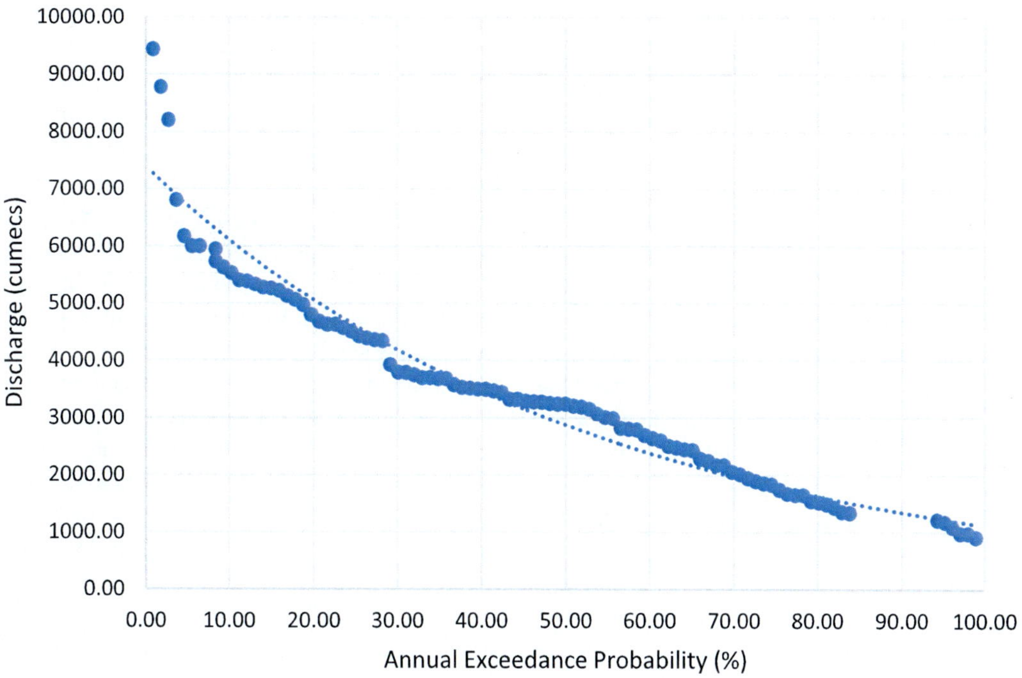

Fig. 4.11 The annual exceedance probability for the Zambezi River peak flow. (Source: Authors)

as recorded at the Big Tree hydrological station. Floods of an annual exceedance probabilities of 0.94% (1957/58), 3.77% (1962/63) and 8.49% (1960/61) occurred during a period of four seasons. As can be noted, high-impact floods have a very low annual exceedance probability. However, during the years of Operation Noah, a series of high-impact floods occurred consecutively imposing a further burden on the governments of Northern and Southern Rhodesia that were already overwhelmed by other response activities. Given that the ideal situation was that the lake would fill up gradually and animals would have a chance to move to higher ground, the authorities in both Zimbabwe and Zambia had not planned any major animal rescue operation along the stretch of the lake. The reality of the floods heightened by the unusual peak flows increased the rate at which the lake levels rose, resulting in many animals drowning or being marooned on temporary islands. This necessitated the unplanned and unbudgeted Operation Noah, which up to date is still one of the largest animal relocation projects in the world. The unusually high flow regimes of the Zambezi River during Operation Noah also led to the dam reaching its maximum levels a year before anticipated, as documented earlier in the literature section.

Scudder (2005) notes that the construction of Lake Kariba drastically altered the Zambezi natural regime. The flood recession agricultural practices downstream from the Zambian side as well as offshore fisheries in Mozambique were negatively affected. The author indicates that there was a failure to properly draw down the Kariba and even the Cahora Bassa reservoirs before the increased rainfall between 1999 and 2000 and 2000 and 2001 rainy seasons. This resulted in downstream loss of life, crops, and village and urban infrastructure in February–March 2000 and 2001.

Our assumption that everybody could have been taken by surprise is confirmed by Scudder (2005: 19) who wrote: "Because Kariba Lake was the first major reservoir formed in the tropics, there was little awareness even among scientists as to what actions should be taken before flooding or the nature of the environmental impacts that would follow inundation". Magadza (2006) concurs. In his view, the lake was designed during a period when human rights, social justice and equity were secondary matters. The same applied to environmental management issues. Hence, most of the management issues that arose over time were not foreseen (Magadza, 2010). Concluding his analysis, Hughes (2006) said unlike the Biblical flood, Kariba's waters did not subside. Rather, the floodwaters stabilised. Furthermore, Tischler (2013: 198) recorded that in February 1959, "construction developed into a fierce race against time following the record floods of 1958".

Other unforeseen hydrological aspects came to a head after more than 50 years of operating the Kariba Dam. Darbourn (2015) highlighted that there was a need to rehabilitate the plunge pool. The spillway torrents excavated a huge carven in the Zambezi River bed, which was ten times bigger and deeper compared to the original design. This risk was threatening the wall foundations. Work on the plunge pool started in September 2015. Additional work was scheduled for the spillway refurbishments. Both projects were projected to cost US$294.4 million.

Of equal significance is the tendency of very high peaks and flows, which present challenges for wildlife and other human activities in Lake Kariba. Dube and Nhamo (2020) already alluded to this by pointing out that due to these often deeps and steeps, animals were being trapped on islands particularly on the Zimbabwean side, which demanded expensive rescue operations. A case in point are rescue operations in 2010, 2016 and 2017 on islands near Matusadonha National Park and Bumi Hills, although the rescue operations were occurring on a much smaller scale. Even on smaller scales these operations tend to be resource intensive, which raises serious concerns given the financial and resource constraints obtaining in Zimbabwe.

4.5 Conclusion

The unusual hydrological events on the Zambezi River during Operation Noah influenced the fast rate at which the levels of the lake were raising. The large number of people and animals needing to be resettled and rescued meant the total amount of resources required was equally huge. Due to high-impact flood events recurring during this period, the levels of water rose faster than the rate at which the animals and probably also the Tonga people could escape. To this end, many animals drowned and/or were marooned on islands and needed to be rescued. With a total count of the known islands sitting at 500, this was a vast area to cover under the circumstances. A revisit of Operation Noah shows that the hydrology also played a big role in the impact of the operation and what could be achieved. It imposed time constraints on the rescuers and consequently affected the strategies that could be used.

References

Alibekov, A. B., Dmitriev, D. S., & Koval'chuk, I. V. (2019). Minimizing the negative impact on wildlife during filling of the lower Bureya HPP reservoir in the Amur Obl. *Power Technology and Engineering, 53*(2), 184–190. https://doi.org/10.1007/s10749-019-01058-0.

Bhat, M. S., Alam, A., Ahmad, B., Kotlia, B. S., Farooq, H., Taloor, A. K., & Ahmad, S. (2018). Flood frequency analysis of river Jhelum in Kashmir basin. *Quaternary International, 507*, 288–294. https://doi.org/10.1016/j.quaint.2018.09.039.

Chow, V. T., Maidment, D. R., & Mays, L. W. (1988). *Applied hydrology*. London: McGraw-Hill International.

Critchley, A. A. (1959). Operation Noah. *Oryx*, 100–107. https://doi.org/10.1017/S0030605300000430.

Darbourn, K. (2015). *Impact of the failure of the Kariba dam*. Pretoria: Institute of Risk Management South Africa.

Dube, K. and Nhamo, G. (2020). Vulnerability of nature-based tourism to climate variability and change: Case of Kariba resort town, Zimbabwe. *Journal of Outdoor Recreation and Tourism, 29*. https://doi.org/10.1016/j.jort.2020.100281.

Goring, D. G., Pearson, C., & Kingsland, S. (1997, July). *Extreme sea levels on the Mt Maunganui shoreline (Moturiki Island)*. Prepared for Environment Bay of Plenty NIWA Client Report No. 97/32. NIWA.

Habthooen, A. M., Lock, J. A., & J.A. (1960). The rescue of rhinoceroses at Kariba Dam. *Oryx*, 352–355. https://doi.org/10.1017/S003060530000079X.

Harnish, A., Cliggett, L., & Scudder, T. (2019). Rivers and roads: A political ecology of displacement, development, and chronic liminality in Zambia's Gwembe Valley. *Economic Anthropology: Water and Economy, 6*, 250–263. https://doi.org/10.1002/sea2.12151.

Holmes, R. R. (2014, January 14). *Floods: Recurrence intervals and 100-year floods*. USGS Website. Retrieved January, 2020, from http://www.water.usgs.gov/edu/

Hughes, D. M. (2006). Whites and water: How Euro-Africans made nature at Kariba Dam. *Journal of Southern African Studies, 32*(4), 823–838. https://doi.org/10.1080/03057070600996846.

Hughes, D. M. (2010). *Whiteness in Zimbabwe: Race, landscape and the problem of belonging*. London: Palgrave Macmillan.

Magadza, C. H. D. (2005). *Kariba reservoir: Experience and lessons learned brief.* Harare: Lake Kariba Research Station University of Zimbabwe.

Magadza, C. H. D. (2006). Kariba reservoir: Experience and lessons learned. *Lakes & Reservoirs: Research and Management, 11*, 271–286.

Magadza, C. H. D. (2010). Environmental state of Lake Kariba and Zambezi River Valley: Lessons learned and not learned. *Lakes & Reservoirs: Research and Management, 15*, 167–192.

Mashingaidze, T. M. (2013). Beyond the Kariba dam induced displacements: The Zimbabwean Tonga's struggles for restitution, 1990s–2000s. *International Journal on Minority and Group Rights, 20*, 381–404.

McGregor, J. A. (2005). Crocodile crimes: People versus wildlife and the politics of postcolonial conservation on Lake Kariba, Zimbabwe. *Goeforum, 36*, 353–368.

Nyandoro, M. (2014). Large dams: Harmony or acrimony with environmental resource management paradigms in Zimbabwe and Lesotho. *Bhatter College Journal of Multidisciplinary Studies, 4*, 50–67.

Ohlsson, L. (1995) *Water and security in Southern Africa*. Publications on Water Resources: No. 1. SIDA: Department for Natural Resources and the Environment.

Scudder, T. (2005). *The Kariba case study*. Pasadena: Division of the Humanities and Social Sciences, California Institute of Technology.

Smithers, R. H. N. (1959). The Kariba Lake. *Oryx*, 21–28. https://doi.org/10.1017/S0030605300000065.

The Holy Bible. (2011). *Noah and the flood*. New International Version®, NIV® Copyright ©1973, 1978, 1984, 2011 by Biblica, Inc.® Used by permission.

Tischler, J. (2013). *Light and power for a multiracial nation the Kariba dam scheme in the Central African Federation*. London: Palgrave Macmillan.

WCD (World Commission on Dams). (2000). *Kariba Dam: Zambia and Zimbabwe*. Cape Town: World Commission on Dams.

ZRA (Zambezi River Authority). (1996). *Kariba dam's operation Noah re-launched*. Kariba: Zambezi River Authority.

5 Victoria Falls Water Flow Regimes: A Tale of Two Half Centuries

Godwell Nhamo, Kaitano Dube, and David Chikodzi

Abstract

The so-called drying Victoria Falls received considerable global media attention from October to December 2019. The debate surrounding the falls have split stakeholders, with strong pushback from the tourism sector. This paper taps into the arguments presented in the debate as well as into two earlier academic publications that used 40 years of data focusing on climate change and variability at Victoria Falls published in 2018 and 2019, respectively. The objective of this paper is to undertake a time-series analysis of the Victoria Falls water flow regimes, with specific reference to two half centuries marked by 1919/20–1968/69 (Half Century I) and 1969/70–2018/19 (Half Century II), thereby using 100 years of available data. Drawing from the analysis of the hydrologic data sets for Victoria Falls, the key finding is that Half Century I had evidently higher flows, compared to Half Century II. In addition, daily mean flow regimes for Half Century II were declining, with marked negative trends for January, August, September, October, November and December. From these months, several Mann-Kendall tests confirmed significant declines for October, November and December. The study recommends a basin-wide rethink of how best the deteriorating conditions of the Victoria Falls water flow regimes can be addressed in the context of climate change and anthropogenic factors leading to reduced water flow regimes in Half Century II. Further studies to determine the influence of land use changes in the upper Zambezi River catchment on water flow regimes are recommended.

Keywords

Victoria falls · Flows · Tourism · Stakeholders · Half centuries · Zambezi River

5.1 Introduction

Water flow regimes are the mainstay of water-based tourism attractions such as the Niagara Falls in Canada and the Victoria Falls in Zimbabwe. With known historical and observable peak and low flow regimes, debates are ongoing to determine if the fluctuations in water flow regimes are regular seasonal or long term, leading to permanent changes in flow regimes

G. Nhamo (✉) · D. Chikodzi
Institute for Corporate Citizenship,
University of South Africa, Pretoria, South Africa
e-mail: nhamog@unisa.ac.za; chikod@unisa.ac.za

K. Dube
Vaal University of Technology,
Vanderbijlpark, South Africa
e-mail: kaitanod@vut.ac.za

G. Nhamo, L. Chapungu (eds.), *The Increasing Risk of Floods and Tornadoes in Southern Africa*,
Sustainable Development Goals Series, https://doi.org/10.1007/978-3-030-74192-1_5

(Chatterjee et al., 2018; Cole, 2015; Klint et al., 2012). In Hudson's (2002: 440) words, "When visiting waterfalls, tourists are often disappointed to find the flow much less than they were led to expect by published descriptions and images". The author further indicates that many believe waterfalls are at their best when the discharge is great, especially after significant downpours. These are usually the views pitched in tourist guidebooks to such places. Hudson (1999) also documents the tragedy of Jamaica's Dunn's River Falls. Receiving approximately one million visitors annually, the waterfall attracts huge tourist volumes, commercial development and environmental degradation. To this list, and with our modern knowledge, we can add climate change and variability, and increased water demand for agriculture and domestic use due to growing populations (Dube & Nhamo, 2018).

Two schools of thought emerge from the growing discourse regarding water flow regimes on waterfalls: (1) those that strongly believe the water flow regime fluctuations are seasonal and temporary variations, and (2) those that strongly believe the flow regime changes are of a permanent, rather than seasonal and temporary, nature and that they are caused by climate change and anthropogenic activities. Given that tourists wish to see water flow regimes at their peak (Dube & Nhamo, 2019; Hudson, 2002), there will ever be increasing concerns if water flow regimes at waterfalls and within their catchment areas around the world decrease to unprecedented and as yet unknown low levels (Niagara Region, 2019; Hudson, 2013). It comes as no surprise, therefore, that the media got attracted regarding the state of the Victoria Falls sparked by Zambian President Edgar Lungu's tweet in which he bemoans the "drying" Victoria Falls in October 2019 and a British Broadcasting Corporation's Hard Talk feature on the same (Baynes, 2019; Allen, 2019; Connolly, 2019). In fact, part of the October 2019 tweet from President Lungu reads, "These pictures of the Victoria Falls are a stark reminder of what climate change is doing to our environment and our livelihood. It is with no doubt that developing countries like #Zambia are the most impacted by climate change and the least able to afford its consequences" (Connolly, 2019: Online).

Everywhere, media headlines popped in response to the Zambian President's and other commentaries. Herein are some of the headline stories. "The False Death of Victoria Falls" (Admin, 2019). "The Victoria Falls are not dry" O'Dea, 2019), "Victoria Falls shrink to a trickle, feeding climate change fears" (Hutchings & Cocks, 2019), "Victoria Falls, one of the natural wonders of the world, is running dry due to climate change" (Matiashe, 2019), "Victoria Falls Dries Drastically After Worst Drought in a Century" (Rosane, 2019), "Victoria Falls 'bone dry'? How drought and dry season are hitting the world's largest waterfall" (Dube, 2019) and "Victoria Falls far from drying" (The Editor, 2019). This resulted in a counter campaign by travel agencies, tourists and some citizens on Twitter, thus, "#VictoriaFallsisnotdry" (The Victoria Falls are not dry) (Admin, 2019). The campaign published daily photos and videos showing the water flows at the famous devil's pool of the Victoria Falls. This is the point where tourists can swim on the same edge of the waterfalls.

As early as 1987, debates concerning the possible impacts of climate change and increasing temperature on the Great Lakes basin water flow regimes that hosts the Niagara Falls emerged (Sanderson & Wong, 1987). The Union of Concerned Scientists and The Ecological Society of America (UCS-ESA) identified the Great Lakes region shared by Canada and the USA as hosting up to 60 million people then (UCS-ESA, 2003). The USC-ESA noted four impacts on the region resulting from the changing climate, namely, (1) winters that were getting shorter, (2) annual average temperatures getting warmer, (3) duration of lake ice cover decreasing as air and water temperatures rose and (4) heavy rainstorms that were becoming more common. In line with the focus of this paper, earlier ice breaking and peaks in spring run-off was likely to change the timing of stream flows, leading to flooding occurring more frequently.

Reductions in summer water levels was likely to diminish underground stream recharge causing smaller streams to dry up. Although not a focus of this paper, Buttle et al. (2004) extend the impacts of climate change to hydropower generation in the region (New York Power Authority, 2005). Colombo et al. (2007) predicted less rainfall in Ontario. Wilcox et al. (2007) attribute the changes in water levels in the Great Lakes to both natural and anthropogenic factors that operate on timescales ranging from hours to millennia. Primary natural factors include the amount of inflow and outflow characteristics. Human factors include diversions into and out of the basin, dredging of outlet channels and regulation of outflows. Seasonal (annual) fluctuations follow the hydrologic cycle that can also be disturbed by climatic changes. Drawing from historical records, the higher water levels come in spring and early summer, with lower levels in the remainder of the year. The authors note that the highest lake levels usually occur in June on Lakes Ontario and Erie, July on Lake Michigan-Huron and August on Lake Superior. On the other hand, the lowest lake levels usually occur in December on Lake Ontario, February on Lakes Erie and Michigan-Huron and March on Lake Su. While monthly mean flow fluctuations are relatively small, fluctuations were observed over long-term periods (multi-year, decadal and longer) and these were attributed to climate change.

The focus of this paper is to investigate the Victoria Falls water flow regimes, comparing two half centuries marked by 1919/20–1968/69 (Half Century I) and 1969/70–2018/19 (Half Century II). The paper is borne from sustained global concerns over the possibility of a drying Victoria Falls at some anticipated point in the near future. Given that the waterfalls are the primary attraction for tourists, emphasis in flow regimes is on working with daily flow data. The objectives of the paper are: (1) to determine the nature of the water flow regimes between the two half centuries; (2) to establish if the trends in water flow regimes are of a temporary or more permanent nature and (3) to provide some policy recommendations regarding possible measures that can be instituted by governments and stakeholders sharing the waterfall and the Zambezi River Basin that hosts it.

5.2 Literature Survey

As noted in the introduction, water flow regime fluctuations that are central to key tourist and recreational attractions like Niagara, Victoria and other global falls raise worldwide attention. In fact, waterfalls have received significant attention from academics since the 1980s for both scientific and aesthetic interest related to their superlative beauty (Hudson, 2013). The interest has even increased given the consensus on the climate change phenomenon in 2007 with more studies emerging on climate change in general and waterfalls and their recreational value specifically (Wang et al., 2020; Shao et al., 2018; Haghe, 2011). In this section, we review literature from such areas and other river basins as a basis for analysing and interpreting the key findings that will come from the Victoria Falls scenario. Hence, run-off, precipitation and stream flow regimes remain central to this work (Higashino & Stefan, 2019; Falchetta et al., 2019; Byun et al., 2019).

Wang et al. (2020) clearly identify hydrological process changes in the Yangtze River in China, particularly stream flow, to have occurred due to natural and human-induced activity over the past decades. The authors analysed daily flows for 60 years with results revealing that dam constructions inevitably alter stream flows, especially downstream. The Yangtze River was also subject to further study by Cheng et al. (2019) who compared the water level regimes of Dongting Lake and Poyand Lake both in middle reach. These are China's two largest freshwater lakes. The authors selected these sites as they are being affected significantly by both climate change and anthropogenic activities. Their work analysed hydrological data from 1955–2016 and 1953–2014 for the two lakes, respectively. The data sets were divided into two periods: (1) the pre-impact period and (2) the post-impact period. The change years were identified as 1979 for Dongting Lake and 2003 for Poyang Lake. The overall degree of hydrological alteration stood at 52.6% and 38.2%, respectively. This showed water level regimes that experienced moderate and low alteration. While water levels at Dongting were responding to increased rainfall and dam regulation, levels at Poyang responded to decreased rainfall, dam regulation and sediment erosion and deposition. The aspects of damming and hydrological alterations also come out strongly in Spanish rivers (De Jalón et al., 2019).

Using the Mann-Kendall, cumulative rank difference and the double mass curve methods, Pirnia et al. (2019) studied the Tajan River basin in Iran using data from 1971 to 2010. The results showed that changes in run-off in both wet and dry seasons after 1999 had significant respective decreasing and increasing trends at 0.01 confidence level. This was mainly attributed to dam construction along the river. However, evidence also emerged that prior to dam construction (1991–1998), climate change variability and human activities contributed 57.76% and 42.24%, respectively, to run-off reduction in wet season. The percentage share changed to 24.68% and 75.32%, respectively, in the post-dam construction period of 1999–2010. In fact, run-off decreased by 116.55 mm. In the same period and in the dry season, climate variability contributed 30.68% and anthropogenic activities contributed 130.68% to run-off increase of 41.45 mm. Hence, run-off changes for both seasons were mainly due to anthropogenic activities associated with damming, especially to supply water for agriculture. Applying the same Mann-Kendall method with data from 37 gauging stations, Dinpashoh et al. (2019) focused on stream flow regimes in Guian Province in Iran. They found that climate change had modified river flow regimes as:

> The median values of low flow timing slopes (Q5% and Q10%) had significant negative trends. Q5% had the steepest downward trends, implying that low flows originating from groundwater occurred earlier than in the past. In contrast, high flow timings (Q90% and Q95%) showed upward trends. Also, Q50% exhibited upward trends for all 37 stations. The trends suggest that climate change has altered the river flow regime (Dinpashoh et al., 2019: 65).

Another study to determine the impacts of climate change and human activities on the flow regime of the dammed Lancang River in southeast China, applied the Mann-Kendall method, supplemented by remote sensing. The findings from 1980 to 2014 data sets (Han et al., 2019) compared the baseline period of 1980–1986 when no dam was constructed with the period after dam construction. It emerged that there was a significant 6% reduction in mean annual streamflow during 1987–2014, especially after 2008 when the construction of the Nuozhadu hydropower plant in the Mekong River basin started. Climatic change was attributed to 57% streamflow changes in the transition period of 1987–2007, while anthropogenic activities accounted for 95% of changes in stream flow post-impact period of 2008–2014. Furthermore, climatic variation was more significant during the dry season between December and May, with precipitation the most important variable. Anthropogenic activities contributed more in the wet season between June and November.

Climate change induced water flow regimes and lake level variations have preoccupied the Niagara region. To prepare for climate change, the Niagara regional government established a team on Climate Change Mitigation and Adaptation Planning in 2007. This resulted in the high-level Climate Change Action Work Plan of 2008. In 2009, the Niagara region joined the Partners for Climate Protection, which is a Federation of Canadian Municipalities (Niagara Region, 2013). From the climate adaptation space, three generic categories of adaption action emerge, namely, building adaptive capacity, taking adaptive action to prevent or reduce impacts, and building resilience (Penney, 2012). In 2014, the Resilient Buffalo Niagara: Strategies to Respond to Climate Change policy was developed (Grower et al. 2014). This became one of the first initiatives in the country proposing a comprehensive and integrated response to address climate change.

In a report by McDermid et al. (2015) that looked at the "State of climate change science in the Great Lakes Basin" with a focus on climatological, hydrological and ecological effects, the findings confirm earlier observations. However, additional evidence emerged. For example, water levels naturally fluctuate by up to 1.5 m. A long-term peak water level was recorded in the 1980s but has been decreasing ever since. There are also projection that future lake water levels will generally fluctuate around lower mean water levels, with lower water levels being a function of warmer air temperatures, increased evaporation and evapotranspiration, drought and changes in rainfall patterns. Increased temperatures (and by default evapotranspiration rates) were observed in 2017. In fact, Environment and Climate Change Canada (2017) reported a record-breaking warmth across the basin in winter with average temperatures of 1–5 °C above the long-term average. On the other hand, the basin winter was generally wet, with parts of Ontario witnessing more than twice its normal amount of precipitation in April and May, with a record October rainfall in Michigan in the fall. This resulted in a record rise in water levels in Lake Ontario from January to June.

Penney (2012) documents specific climate change issues in the Niagara region that holds 20% of the planet's unfrozen fresh surface water (Gronewold et al., 2013). Penney (2012) reports the following changes that have already been observed, namely, an increase in annual average temperature of about 1.3 °C in 40 years, a growing trend towards more days with temperatures over 30 °C, more heat waves and/or more consecutive hot days, as well as a longer growing season marked by May and September becoming significantly warmer. Other trends include increase in average number of frost-free days (with 10 more per year compared to 1970), small increase in annual precipitation in winter, more rain and less snow in winter, increased summer droughts and dry spells, and an increase in heavy rain events. These trends were predicted to continue into the 2050s. Penney (2012) brings in the impacts of the changes on tourism and recreation, which would be mainly positive. For example, an earlier spring and warmer summer will likely extend the main tourism season. This implies that main outdoor activities, which include falls viewing, touring wineries, beach visits, parks and nature trails as well as water sports, fishing and golf will benefit. However, should there be an inverse in climate trends, declining water levels could negatively affect the falls and dock stations, and increase navigation hazards.

Of interest in the statistics coming up are irregular disruptions within generally lower mean water level fluctuations although 2017 was a record-breaking year. Lake Ontario is the most downstream of the Laurentian Great Lakes system (Niagara Region, 2019). In fact, Gronewold and Rood (2019) noted that the spring record high followed a period of record low levels. The scenario had caused the public and expert discussions to focus on the possibility of a long-term and more permanent low-level scenario due to climate change, diversions from the lakes and dredging. The authors conclude by advocating for the appropriate understanding and communication of drivers leading to water levels variability, especially in light of the past low-level extremes. The outcry over low water and lake levels in the Great Lakes region mirrors similar outcries over the Victoria Falls fluctuations.

Shamseddin and Chaibi (2019) turned their energy to scanning climate change impacts on water resources of the largest African river basins. The work looked at 61 peer-reviewed papers that documented seven river basins, namely, the Nile (30% of the papers), Senegal (25%), Volta (25%), Niger (16%), Congo (13%), Zambezi (16%) and Limpopo (13%). The Nile River basins usually draw attention due to Egypt's dominance in that space (Ahmed et al., 2019). Overall, upstream country activities that involve damming and other abstractions, coupled with droughts, caused significant changes in flow regimes downstream. Coming to the river basins under consideration, Shamseddin and Chaibi (2019) categorised the climate-change-related trends for rainfall, stream flow and run-off into three trends reflecting, namely, decreasing, increasing and uncertain trends. There were clear agreements in decreasing trends for run-off for the Congo, Niger and Senegal, while the Zambezi trends were split equally between decreasing and uncertain, with no response on increasing trends. As for rainfall, there were clear decreasing trends for Senegal, Niger and Zambezi compared to increasing trends. Lastly, there were also clear decreasing trends for Senegal and split presentations on Zambezi in terms of stream flows. The Congo studies agreed on clear increased flow. Another element investigated by the authors was the inclusion of land use considerations in the climate change studies to which the Zambezi basin studies had the least, while studies for the Niger and Congo basins clearly included land use. In an earlier study, Beck and Bernauer (2011) revealed that the abundant water resources in many parts of the Zambezi River basin were unlikely to last. This was attributed more to the increasing population, economic growth, increased irrigated agriculture and inter-basin water transfers than to climate change. This scenario would lead to drastic reductions in run-off during dry seasons at key locations, an element that requires effective governance mechanisms for water allocations within the basin. It further called for effective mechanisms to deal with flow variability, a point raised earlier by Palmer et al. (2008). However, work by Shao et al. (2018) on the Hailiutu River basin in China determined that the impact of climate variability on the streamflow was more pronounced compared to land use change. The next section documents the main methods used to generate data and the analysis thereof.

5.3 Materials and Methods

This research investigates the Victoria Falls water flow regimes by comparing two half centuries marked by 1919/20–1968/69 (Half Century I) and 1969/70–2018/19 (Half Century II). To this end, the objectives of the paper are to: (1) determine the nature of the water flow regimes between the two half centuries; (2) establish if the trends in water flow regimes are of a temporary or more permanent nature and (3) provide some policy recommendations regarding possible measures from governments and stakeholders sharing the waterfall and the Zambezi River Basin that hosts it.

The Zambezi River Authority (ZRA), which is the custodian of the portion of the Zambezi River under study shared by both Zambia and Zimbabwe, generated the time-series run-off data. The data includes annual, monthly and daily mean flows in Cumecs measured at the Big Tree hydrologic station just before the Victoria Falls. The use of the outlined data sets in river and falls flow regimes is common practice (Glas et al., 2019; Cheng et al., 2019; Choudhury et al., 2019). While data sets for annual and monthly means spread over the entire 100 years of interest, daily flows leave out 5 years from 1919/20 to 1924/25. Figure 5.1 presents the location of the study area and the Big Tree flow gauging station.

The analysis started by plotting simple annual mean time-series trends comparing the two half centuries. In addition, decadal means computations and detailed daily means across the two half centuries were also done. Other analysis included decadal comparisons between the two half centuries. To determine if the observed trends in the time-series data were statistically significant or just changing within the normal cycle of events, the Mann-Kendall method was applied with 37 such tests done. The Mann-Kendall test is a non-parametric test for the detection of trends in time-series data. Auto-correlation was also done. Auto-correlation can be defined as the correlation of a time series with its future and past values. This normally occurs because physical time series such as run-off frequently have inertia or carry-over effects, which can complicate the application of statistical tests by reducing the number of independent observations and increasing the chances of detecting significant trends when in fact they are absent. Before the Mann-Kendall test, data was first tested for serial correlation. The highlighted analysis methods are common when studying flow regimes as these have been applied by Han et al. (2019), Dube and Nhamo (2018, 2019), Pirnia et al. (2019) and Dinpashoh et al. (2019).

During detailed analysis of the daily flows, the authors were interested in determining trends that included establishing if there could be movements in peak and low flows and the months such were identified. Furthermore, the analysis teased out the daily maximum, minimum and mean flows. The key findings are presented in the next section.

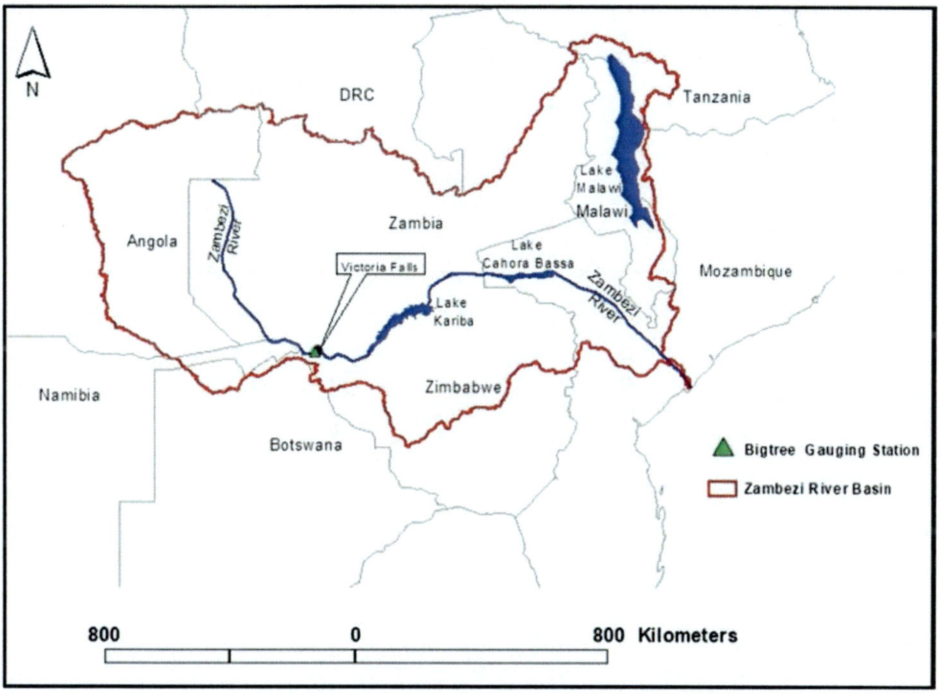

Fig. 5.1 Locations of the study area and Big Tree flow gauging station. (Source: Authors)

5.4 Presentation of Data and Discussion of Findings

This section consists of three sub-sections. These sub-sections include annual and decadal mean flow analysis, daily peak and low flow migrations, and daily mean flow regime comparisons. Each of these sections will be considered in much depth in the following paragraphs.

5.4.1 Annual and Decadal Mean Flow Analysis

To get a general view concerning the annual mean flows from the two half centuries, a trend analysis was done (Fig. 5.2). It emerged that there was a general upward trend in flows in Half Century I, while Half Century II shows a decline in flow regimes. While the peak in Half Century II came in 1977/78 at 1889 Cumecs, there were three peaks above this value in Half Century I for 1957/58 at 2328 Cumecs (all-time high for the century and entire period with data from 1907/08), 1968/69 with 2137 Cumecs and 1962/63 with 2112 Cumecs. The average flow for the half centuries were 1152 Cumecs for Half Century I and 1060 Cumecs for Half Century II.

The decadal comparison, another common analysis in flow regimes and climate change and variability-related studies, was done for the two half centuries. Apart from the first set of decades 1919/20 to 1929/30 and 1969/70 to 1979/80 where Half Century II had a higher mean, the remaining four sets of decades recorded higher means for Half Century I. In fact, the last two decades for Half Century I recorded volumes higher than the peak of Half Century II, which stood at 1309 Cumecs, compared to 1496 Cumecs and 1485 for the last two decades, respectively. Details are shown in Fig. 5.3.

To conclude the analysis in this section, the five lowest annual mean records were determined for the two half centuries. On average, Half Century I registered 572 Cumecs in the selected 5 years compared to 501 for the 5 years in Half Century II. For Half Century I the lowest annual mean flow regimes were in the 1948/49, 1921/22,

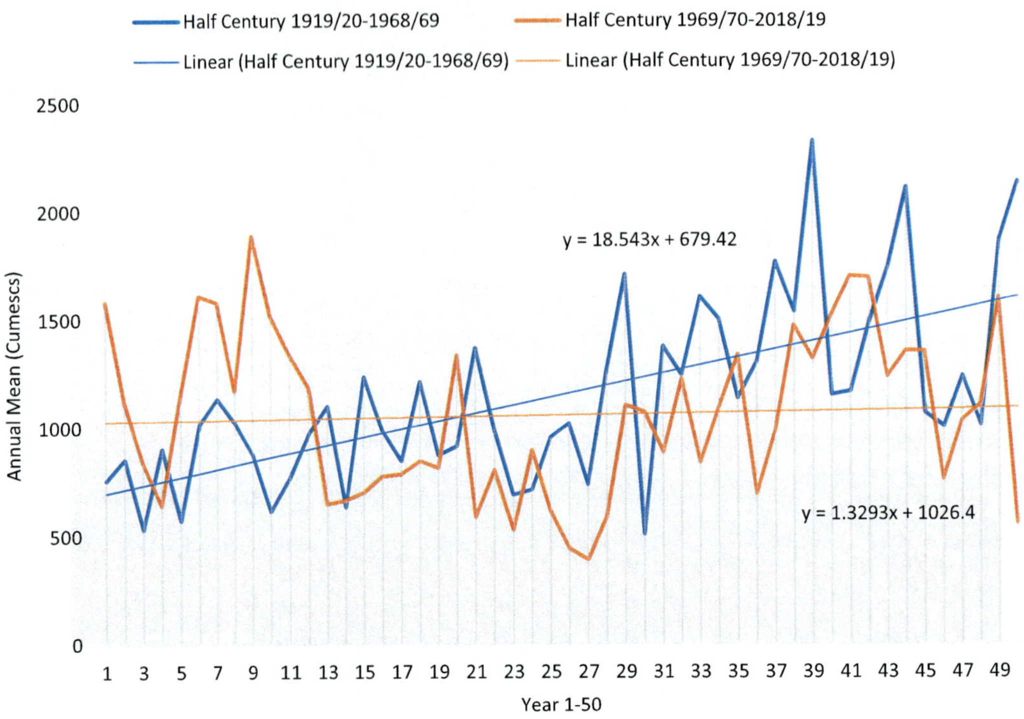

Fig. 5.2 Tale of two half centuries annual mean flows. (Source: Authors)

1923/24, 1928/29 and 1932/33 seasons reported as 508, 532, 572, 614 and 634 Cumecs in that order. As for Half Century II, the lowest annual mean flow regimes were in 1995/96, 1994/95, 1991/92, 2018/19 and 1989/90 seasons and were reported as 390, 442, 528, 557 and 588 Cumecs in that order. A replica analysis was made for the peak annual mean flows. For Half Century I the peak annual mean flow regimes were in the 1957/58, 1968/69, 1962/63, 1967/68 and 1955/56 seasons and were reported as 2328, 2137, 2112, 1864 and 1769 Cumecs in that order. As for Half Century II, the peak annual mean flow regimes were in the 1977/78, 2009/10, 2010/11, 1974/75 and 2017/18 seasons and were reported as 1889, 1702, 1697, 1610 and 1604 Cumecs in that order. On average, Half Century I recorded 2042 Cumecs in the 5 years under consideration compared to 1700 Cumecs for the 5 years in Half Century II. The differences between the two half centuries are huge and significant and it generally points to a wetter Half Century I and drier Half Century II.

5.4.2 Peak and Lows Daily Flow Migrations

Since tourists are interested in viewing the waterfalls at their peak flow as alluded to earlier (Hudson, 2002; Dube & Nhamo, 2018), the daily flows across all 12 months of the year for the 50 years were analysed. A total to 540 months were analysed for Half Century I and 600 months for Half Century II. The first analysis looked at the daily maximum flows (Fig. 5.4). Results show a decline in mean peak flows for March from 20% in Half Century I to 13% in Half Century II. There was a similar, but drastic reduction in the percentage of mean peak daily flows for April. The percentage decreased from 58% registered in Half Century I to a mere 37% recorded for Half Century II. The gain went to peak mean daily flows for May for Half Century II. There was a 227.27% increase with a 22% reported for May peak daily flows in Half Century I, compared to 50% in Half Century II. The significant migration of mean peak daily flows

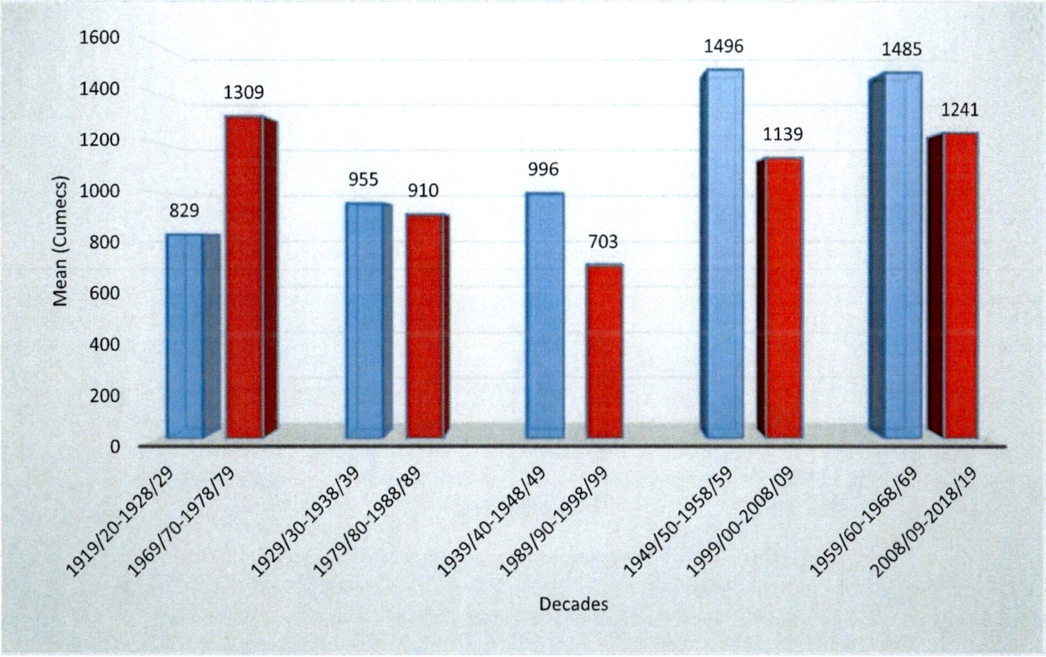

Fig. 5.3 Decadal mean flow comparisons. (Source: Authors)

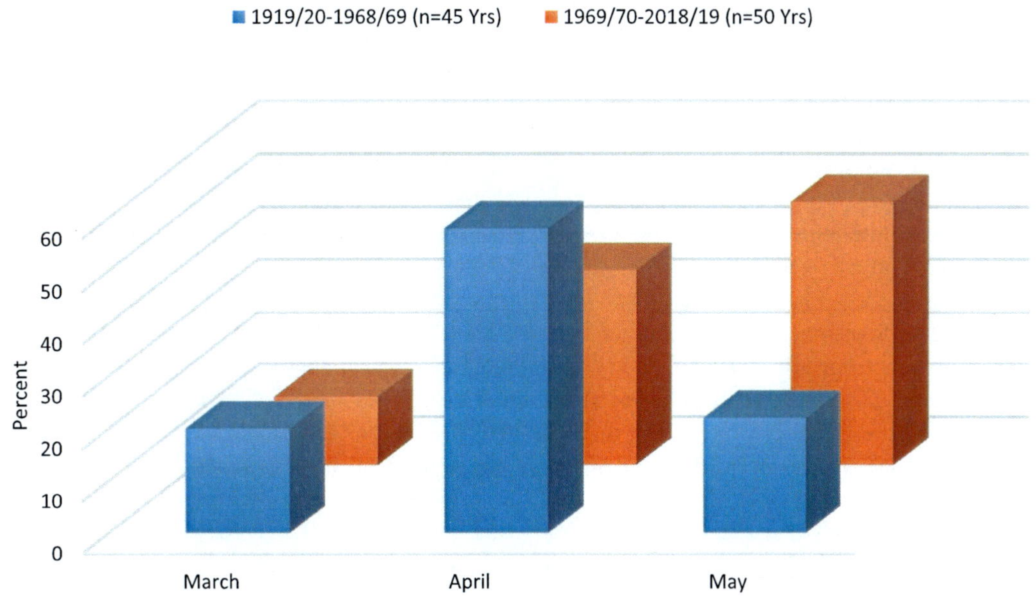

Fig. 5.4 Percentage daily maximum flows for March April and May. (Source: Authors)

from April to May could be linked to the delayed onset of the rains as observed by Dube and Nhamo (2018, 2019). This observation is significant given that the Victoria Falls traditionally sells April as the month of peak daily mean flows. For the record, the peak daily mean flows were recorded in 1957/58 and stood at 9436 Cumecs. This figure was recorded consecutively for 5 days from 5–9 March 1958 for Half Century I. Hence, tourists who visited Victoria Falls on these

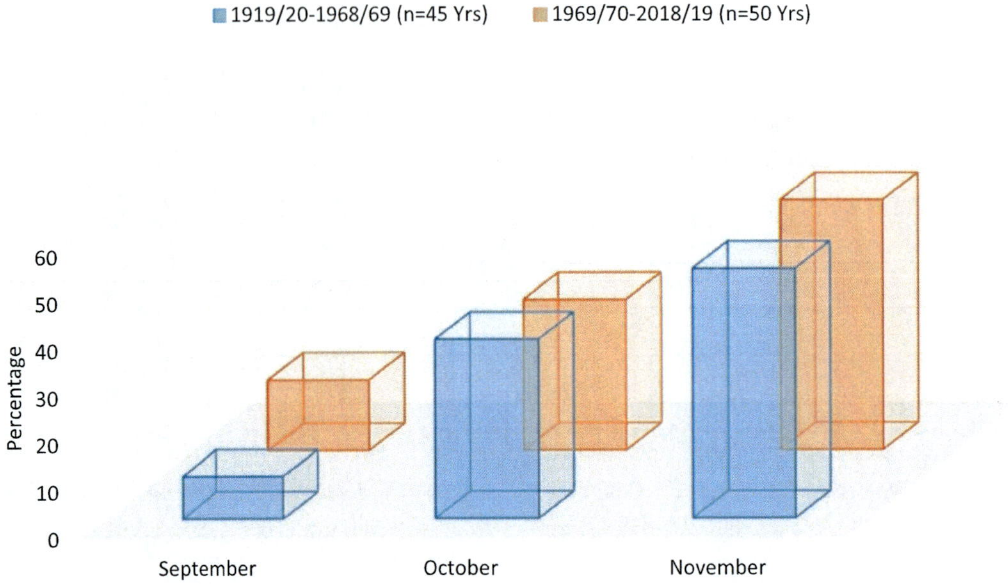

Fig. 5.5 Percentage daily minimum flows for September, October and November. (Source: Authors)

5 days, had the most spectacular experience on record since records on daily flow regimes started in 1924/25. The peak daily flow was recorded at 6172 Cumecs and this was done on two occasions on 26 March and 1 May 1978 for Half Century II. Overall, the peak daily mean flows for the two half centuries show a huge difference of 3264 Cumecs.

Trends for minimum daily mean flows were also plotted and these are shown in Fig. 5.5. What emerges from Fig. 5.5 is that more lows are coming in September in Half Century II as these increased from 9% in Half Century I to 15%. There are also less lows in October in Half Century II as these reduced from 38% in Half Century I to 32%. However, the data and analysis shows no change for lows in November for the two half centuries that share an exact 53% apiece. Given that, October has been known as the month that shocked Victoria Falls visitors with its low flows (Baynes, 2019; Connolly, 2019). Data and the analysis done present yet another picture. The month bringing up more lows in both half centuries is November. Additional revelations will come from the month-by-month daily mean analysis, especially for December.

For the record, the daily mean lowest flows were recorded in the 1925/26 and 1996/97 for Half Century I and Half Century II, respectively. The daily mean flows stood at 149 and 92 Cumecs, respectively, and these figures were recorded on 10 November 1925 and for 2 days on 12 and 13 November 1996. These figures are in sharp contrast and further confirm the severity of reductions in daily flows at Victoria Falls in Half Century II. The 92 Cumecs is currently the all-time low from available daily data recorded since 1924/25. However, as weather conditions remained dry in southern Africa, with regional and global outcries of a drought in 2019/2020 that is already being toted to surpass all those experienced in the region since 1981 (OCHA, 2020). There remains a real chance that the all-time low record of 92 Cumecs could be broken.

5.4.3 Daily Mean Flow Regime Comparisons

Since tourists visit the Victoria Falls on a daily basis, a month-by-month comparison between the half centuries was deemed necessary. This permitted the authors to have a holistic picture in preparation of comprehensive policy and further research recommendations. The analysis was categorised into two segments signalling the beginning and end of both wet and dry seasons. The wet (peak) season commences in November ending in April, while the dry season is marked by May ending in October. The trends for the wet season are shown in Fig. 5.6a–f. Generally, the 6 months all show declining trends in daily mean flow regimes for Half Century II compared to Half Century I. Three of the months, namely, November, December and January, present negative trends, with the highest negative decline recorded in December. Since December is becoming drier, with reduced daily mean flow regimes, this has an additional negative bearing on January that is also witnessing negative daily mean flows. The minimum disruptions in daily flow regimes seem to fall in February. With these trends, it comes as no surprise that March and April flows have witnessed significantly reduced flows in Half Century II. The delayed onset of the rain season and increasing temperatures in the Victoria Falls area have been documented by Dube and Nhamo (2018, 2019).

The next set of trends focus on the dry season marked in this work by the period May to October. The emerging picture is shown in Fig. 5.7a–f. Overall, the 6 months all show declining trends in daily mean flow regimes for Half Century II compared to Half Century I. However, August, September and October all present negative trends, with the highest negative decline recorded in October. The declining and negative daily flow regimes for these 3 months leading to the wet season imply delayed run-off as the entire Zambezi River basin will be dry, thereby absorbing more. The minimum disruptions in daily flow regimes seem to fall in June with similar trends for May and July. The 3 months from May to July generally have the lowest temperatures in southern Africa and for the study area as well. Hence, part of the decline in fluctuations in daily mean flows (although at reduced levels) could be a response to these low temperatures. With rapid deterioration in daily flows from August through to October, tourists are likely to be quick to notice the declines, especially in October, thereby generating much noise on social media platforms such as that witnessed from Zambian President Edgar Lungu in October 2019 and the tweets that followed.

To sum up, the overall trends indicate negative declines for 6 months in Half Century II.

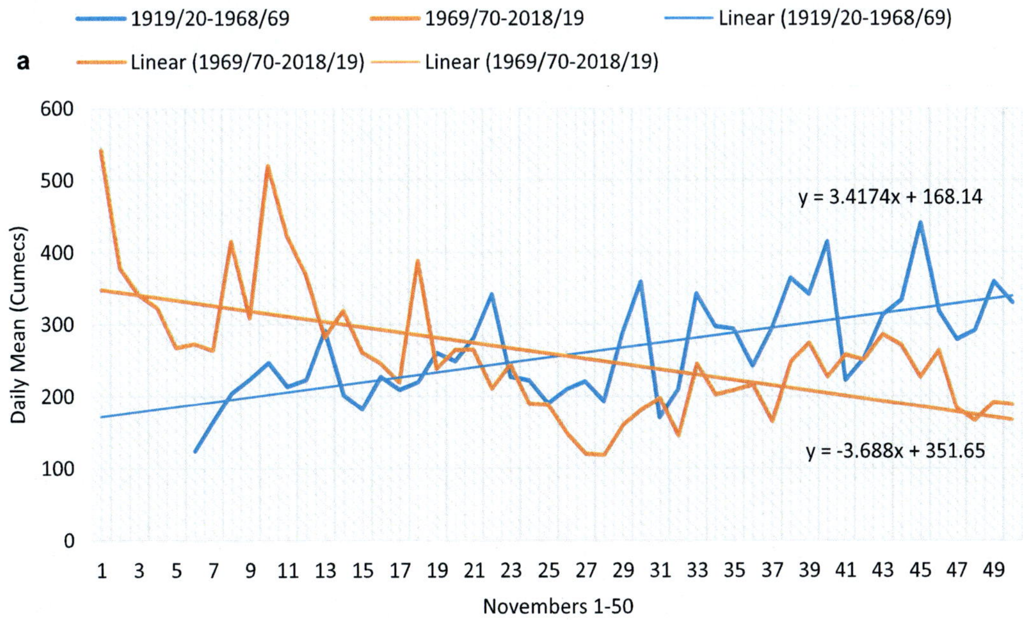

Fig. 5.6 Daily mean flows for (**a**) November, (**b**) December, (**c**) January, (**d**) February, (**e**) March, (**f**) April. (Source: Authors)

b

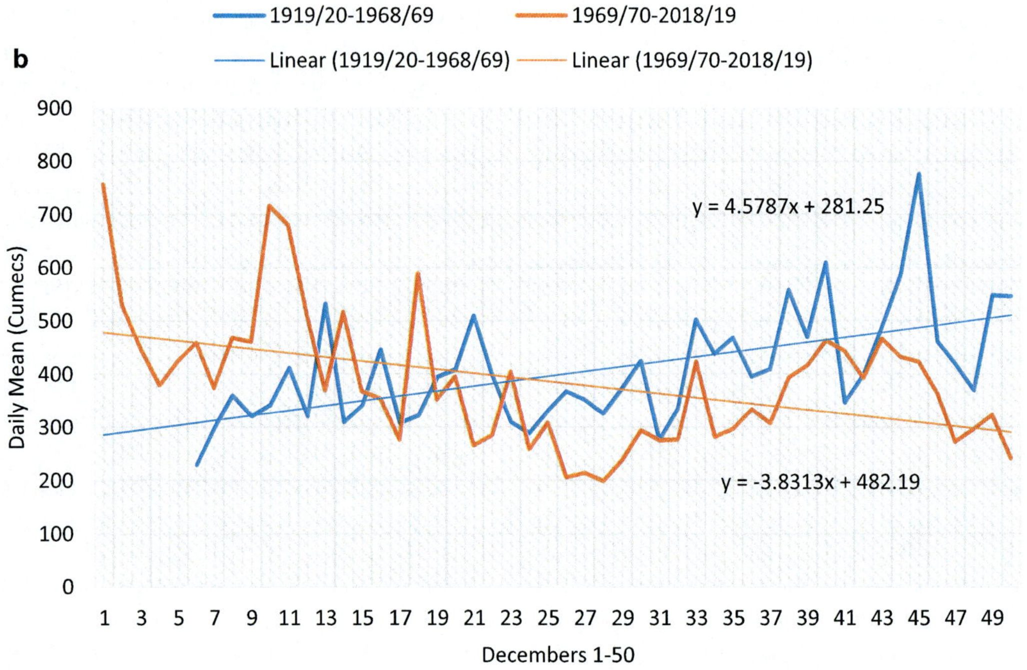

c

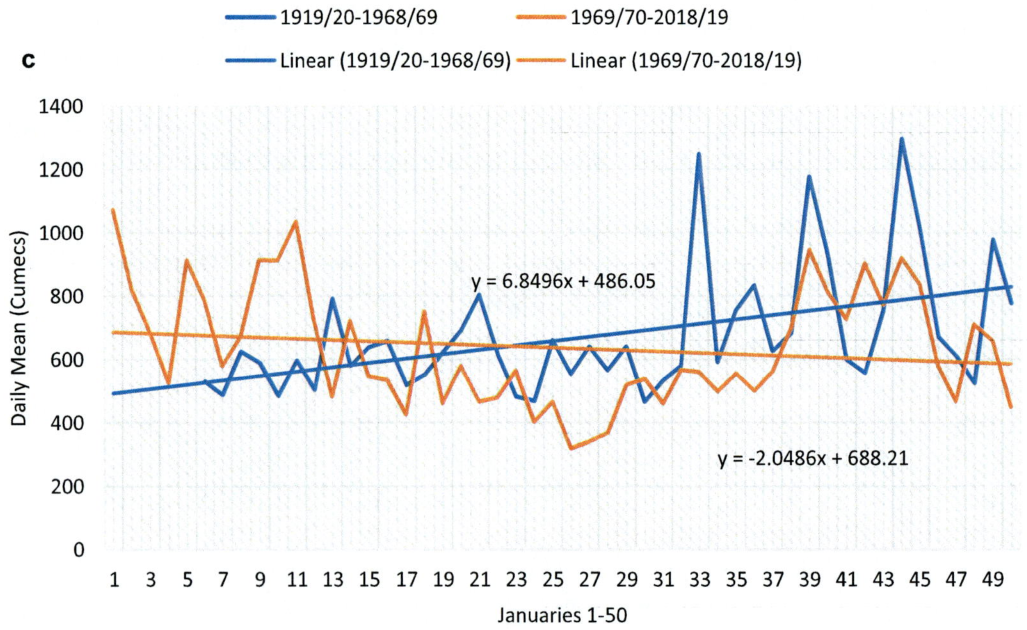

Fig. 5.6 (continued)

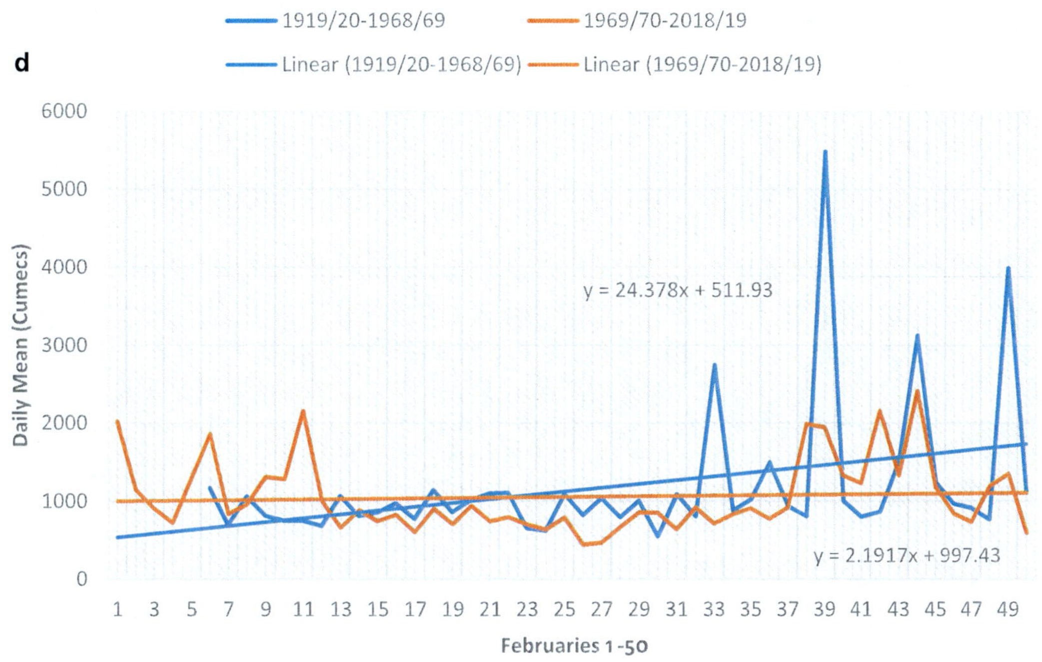

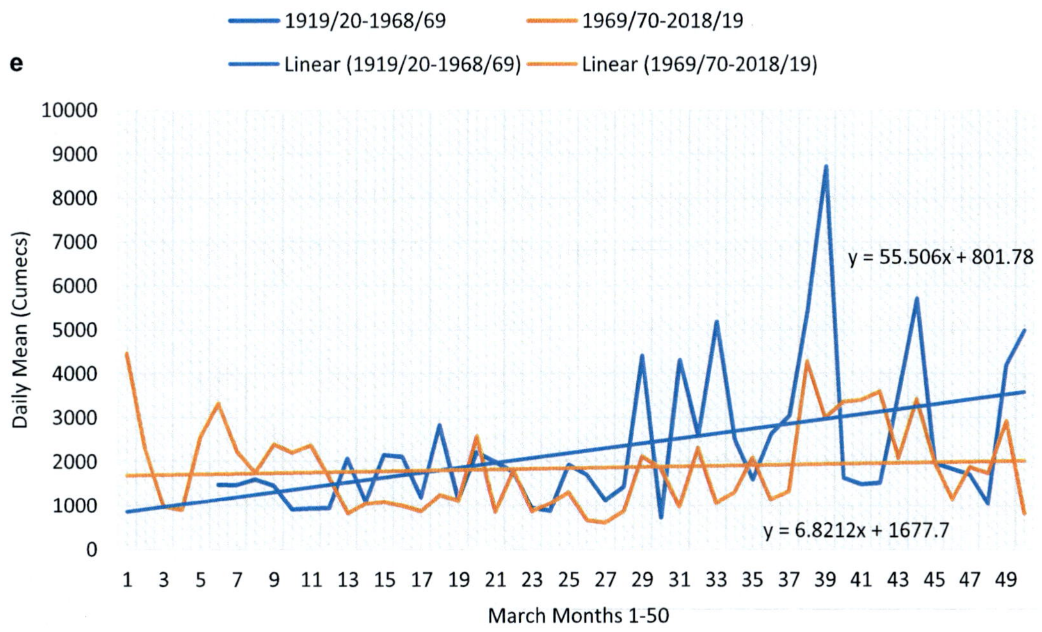

Fig. 5.6 (continued)

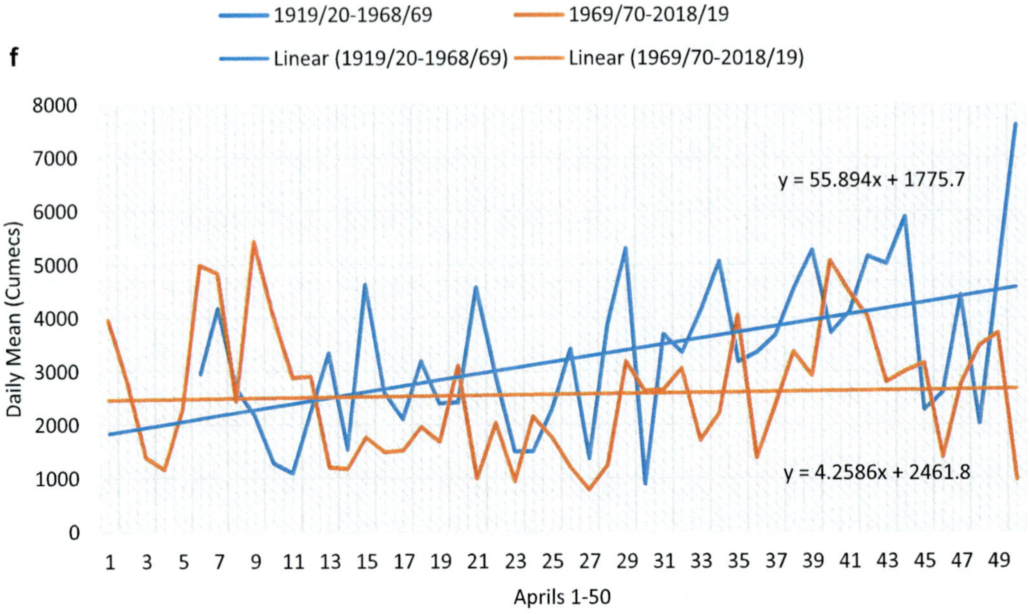

Fig. 5.6 (continued)

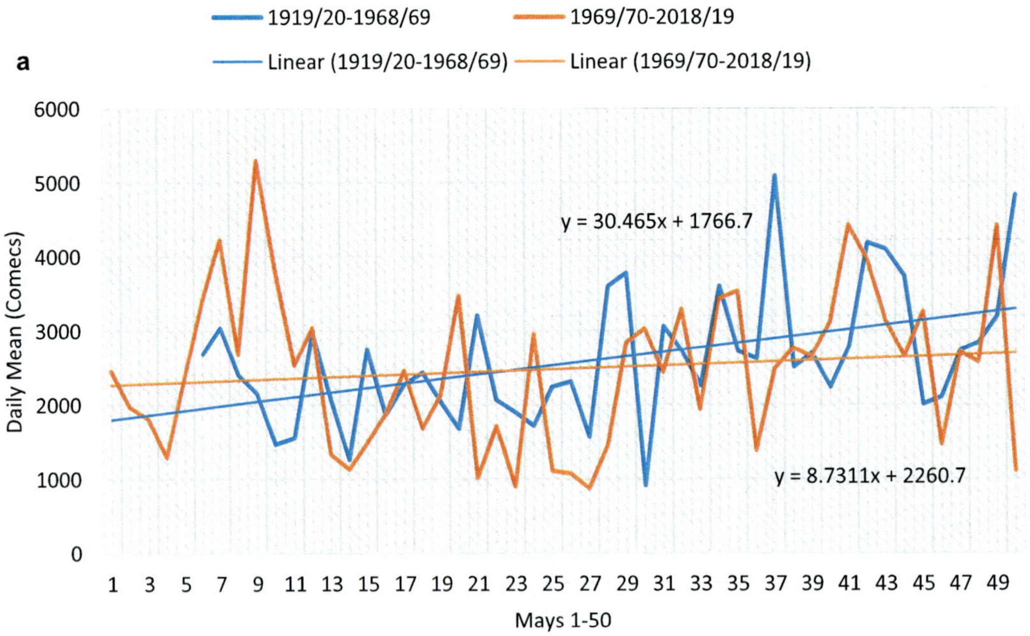

Fig. 5.7 Daily mean flows for (**a**) May, (**b**) June, (**c**) July, (**d**) August, (**e**) September, (**f**) October. (Source: Authors)

b

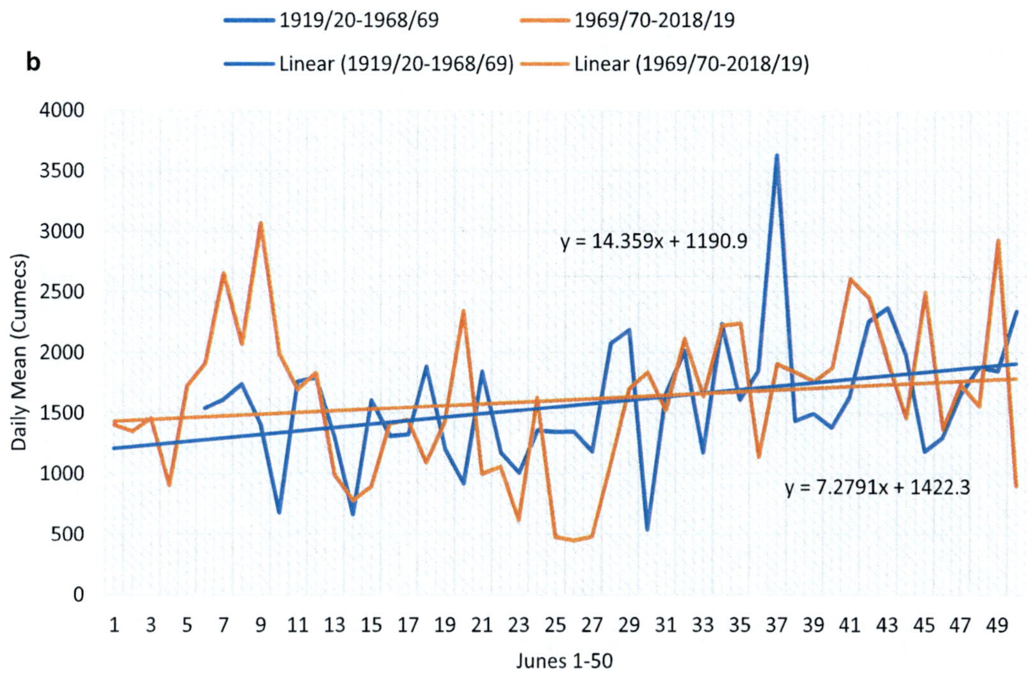

c

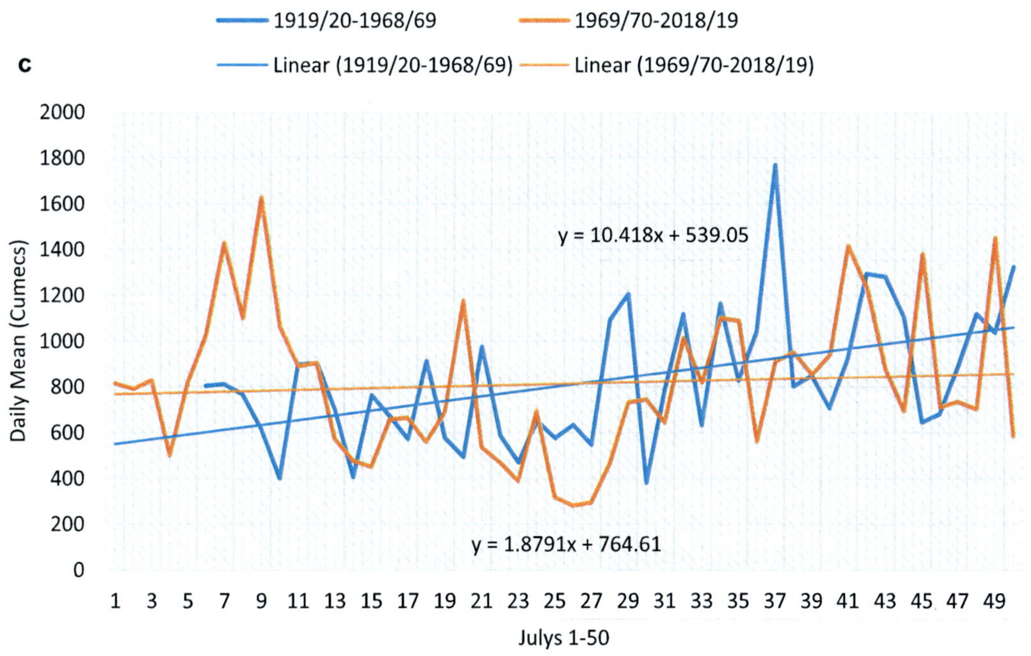

Fig. 5.7 (continued)

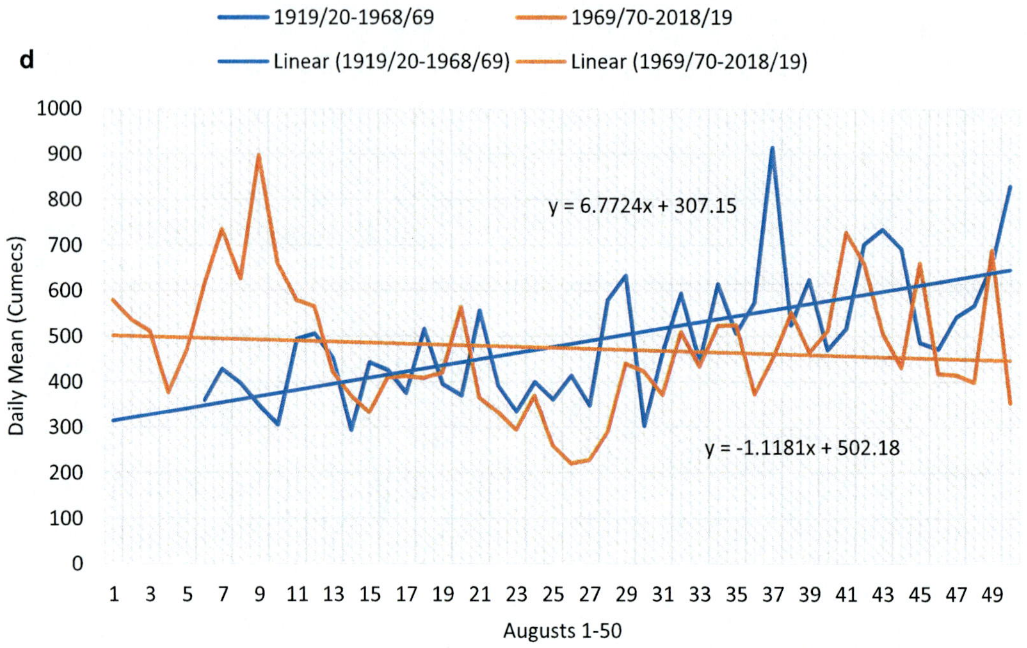

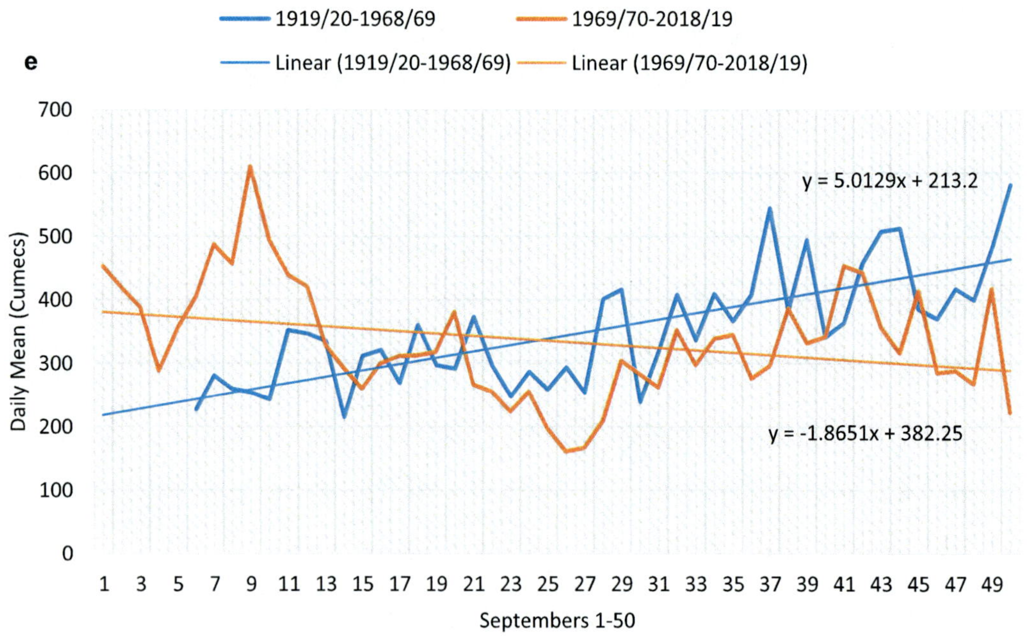

Fig. 5.7 (continued)

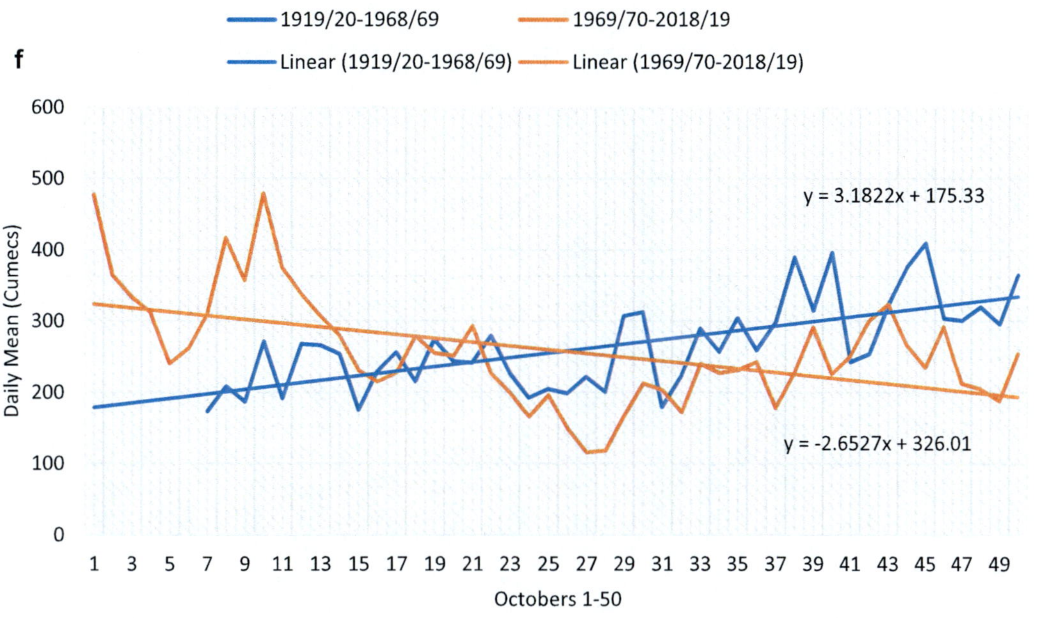

Fig. 5.7 (continued)

These months are January, August, September, October, November and December. Although still on reduced daily mean flow regimes compared to Half Century I, there are traces of an increasing trend for May, June and July. However, these are insignificant. Furthermore, February is the only month with minimal variance in daily mean flow regimes. As reflected in the literature section, flow regime changes during the dry seasons are mainly due to climate change as opposed to anthropogenic activities that include, especially, water abstractions for agriculture in the catchment area. Given that there will be limited agricultural activity in the Zambezi River basin in winter, coupled with cooler temperatures, there is likely to be limited variations in daily flow regimes. Studies by Dube and Nhamo (2018, 2019) further show increasing temperature trends that could result in increased evaporation. The daily mean trends for the two half centuries are presented in Fig. 5.8.

The daily mean trends for the two half centuries confirm a wetter Half Century I in comparison to a drier Half Century II. All other months, apart from June, show higher daily mean flows for Half Century I. There are also huge differences in daily mean flows for March and April, with drastic reductions in Half Century II. Once more, the confirmed drastic reductions in daily mean flows for peak months of March and April in Half Century II imply that tourists are generally experiencing less in terms of peak volume flows than before. Overall, to check the significance of trends across the months and years, 37 Mann-Kendall test were undertaken. Table 5.1 presents a summary of the results.

From the 37 Mann-Kendall tests done (one of which was for the entire 100 years on mean annual flow), all the mean daily water flow regimes for the 12 months for Half Century I, recorded significant increases. This is in contrast to half (six) of the months for Half Century II that recorded insignificant increases in water flow regimes. On the other hand, of the 6 months on the decline in Half Century II, three of these, October, November and December, recorded significant declines. Overall, and considering the entire 100 years, 7 months show a declining trend although not significant.

Emerging findings confirm similar trends elsewhere. In Sri Lanka, Dissanayaka and

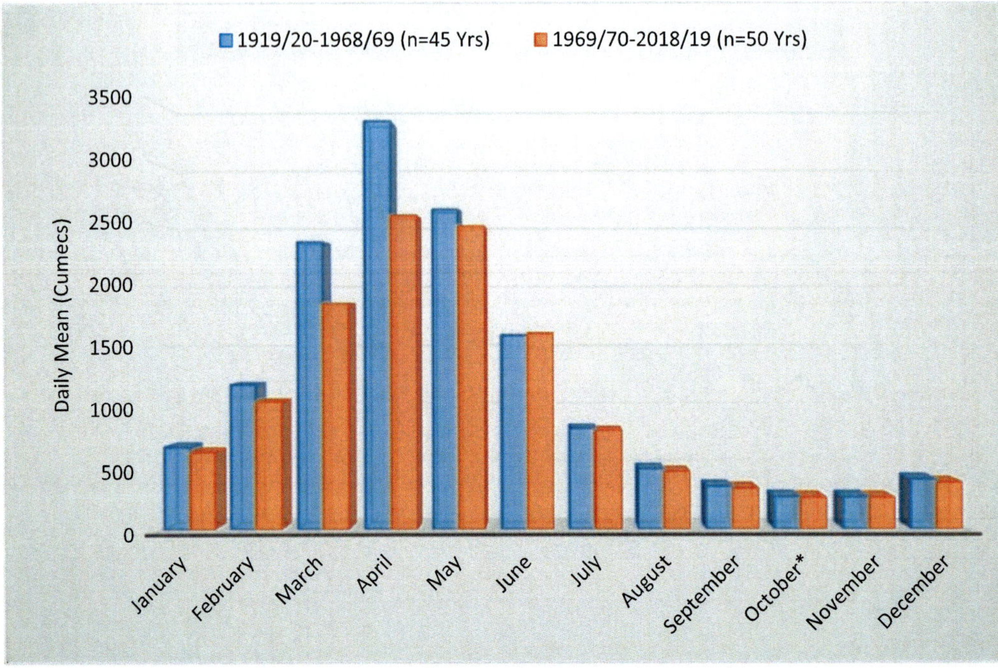

Fig. 5.8 Daily mean trends for the two half centuries. *October's mean comes out of 44 years. (Source: Authors)

Rajapakse (2019) focused on long-term rainfall trends and climate extremes in Kelani River basin. Their findings suggest that temperature and rainfall extremes were increasing, while annual mean precipitation in the catchment was decreasing. The northeastern part of India has been rapidly drying since 1979 and this phenomenon has attracted academic and other interests. The region is home to the Himalayan mountain range. Choudhury et al. (2019) raised a question as to whether this rapid drying is due to climate change or natural variability. Northeast India is considered the wettest area on the planet, yet it has witnessed a decrease of about 355 mm in summer monsoon rainfall in the 36 years between 1979 and 2014. The authors concluded that the decrease in rainfall is the result of the strong interdecadal variability of the subtropical Pacific Ocean. In fact, some of the places in Northeast India that used to receive an annual rainfall of more than 3000 mm during the monsoon period reveal drying of a maximum of 1000 mm (20–35% of the mean). Although there are small pockets where precipitation has increased, the decrease is widespread. Modelling work done to assess climate change impacts on river flow regimes in the 11 sub-basins of the Tonle Sap Lake Basin of Cambodia for three time periods up to the 2030s, 2060s and 2090s revealed the likelihood of decreased flows in both wet and dry seasonal flows (Oeurng et al., 2019). The projected reductions in flows range from 9% to 29%, 10% to 35% and 7% to 41% for the three periods in that order.

Glas et al. (2019) focused on the historical changes in New York State stream flow regimes using 56 years of daily flow data from a network of 97 gauging stations. The results showed that stream flow is strongly related to precipitation, and that across the state, winter-spring centre-volume date has migrated to earlier periods accompanied with increasing January stream flows. In Turkey, Balov and Altunkaynak (2019) researched the Melen and Munzur catchments. From their model, they found that the total water

Table 5.1 Mann-Kendall test at alpha 0.05

Period	Trend line equation	*P* value	Trend description
January 1919/20–1968/69	$y = 6.8496x + 486.05$	0.003	Significantly increasing
January 1969/70–2018/19	$y = -2.0486x + 688.21$	0.432	Declining but not significant
January Composite 1919/20–2018/19	$y = -0.3233x + 672.49$	0.738	Declining but not significant
February 1919/20–1968/69	$y = 24.378x + 511.93$	0.043	Significantly increasing
February 1969/70–2018/19	$y = 2.1917x + 997.43$	0.854	Increasing but not significant
February Composite 1919/20–2018/19	$y = 0.7402x + 1084.5$	0.893	Increasing but not significant
March 1919/20–1968/69	$y = 55.506x + 801.78$	0.004	Significantly increasing
March 1969/70–2018/19	$y = 6.8212x + 1677.7$	0.503	Increasing but not significant
March Composite 1919/20–2018/19	$y = -1.5092x + 2168.8$	0.959	Declining but not significant
April 1919/20–1968/69	$y = 55.894x + 1775.7$	0.001	Significantly increasing
April 1969/70–2018/19	$y = 4.2586x + 2461.8$	0.340	Increasing but not significant
April Composite 1919/20–2018/19	$y = -5.7387x + 3207.6$	0.403	Declining but not significant
May 1919/20–1968/69	$y = 30.465x + 1766.7$	0.005	Significantly increasing
May 1969/70–2018/19	$y = 8.7311x + 2260.7$	0.270	Increasing but not significant
May Composite 1919/20–2018/19	$y = 2.5306x + 2426.3$	0.429	Increasing but not significant
June 1919/20–1968/69	$y = 14.359x + 1190.9$	0.019	Significantly increasing
June 1969/70–2018/19	$y = 7.2791x + 1422.3$	0.146	Increasing but not significant
June Composite 1919/20–2018/19	$y = 2.8685x + 1465.3$	0.115	Increasing but not significant
July 1919/20–1968/69	$y = 10.418x + 539.05$	0.003	Significantly increasing
July 1969/70–2018/19	$y = 1.8791x + 764.61$	0.493	Increasing but not significant
July Composite 1919/20–2018/19	$y = 1.1166x + 768.31$	0.338	Increasing but not significant
August 1919/20–1968/69	$y = 6.7724x + 307.15$	<0.0001	Significantly increasing
August 1969/70–2018/19	$y = -1.1181x + 502.18$	0.514	Declining but not significant
August Composite 1919/20–2018/19	$y = 0.1134x + 480.56$	0.550	Increasing but not significant
September 1919/20–1968/69	$y = 5.0129x + 213.2$	<0.0001	Significantly increasing
September 1969/70–2018/19	$y = -1.8651x + 382.25$	0.079	Declining but not significant
September Composite 1919/20–2018/19	$y = -0.1176x + 350.46$	0.787	Declining but not significant
October 1919/20–1968/69	$y = 3.1822x + 175.33$	<0.0001	Significantly increasing
October 1969/70–2018/19	$y = -2.6527x + 326.01$	0.002	Significantly declining
October Composite 1919/20–2018/19	$y = -0.2629x + 275.26$	0.592	Declining but not significant
November 1919/20–1968/69	$y = 3.4174x + 168.14$	<0.0001	Significantly increasing
November 1969/70–2018/19	$y = -3.688x + 351.65$	<0.0001	Significantly declining
November Composite 1919/20–2018/19	$y = -0.3749x + 279.81$	0.368	Declining but not significant
December 1919/20–1968/69	$y = 4.5787x + 281.25$	<0.0001	Significantly increasing
December 1969/70–2018/19	$y = -3.8313x + 482.19$	0.009	Significantly declining
December Composite 1919/20–2018/19	$y = -0.5949x + 426.35$	0.368	Declining but not significant
Overall 1919/20–2018/19	$y = -0.122x + 1133.2$	0.814	Declining but not significant

Source: Authors

volume would decrease in response to precipitation diminution and temperature increase during the mid-period of 2041–2070 and late future period of 2071–2100 in both watersheds. The predicted decrease in run-off volume was 15.42% and 26.65% on average for the Melen and Munzur rivers, respectively, with no change in monthly distribution of the run-off during the current century.

5.5 Conclusion

What is clear from this work is a tale of two half centuries in terms of the Victoria Falls water flow regimes. Half Century I (1919/20–1968/69) had higher flows compared to Half Century II (1969/70–2018/19). These trends were confirmed by several analyses that included annual means, decadal means and overall daily means. Although our main focus was not to determine the causes of the changes in flow regimes, recent studies confirmed that these are linked to both climate change and variability, as well as anthropogenic activities. It also emerged that of the 6 months on the decline in Half Century II, three of these, October, November and December, recorded significant declines. Overall, and considering the entire 100 years, 7 months show a declining trend although not significant. In addition, peak flows have been shifting from the commonly publicised April to May for Half Century II. However, the results show that the water flow regime fluctuations are more of a permanent, rather than temporary, nature. To this end, the results challenge the traditional and the much preferred climate variability narrative. This narrative remains a convenient truth as selected stakeholders such as tour operators, tourists and some politicians wish to delay the inevitable. The facts based on the two half centuries comparison approach confirm more of a permanent lower flow regimes equilibrium in Half Century II. Should the population increase and other anthropogenic development activities such as industrialisation, expanding irrigation, agriculture and domestic use of water upstream of the Victoria Falls be brought in, there is a real possibility of a completely dry Victoria Falls in the dry season, particularly in October, November and December. We therefore raise one policy recommendation through a rhetoric question: Is it not time that responsible Victoria Falls stakeholders accept the reality of a possibility of the drying falls, especially in the dry season, and put in place appropriate adaptation and resilience-building measures. Time may be running out!

References

Admin. (2019). *The false death of Victoria falls*. Retrieved January 2, 2020, from https://www.fr24news.com/2019/12/the-false-death-of-victoria-falls-the-society.html

Ahmed, Y., Al-Faraj, F., Scholz, M., & Soliman, A. (2019). Assessment of upstream human intervention coupled with climate change impact for a transboundary river flow regime: Nile river basin. *Water Resources Management, 33*, 2485–2500. https://doi.org/10.1007/s11269-019-02256-1.

Allen, F. (2019). *Shocking pictures show Victoria Falls dried up after devastating drought leaves it with lowest water level in 25 years*. Retrieved January 2, 2020, from https://www.thesun.co.uk/news/10496407/victoria-falls-dried-up-devastating-drought-pictures/

Balov, M. N., & Altunkaynak, A. (2019). The impacts of climate change on the runoff volume of Melen and Munzur Rivers in Turkey based on calibration of WASMOD model with multiobjective genetic algorithm. *Meteorology and Atmospheric Physics, 132*, 85–98. https://doi.org/10.1007/s00703-019-00676-7.

Baynes, C. (2019). *Victoria Falls: One of world's natural wonders at risk of drying up because of climate change, warns Zambian president*. Retrieved January 2, 2020, from https://www.independent.co.uk/news/world/africa/victoria-falls-climate-change-crisis-drought-zambia-zimbabwe-natural-wonders-a9231196.html

Beck, L., & Bernauer, T. (2011). How will combined changes in water demand and climate affect water availability in the Zambezi River basin? *Global Environmental Change, 21*, 1061–1072. https://doi.org/10.1016/j.gloenvcha.2011.04.001.

Buttle, J., Muir, T., & Frain, J. (2004). Economic impacts of climate change on the Canadian Great Lakes HydroElectric power producers: A supply analysis. *Canadian Water Resources Journal, 29*(2), 89–110. https://doi.org/10.4296/cwrj089.

Byun, K., Chiu, C. M., & Hamlet, F. (2019). Effects of 21st century climate change on seasonal flow regimes and hydrologic extremes over the Midwest and Great Lakes region of the US. *Science of the Total Environment, 650*, 1261–1277. https://doi.org/10.1016/j.scitotenv.2018.09.063.

Chatterjee, S., Daniels, M. D., Sheshukov, A. Y., & Gao, J. (2018). Projected climate change impacts on hydrologic flow regimes in the Great Plains of Kansas. *River Resilience Application, 34*, 195–206. https://doi.org/10.1002/rra.3249.

Cheng, J., Xu, L., Feng, W., Fan, H., & Jiang, J. (2019). Changes in water level regimes in China's two largest freshwater lakes: Characterization and implication. *Water, 11*, 917. https://doi.org/10.3390/w11050917.

Choudhury, B. A., Saha, S. K., Konwar, M., Sujith, K., & Deshamukhya, A. (2019). Rapid drying of Northeast India in the last three decades: Climate change or natural variability? *Journal of Geophysical Research: Atmospheres, 124*, 227–237. https://doi.org/10.1029/2018JD029625.

Cole, E. (2015). Impetuous Torrents: Scottish Waterfalls in Travellers' Narratives, 1769–1830. *Scottish Geographical Journal, 131*(1), 49–66. https://doi.org/10.1080/14702541.2014.988287.

Colombo, S. J., McKenney, D. W., Lawrence, K. M., & Gray, P. A. (2007). *Climate change projections for Ontario: Practical information for policymakers and planners*. Marie: Ontario Ministry of Natural Resources.

Connolly, L. (2019). *World-famous Victoria falls all but dries up in devastating drought*. Retrieved January 2, 2020, from https://www.unilad.co.uk/news/world-famous-victoria-falls-all-but-dries-up-in-devastating-drought/

De Jalón, S. G., Del Tánago, M. G., & De Jalón, D. G. (2019). A new approach for assessing natural patterns of flow variability and hydrological alterations: The case of the Spanish rivers. *Journal of Environmental Management, 233*, 200–210. https://doi.org/10.1016/j.jenvman.2018.12.049.

Dinpashoh, Y., Singh, V. P., Biazar, S. M., & Kavehkar, S. (2019). Impact of climate change on streamflow timing (case study: Guilan Province). *Theoretical and Applied Climatology, 138*, 65–76. https://doi.org/10.1007/s00704-019-02810-2.

Dissanayaka, K. D. C. R., & Rajapakse, R. L. H. L. (2019). Long-term precipitation trends and climate extremes in the Kelani River basin, Sri Lanka, and their impact on streamflow variability under climate change. *Paddy and Water Environment, 17*, 281–289. https://doi.org/10.1007/s10333-019-00721-6.

Dube, K., & Nhamo, G. (2018). Climate variability, change and potential impacts on tourism: Evidence from the Zambian side of the Victoria Falls. *Environmental Science and Policy, 84*, 113–123. https://doi.org/10.1016/j.envsci.2018.03.009.

Dube, K., & Nhamo, G. (2019). Climate change and potential impacts on tourism: Evidence from the Zimbabwean side of the Victoria Falls. *Environment, Development and Sustainability, 21*, 2025–2041. https://doi.org/10.1007/s10668-018-0118-y.

Dube, T. (2019). *Victoria Falls 'bone dry'? How drought and dry season are hitting the world's largest waterfall.* Retrieved January 2, 2020, from https://factcheck.afp.com/victoria-falls-bone-dry-how-drought-and-dry-season-are-hitting-worlds-largest-waterfall

Environment and Climate Change Canada. (2017). *2017 Annual Climate Trends and Impacts Summary for the Great Lakes Basin.* Retrieved January 5, 2020, from www.glisa.umich.edu/resources/annual-climate-summary

Falchetta, G., Gernaat, D., Hunt, J., & Sterl, S. (2019). Hydropower dependency and climate change in sub-Saharan Africa: A nexus framework and evidence-based review. *Journal of Cleaner Production, 231*, 1399–1417. https://doi.org/10.1016/j.jclepro.2019.05.263.

Glas, R., Burns, D., & Lautz, L. (2019). Historical changes in New York State streamflow: Attribution of temporal shifts and spatial patterns from 1961 to 2016. *Journal of Hydrology, 574*, 308–323. https://doi.org/10.1016/j.jhydrol.2019.04.060.

Gronewold, A. D., & Rood, R. B. (2019). Recent water level changes across Earth's largest lake system and implications for future variability. *Journal of Great Lakes Research, 45*, 1–3. https://doi.org/10.1016/j.jglr.2018.10.012.

Gronewold, D. W., Fortin, V., Lofgren, B., Clites, A., Stow, C. A., & Quinn, F. (2013). Coasts, water levels, and climate change: A Great Lakes perspective. *Climatic Change, 120*, 697–711. https://doi.org/10.1007/s10584-013-0840-2.

Grower, H., Smith, M., Cao, Y. and Yang, Y. (2014). *Resilient Buffalo Niagara: Strategies to respond to climate change.* Buffalo: University at Buffalo School of Architecture and Planning.

Haghe, J. P. (2011). Do waterfalls have value in themselves? A metamorphosis in the values of the Gimel waterfall in France. *Policy and Society, 30*(4), 249–256. https://doi.org/10.1016/j.polsoc.2011.10.008.

Han, Z., Long, D., Fang, Y., Hou, A., & Hong, Y. (2019). Impacts of climate change and human activities on the flow regime of the dammed Lancang River in Southwest China. *Journal of Hydrology, 570*, 96–105. https://doi.org/10.1016/j.jhydrol.2018.12.048.

Higashino, M., & Stefan, H. G. (2019). Variability and change of precipitation and flood discharge in a Japanese river basin. *Journal of Hydrology: Regional Studies, 21*, 68–79. https://doi.org/10.1016/j.ejrh.2018.12.003.

Hudson, B. J. (1999). Fall of beauty: The story of a Jamaican waterfall—A tragedy in three acts. *Tourism Geographies, 1*(3), 343–357. https://doi.org/10.1080/14616689908721326.

Hudson, B. J. (2002). Best after rain: Waterfall discharge and the tourist experience. *Tourism Geographies, 4*(4), 440–456. https://doi.org/10.1080/14616680210158164.

Hudson, B. J. (2013). Waterfalls, science and aesthetics. *Journal of Cultural Geography, 30*(3), 356–379. https://doi.org/10.1080/08873631.2013.828482.

Hutchings, M., & Cocks, T. (2019). *Victoria Falls shrink to a trickle, feeding climate change fears.* Retrieved January 2, 2020, from https://www.moneyweb.co.za/news-fast-news/victoria-falls-shrink-to-a-trickle-feeding-climate-change-fears/

Klint, L. M., Jiang, M., Law, L., Delacy, T., Filep, S., Calgaro, E., Dominey-Howes, D., & Harrison, D. (2012). Dive tourism in Luganville, Vanuatu: Shocks, stressors, and vulnerability to climate change. *Tourism in Marine Environments, 8*(1/2), 91–109. https://doi.org/10.3727/154427312X13262430524225.

Matiashe, F. S. (2019). *Victoria Falls, one of the natural wonders of the world, is running dry due to climate change.* Retrieved January 2, 2020, from https://qz.com/africa/1765688/photos-of-victoria-falls-running-dry-in-zimbabwe-and-zambia/

McDermid, J. L., Dickin, S. K., Winsborough, C. L., Switzman, H., Barr, S., Gleeson, J. A., Krantzberg, G., & Gray, P. A. (2015). *State of climate change science in the Great Lakes Basin: A focus on climatological, hydrological and ecological effects.* Ontario: Ontario Climate Consortium and Ontario Ministry of Natural, Resources and Forestry.

New York Power Authority. (2005). *Niagara River water level and flow fluctuation study final report.* New York: New York Power Authority.

Niagara Region. (2013). *Corporate climate change action plan: Partners for climate protection: Milestones 2 and 3.* Niagara: Niagara Region.

Niagara Region. (2019). *New Niagara Official plan: Natural Environment Work Program—Watershed Planning Discussion Paper*. Niagara: Niagara Region.

OCHA (United Nations Office for the Coordination of Humanitarian Affairs). (2020). *Zambia: Prolonged drought increases food insecurity*. Retrieved January 9, 2020, from https://www.unocha.org/story/zambia-prolonged-drought-increases-food-insecurity

O'Dea, M. (2019). *Is Victoria Falls drying up?* Retrieved January 2, 2020, from https://www.lonelyplanet.com/articles/victoria-falls-drought-climate-change

Oeurng, C., Cochrane, T. A., Chung, S., Kondolf, M. G., Piman, T., & Arias, M. E. (2019). Assessing climate change impacts on river flows in the Tonle Sap Lake Basin, Cambodia. *Water, 11*, 618. https://doi.org/10.3390/w11030618.

Palmer, M. A., Liermann, C. A. R., Nilsson, C., Flörke, M., Alcamo, J., Lake, P. S., & Bond, N. (2008). Climate change and the world's river basins: Anticipating management options. *Frontal Ecology Environment, 6*(2), 81–89. https://doi.org/10.1890/060148.

Penney, J. (2012). *Adapting to climate change: Challenges for Niagara*. Niagara: Niagara Region.

Pirnia, A., Golshan, M., Darabi, H., Adamowski, J., & Rozbeh, S. (2019). Using the Mann-Kendall test and double mass curve method to explore stream flow changes in response to climate and human activities. *Journal of Water and Climate Change, 10*(4), 725–742.

Rosane, O. (2019). *Victoria Falls dries drastically after worst drought in a century*. Retrieved January 2, 2020, from https://www.ecowatch.com/victoria-falls-drought-2641553066.html

Sanderson, M., & Wong, L. (1987). The influence of climate change and climatic variability on the Hydrologie Regime and water resources. Proceedings of the Vancouver Symposium, August 1987. *IAHS, 168*, 477–487.

Shamseddin, A. M., & Chaibi, T. (2019). Scanning climate change impacts on water resources of the largest African river basins. *International Journal of River Basin Management, 18*, 33–38. https://doi.org/10.1080/15715124.2019.1576699.

Shao, G., Guan, Y., Zhang, D., Yu, B., & Zhu, J. (2018). The Impacts of Climate Variability and Land Use Change on Streamflow in the Hailiutu River Basin. *Water, 10*, 814. https://doi.org/10.3390/w10060814.

The Editor. (2019). *Editorial comment: The Victoria Falls far from dying*. Retrieved January 2, 2020, from https://www.herald.co.zw/editorial-comment-the-victoria-falls-far-from-dying/

The Union of Concerned Scientists and The Ecological Society of America (UCS-ESA). (2003). *Confronting climate change in the Great Lakes Region: Impacts on our communities and ecosystems*. New York: UCS-ESA.

Wang, Y., Tao, T., Sheng, D., Zhou, Y., Wang, D., Shi, X., Wu, J., & Ma, X. (2020). Quantifying the change in streamflow complexity in the Yangtze River. *Environmental Research, 180*, 108833. https://doi.org/10.1016/j.envres.2019.108833.

Wilcox, D. A., Thompson, T. A., Booth, R. K., & Nicholas, J. R. (2007). *Lake-level variability and water availability in the Great Lakes*. Reston: U.S. Department of the Interior/U.S. Geological Survey.

Part III

Urban Flooding and Insurance Matters

6 Hotel Management Under Increasing and More Intense Floods: A Focus on The Centurion Hotel, South Africa

Godwell Nhamo and David Chikodzi

Abstract

Hotel management is becoming more difficult in the face of increasing and repeated natural and human-induced disasters, particularly floods, cyclones and earthquakes. Hotel managers should come up with disaster risk reduction (DRR) and management plans that include early warning, search and rescue (evacuations) and relief and recovery. While there are numerous incidences of hotel infrastructure being razed to the ground, many hotels remain operational, but under the risk of repeat events. Through repeated direct observations, interviews and Geographical Information Systems, this chapter investigates the extent and potential causes of repeated flooding of The Centurion Hotel, South Africa. It further examines the DRR and management protocols put in place by the hotel and their effectiveness. The Centurion Hotel has been a victim of repeated flooding, with the most recent event in December 2019 being the worst, forcing it to close twice in 2020 affecting 84 employees. The flooding is aggravated by the heavily silted Centurion Lake and land use changes in the upper catchment of the Hennops River stretching across the three municipal jurisdictions of Ekurhuleni, Johannesburg and City of Tshwane metropoles. The results of this study further revealed that the hotel's design originally took into account the possibility of flooding, although the flood lines have since shifted and high-impact floods have become more frequent. The hotel also put into place measures to counter the effects of flooding although these were overwhelmed by the 2019 floods. The hotel has a sound standard operating procedure for the evacuation of guests in the event of a disaster. Given that such extreme weather events will only increase in the future, we recommend that The Centurion Hotel continues adapting and building back better and smarter with lessons from each flood event, including carefully planning for possible seasonal operations in the future.

Keywords

Hotels · Disaster management · Climate change · Floods · Building back better

6.1 Introduction

The hotel industry remains part of the broader tourism sector that is now at risk from both human-induced and natural disasters like floods, storms (Vila et al., 2019), earthquakes, tornadoes

G. Nhamo (✉) · D. Chikodzi
Institute for Corporate Citizenship,
University of South Africa, Pretoria, South Africa
e-mail: nhamog@unisa.ac.za; chikod@unisa.ac.za

G. Nhamo, L. Chapungu (eds.), *The Increasing Risk of Floods and Tornadoes in Southern Africa*, Sustainable Development Goals Series, https://doi.org/10.1007/978-3-030-74192-1_6

and cyclones (Ritchie, 2008; Brown et al., 2017). The hotel sector's vulnerabilities are multi-faceted, including those from climate change (Nhep et al., 2018). Hotel buildings can be severely damaged or wiped out physically by natural disasters as happened in Phuket in Thailand during the 2004 Indian Ocean Tsunami (Tang et al., 2019). If the hotels survive these catastrophes, tourists may develop negative perceptions of the establishment leading to reduced guest volumes (Ivkov et al., 2019). In Phuket, several tourism facilities went bankrupt or into a slump, because they started their business again immediately after the reconstruction, yet tourists did not return.

Understanding the generic disaster risk reduction and management (DRR&M) cycle remains fundamental in the hotel sector (Tsai & Chen, 2011). Such a DRR&M cycle is marked by disaster management readiness (sometimes called preparedness), search and rescue, and relief and recovery. However, the DRR&M cycle and model that lead to building disaster resilience[1] seem to be separate activities from commercial concerns (Brown et al., 2017). A case in point is that while managing a profitable hotel business, managers seem not to prioritise planning for continued operations following a disaster.

Both the hotel management and its staff should be up to date with procedures and processes at each one of these stages. While the hotel sector may wish to understand and mitigate the impact of these disasters, there remains a gap in terms of information, and the chaotic nature and uncertainty of the extreme weather events make it even more difficult (Brown et al., 2019). Hence, a holistic DRR&M framework is needed if the tourism industry in general and the hotel industry in particular is to remain a viable business venture. As predicted, both anthropogenic and climate-related disasters, especially floods, are likely to continue on the upward trend and become harsher (Tsai et al., 2016; Sydnor-Bousso et al., 2011).

Among the loss modification options, spreading the financial burden through relief and insurance could be included (Ritchie, 2008). However, given the escalating incidences of extreme weather events and uncertainty surrounding these events, there is a likelihood that insurance premiums will skyrocket and continue to do so soon. This is likely to drive more and more players to either under-insure or avoid these premiums completely (Arseculeratne et al., 2018). There is also the other side, where insurers will avoid covering high-risk properties. While event modification to suppress the impact of extreme weather events like floods through barrier erections, building back better (BBB) and retrofits, is an option, it remains limited, especially for old properties (Brown et al., 2017). The estimation of the disaster impact on a particular hotel in terms of its frequency, scope and vulnerability is paramount (Tsai & Chen, 2011). On studying hotel resilience to disasters, Sydnor-Bousso et al. (2011) document that the presence of roads and public buildings of a community, higher levels of religious associations in a community, and having larger numbers of college-educated persons provide at least 24% of the explanation of how the hospitality industry can lessen the negative effects of a disaster.

The Centurion hotel has experienced several flood events. However, there remains a gap in knowledge about the causes of repeated flooding at the hotel. Furthermore, the effectiveness of the hotel's DRR and management controls have not been interrogated. Given the foregoing, we set two research objectives in this chapter: (1) to determine the extent, nature and possible causes of the repeated flooding of The Centurion Hotel; and (2) to investigate and document the Centurion Hotel DRR and management protocols and their effectiveness. Flood risk may not be eliminated, but the hotel can adapt to keep the business going. For The Centurion Hotel, physical, hydro-meteorological and social perceptions remain real. The rest of the chapter includes the literature review followed by the materials and methods. The results of the study are then presented and discussed, prior to the conclusion.

[1] "Disaster resilience of hotels describes the organisations' ability to withstand and recover from a disaster" (Brown et al., 2019: 108).

6.2 Literature Review

> The hotel sector's around the clock, 365 days a year model, elevates the importance of disaster resiliency. Guests will always be present, as will staff. Disaster resiliency for hotels may translate into lives saved, as well as business reputation (Brown et al., 2017: 367).

The above statement makes it abundantly clear that it cannot be business-as-usual in the hotel industry in the era of DRR&M. In focusing on the impact of disaster events on tourism development in Indonesia from 1998 to 2016, Rindrasih et al. (2019) found that disasters affect the performance of the tourism industry differently in terms of the scale of destruction, the location of disasters and the types of disasters. In Taiwan, a country prone to natural disasters triggered by earthquakes and typhoons (Tsai et al., 2016), the hotel industry has been on the receiving end (Tsai & Chen, 2011). For example, Typhoons Nari, Toraji (2001), Sinlaku (2008) and Morakot (2009) resulted in severe damage to the local hotel industry. To this end, a mechanism for typhoon and flood-risk assessment and management for the local hotel industry is necessary. Jordanian hotels face similar challenges and the research shows that they are still not well prepared (Sawalha et al., 2013). Lamanna et al. (2012) focus on the effects of disasters on businesses in the tourism industry. The authors emphasise that effective and efficient preparation, speedy response and strategic recovery are crucial for tourism businesses. The authors' focus on hotels in the Greater New Orleans area was examined to determine their resiliency in handling the effects of Hurricane Gustav.

The results reveal that 88% of the hotels closed to the public before Gustav (Lamanna et al., 2012). The hotels further remained closed for an average of 4 days. An estimated 13% of hotels lost power, 17% lost telephone services and 42% lost the minimum base of employees that were required for continued full guest services. Overall, human resource aspects were related to several challenges faced by the hotel industry. Business interruptions remain a huge contributing factor to hotel losses. For example, the Northridge earthquake led to 23% of the total loses in business. Disaster times have resulted in both voluntary and involuntary terminations of employment. At times the hotel and related industries are forced to cut their workforce due to economic losses and some employees may choose to terminate their employment due to trauma. Since the disasters occur close to the hotels as well as far and wide, at times employees are left without resources to return to work (Ibid.) and are even cut off due to damaged transportation and communication infrastructure.

In a study involving hotels in Greater New Orleans, the USA, after Hurricane Gustav, Lamanna et al. (2012) established the following reasons why hotels shut down (n = 24): mandatory evacuation (55%), failure of employees to get to work (13%), lack of guests (11%), not enough employees (8%), disruption of the flow of supplies (3%), disrupted tourism infrastructure (3%), transportation system disruption (3%) and other reasons (4%). Reasons for continued closure were also presented. These included failure of employees to get to work (32%), mandatory evacuation (30%), not enough employees (8%), lack of guests (6%), loss of electricity (6%), disruption of the flow of supplies (2%), damage to the hotel property (2%), unavailability of outsourced services (2%) and other reasons (12%) (Ibid.). In Phuket, the reconstruction period for hotel rooms was around 9.2 months, with the average period for reconstruction and tourism recovery estimated at 20 months (Tang et al., 2019). Given that hotels may close for such long periods and experience low volumes during relief and after re-opening (Choi et al., 2019), attracting old and new guests may demand promotions and incentives including price cuts (Kim et al., 2019). However, hotels not affected may increase prices drastically, even if this may be illegal. This was a scenario in some hotels in North California after Hurricane Florence—the concept of price gouging (Lee & Lee, 2019).

As the disasters threaten the hotel industry, whether natural or anthropogenic, there is no substitute for good disaster preparedness (Nguyen et al., 2017; Nhep et al., 2018). Such readiness should involve formulating plans to deal with disasters and emergencies (Ritchie, 2008) and may even involve having information send to guests prior to visiting disaster-prone areas (Hall et al., 2019). Ritchie (2008) provides ingredients of a disaster and tourism disaster strategy that remain relevant to this day (Table 6.1).

Other strategies aimed at mitigating disaster risk, especially those linked to floods, may include the following (Nguyen et al., 2017): avoiding low-lying, river-side and hillside areas; reducing the amount of decoration and attached facilities on lower floors; having emergency water supply and power generation equipment; avoiding basement parking; storing emergency rescue and life support materials; and locating critical equipment on higher floors (Tsai & Chen, 2011). Among the important equipment could be water and electrical accessories.

During disasters, guests look up to hotel staff to provide guidance (Ritchie, 2008). Hence, if the hotel property is prone to flooding, room information should include this in the hazards checklist. If such a hazard is imminent, the management and/or staff can distribute specific information packs. Additionally, the hotel should keep up-to-date guest records so that authorities can be alerted as to who is in residence, who has departed and where guests plan to go. Management and staff should have been trained on how to handle guests who check in or out prior to the onset of a disaster. The guests' place of origin is a critical part of the hotel records. As some guests will not be familiar with the local conditions and at times have challenges with language, there should be special attention from tour operators and resort managers. Guests will require assistance for

Table 6.1 Potential ingredients of a disaster and tourism disaster strategy

Strategy development	Implementation
• Forming a disaster recovery committee and convening meetings for sharing information • Assessing risk (identifying possible threats/disasters and prioritising in terms of probability of occurrence—real, likely and historical threats. Perhaps stimulated by a definition and classification of potential disasters) • Analysing anticipated short- and long-term impacts • Identifying strategies for avoiding/minimising impacts, critical actions necessary, chain of command for coordination, responsibilities and resources • Preparing and disseminating manual and secure commitment from responsible parties and relevant agencies. Including relevant contact information	• Holding disaster drills, rehearsals and simulations • Developing techniques for training, knowledge transfer and assessments • Formulating memoranda of understanding and mutual aid agreements • Educating the public and others involved in the planning process • Obtaining, positioning and maintaining relevant material resources • Undertaking public educational activities • Establishing informal linkages between involved groups • Thinking and communicating information about future dangers and hazards • Drawing up organisational disaster plans and integrating them with overall community mass emergency plans • Continually updating obsolete materials/strategies
Tourism disaster strategy ingredients	
• Selecting a team leader: a senior person with the authority and ability to command respect (ability to communicate effectively, prioritising and managing multiple tasks, delegating, coordinating and controlling, working cohesively with a crisis management team and making decisions quickly) • Team development: permanent and integral feature of strategic planning; ability to identify and analyse possible crises and develop contingency plans • Contingency plan: including the mechanism for activating the plan, possible crisis, objectives, worst-case scenario, the trigger mechanism • Actions: strategic actions including gathering information and developing relations with other agencies and stakeholders internal and external to the tourism industry such as warning, confirming, mobilising, providing customer information and shelters, dealing with employee concerns, transportation, looting and pre-entry issues • Crisis management command centre: a specific location and facility with relevant communication and other resources for the crisis management team	

Source: Ritchie (2008: 332)

travel and transport coordination, emergency accommodation, food supplies, sanitary facilities and the like. Such evacuation centres should be identified well in advance of the disaster striking. In other instances, these may include normal local emergency gathering areas (in the case of The Centurion Hotel it is the second floor in the case of floods). Hotel management should issue media statements to which journalist have access.

Disaster planning criteria for New Orleans hotels were investigated following Hurricane Gustav (Lamanna et al., 2012). These criteria are similar to those applied by Rittichainuwat (2013) in studying safety features in the aftermath of the 2004 tsunami in Japan. Lamanna et al. (2012) asked questions about whether the hotel had: a written disaster plan; informal discussions during planning; inclusion of preparedness, response, recovery and mitigation—functional approach; a plan focused on a specific business site; a regularly revised plan—at least once per year; regular procedures for training; staff participation disaster drills at least once annually; and staff involved in creating and implementing the plan—a team approach. The findings are shown in Fig. 6.1. While on average the majority of sampled hotels could tick five of the boxes, there were challenges with three, namely annual drills (46%), team approach (54%) and staff training (5%).

Nguyen et al. (2017, 2018) advocate for public-private collaboration for DRR&M in coastal destinations, particularly between the hotel industry and local government (see also Filimonau & De Coteau, 2019). Nguyen et al. (2017) focus on Matsushima, Japan, a popular coastal destination in the Tohoku Region, where the 2011 tsunami had a severe impact. The findings were that hotels can play a major role in working with municipalities towards DRR&M. However, collaboration gaps between the stakeholders remained a key hindrance to realising the adoption of the proposed hotel-based DRR&M model. The collaboration is necessary since most tourist attractions, especially hotels, are located along beachfronts that are perpetually at risk of the rising sea levels and tides like in Colombia. Some common hazards in coastal zones include tsunamis, flooding, earthquakes and landslides (Nhep et al., 2018; Wijayanto et al., 2019). Prohibiting aspects

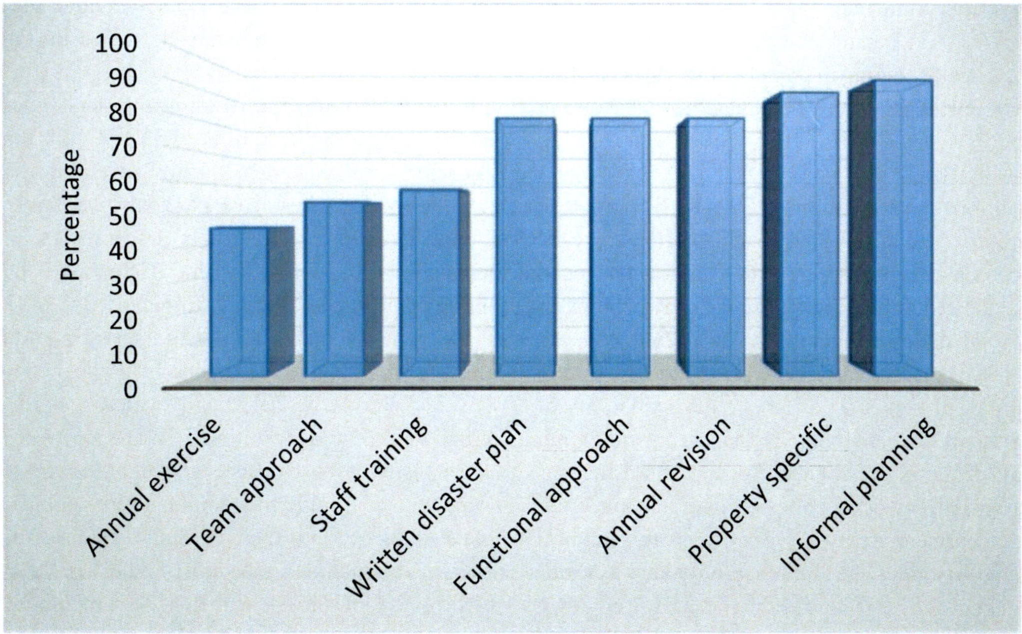

Fig. 6.1 New Orleans hotels disaster planning criteria (n = 24). (Source: Authors; data from Lamanna et al. (2012: 220))

include lack of financial resources, manpower, knowledge and differences in perceived responsibilities. To mitigate against these negative elements, there could be high-quality collaborative agreements that address key stakeholders' needs, consensus making, independent facilitation, need to digest and understand information and make informed decisions, and providing a level platform to influence decisions (Nguyen et al., 2017). Once the hurdles are addressed, collaborative outcomes include the sharing of technical, financial and human resources; reduced stakeholder conflict; and improved resilience towards disasters (Ivkov et al., 2019).

Building resilience and adaptation in Colombian hotels (Nhep et al., 2018) focuses on droughts and limited water supplies, excessive rainfall, beach erosion, storm and storm surges, and increasing temperatures. Given the focus in this publication, we will just touch on how the hotels are adjusting to excessive rainfall. The hotels redesigned and reconstructed their buildings, thus, roofs, guttering, doors, windows and other building features exposed to rain. However, upscale and luxurious hotels embarked on mega changes that included redoing drainage systems for better rainwater flows. However, Nguyen et al. (2018) bring another angle based on the post-2011 tsunami in Japan. They indicate that very few hotels were designated as evacuation centres during the 2011 Great East Japan Earthquake.

In other work, Brown et al. (2018) apply the capital-based approach to evaluate disaster resilience in the hotel industry. The capitals used include economic, social, human, physical, natural and cultural capital (see also Ivkov et al., 2019). Aspects that emerge from the economic capital (Brown et al., 2018) are the availability of resources, diverse income, financial strength and staff economic resilience. Social capital focuses on social networks, cohesion and/or connectedness, trust and ability to work in a group. Cultural capital includes the influence on social systems and knowledge. Natural capital is narrowed down to the environmental risks of the hotel's location and its effects on the environment. Human capital harnesses health, skills, the capacity to adapt, knowledge and business continuity. Lastly, physical capital touches on life safety and also business continuity (Brown et al., 2019).

Overall, new measures not discussed earlier that are brought up (Brown et al., 2018) include the number of rooms and hotel vehicles, access to credit, length of hotel in operation, full-time versus part-time work, creativity and innovation, situational awareness and problem-solving skills, early warning systems, willingness to respond, manual procedures and having multiple suppliers and multiple communication tools and platforms (see also Eugene, 2018; Karam, 2018; Ivkov et al., 2019). In a study in Padang city at risk from flood, earthquake, tsunami and abrasion, Ananda et al. (2018) advocate for optimising the multi-hazard early warning system for contingencies against disasters. The work revealed that affording contingency efforts is affected by the hotel's star rating, with those with lower gradings having more challenges. However, the most important sections in a hotel's contingency plan against coastal hazards are the evacuation route map and clear signs of evacuation directions. In Sri Lanka, Chandrasekera and Hebert (2019) found that 100% of the sampled coastal hotels indicate that their emergency preparedness plans include sections that discuss the procedures to follow during a tsunami.

Eugene identifies the features and challenges facing Chennai that are similar to The Centurion Hotel scenario: "Chennai houses several water bodies—ponds, lakes, rivers, canals, wetlands, storage tanks and water reservoirs but most of them are beleaguered by encroachments and mounting piles of garbage" (Eugene, 2018: 44). In addition, Chennai experienced serval natural disasters including the tsunami in 2004, heavy rains and floods in 2015, and cyclonic storms Vardah in 2016 and Ockhi in 2017. The frequency and intensity of these disasters brought the city to a standstill. The study then selected some popular hotels to investigate their readiness to face the floods and concluded that most could handle it. However, it became clear that an empathetic emotional support structure and the human touch are more critical than rigid sets of rules and/or mandatory protocols. This observation is diagonally opposite to El Shaer's (2018) findings in Egypt where hotels embarked on reactive rather than proactive mechanisms. A number of areas may lead to failure in DRR&M and these include the lack of efficient communication, poor coordination plans, failure to designate a person or persons in charge and lack of training.

6.3 Materials and Methods

The study was conducted at The Centurion Hotel, which is located at number 1001 Lenchen Avenue North in Centurion Central, City of Tshwane, South Africa (Figs. 6.2 and 6.3). The Centurion Hotel, a four-star-rated establishment, is built across the Hennops River at the eastern bank of the Centurion Lake. The hotel, built before the increase in environmental change and extreme weather events such as floods, does not have a ground floor. The Legacy Hotels and Resorts Group owns, manages and develops luxury hotels, game lodges and residences in Africa (The Legacy Hotels & Resorts Group, 2019a). The Group owns 24 hotels and resorts across four countries (Gabon, Ghana, Namibia and South Africa) with the bulk of these hotels located in South Africa. Some of its top hotels include Michelangelo in Sandton and The Commodore in Cape Town (both in South Africa). The Centurion Hotel accommodates 160 rooms comprising 91 twin beds, 44 queen beds, 14 executive king suites, two luxury suites, one accessible room and eight studio suites (The Legacy Hotels & Resorts Group, 2019b).

The methods used include earth observation systems, specifically Geographical Information Systems (GIS), interviews and repeated on-the-ground observation that took place from December 2019 to December 2020. Interviews were granted by key informants including The Legacy Hotels and Resorts Group divisional

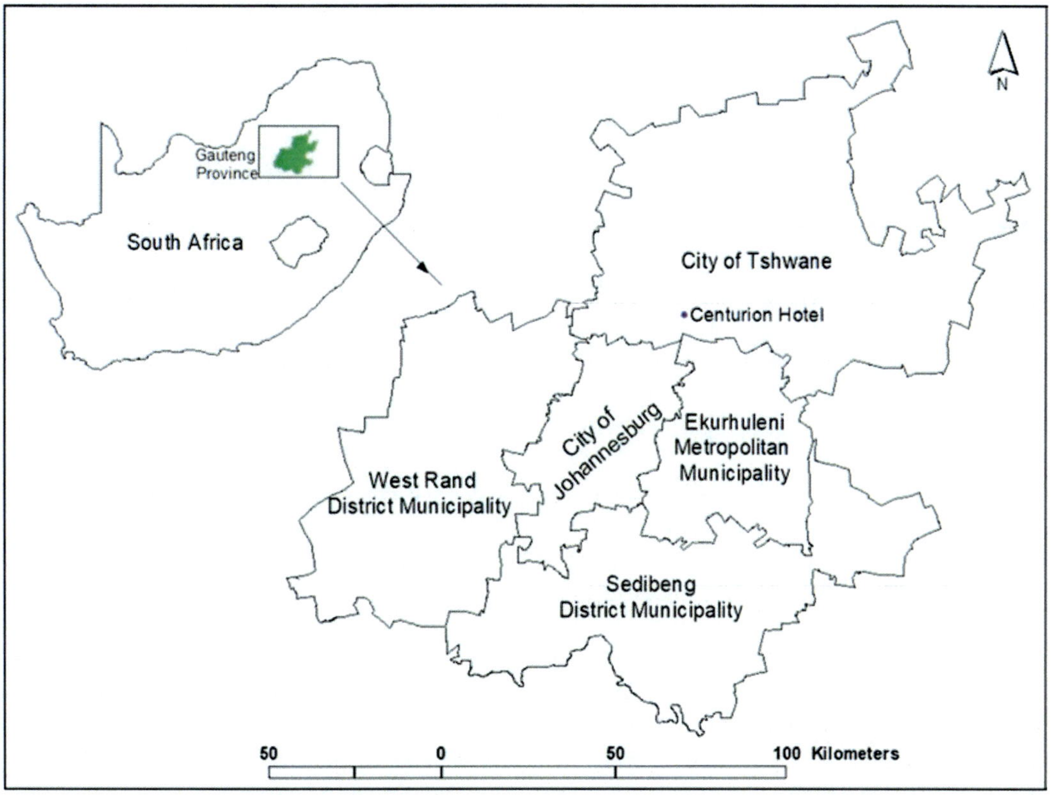

Fig. 6.2 The Centurion Hotel location. (Source: Authors)

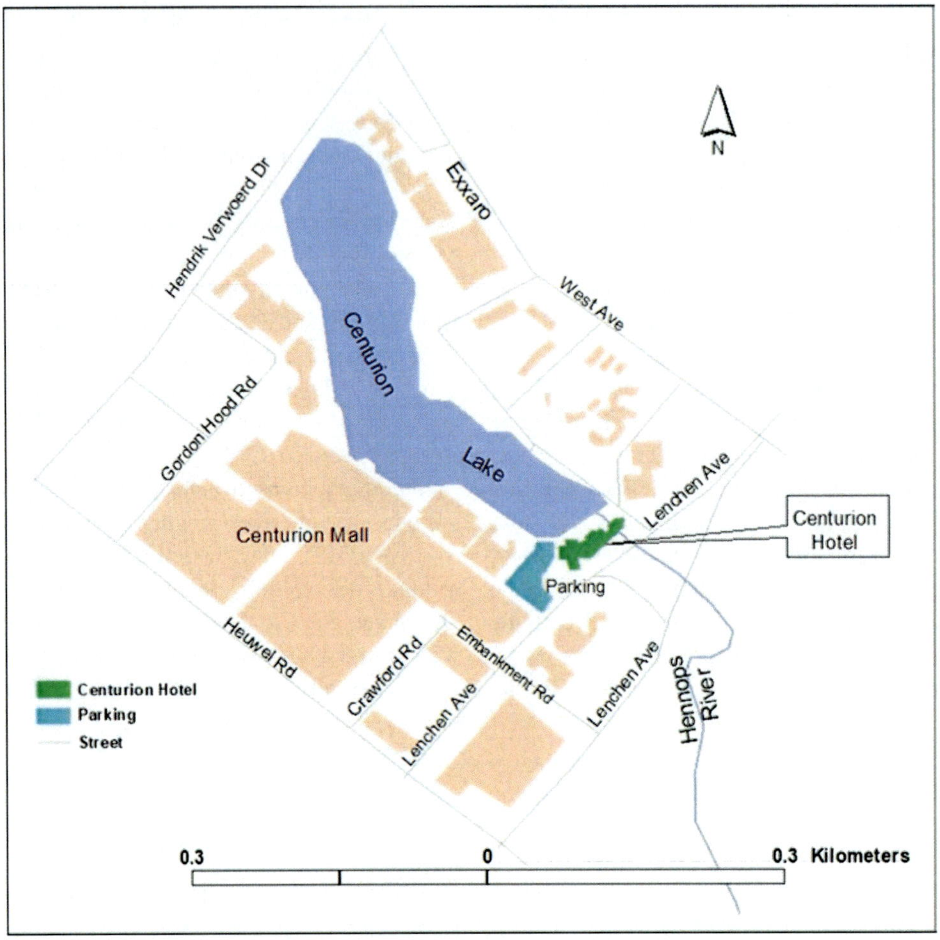

Fig. 6.3 The Centurion Hotel in relation to its vicinity. (Source: Authors)

operations manager, The Centurion Hotel general manager, Centurion Mall security manager and City of Tshwane staff from the Sustainability Division. These methods have been applied by other authors in the recent past, including Rittichainuwat (2013).

A case study design approach was used and the selection of The Centurion Hotel and the inseparable Centurion Lake was purposive. This was done in order to have a deeper understanding of the flood events that have occurred here over the years and also to unpack measures that have been taken to manage the situation. The case study method provides both scope and insight into the phenomena under study (Farquhar, 2013). In-depth interviews with two hotel group key informants were carried out at the hotel site a week after the 9 December 2019 floods. The interviews unpacked the events of the December 2019 floods, the hotel flood management plan, historical floods as well as measures put in place to counter future floods. Up to four follow-up interviews were done with the hotel's general manager during the renovation and re-operational phase of the hotel to track progress on several fronts, including the situation for the 2020/2021 rain season.

Field observations were carried out at the hotel site, Centurion Lake and the upper catchment of the Centurion Lake, mainly feeding into the Hennops River. This was meant to gather evidence of land use practices that were likely to result in an increased risk of flooding on the lower parts of the catchment, including The

Centurion Hotel and Centurion Lake. Many scenes were captured as photographs and used to support claims from interviews and written reports including those from media. High-resolution satellite imagery for the Centurion Lake was obtained from Google Earth and used to show the accumulation of sediments from year 2002 to year 2020.

6.4 Presentation and Discussion of Key Findings

This section presents and discusses the key findings in three subsections. The first subsection focuses on presenting the nature and extent of damage from the 9 December 2019 floods. The second subsection elaborates on the extent and possible causes of the repeated flooding of The Centurion Hotel. The third subsection presents the measures taken by the hotel management to minimise the impact of flooding as well as assess their effectiveness.

6.4.1 The Nature and Extent of Damage to The Centurion Hotel by the 9 December Floods

Following a devastating flood on 9 December 2019, The Centurion Hotel closed for business (The Legacy Hotels & Resorts Group, 2019c). A statement posted by the Group (retrieved on 22 December 2019) read:

> Due to severe flooding at The Centurion Hotel on 9 December, the hotel has experienced devastating damage which requires us to close the property to embark on a safety assessment and upgrade if necessary and complete a refurbishment of the affected areas. We expect to be able to re-open the hotel in March 2020 (The Legacy Hotels & Resorts Group, 2019c online).

This was the extent of the impact of the floods that led to the researchers deciding to document this case study. From the authors' lived experience, it rained non-stop from Thursday, 5 December to Wednesday, 12 December 2019. Images of the devastation following the floods across the Gauteng province, especially the Centurion and Mamelodi area in the Tshwane Metropolitan Municipality, started circulating on social media and traditional media platforms on 9 December. The damage was significant, as can be seen in Figs. 6.4, 6.5, 6.6, 6.7, and 6.8. The hotel eventually reopened its doors on 1 March 2020 as planned.

According to The Centurion Hotel general manager, the 9 December floods were the worst in living memory. This view was also supported by the Centurion Mall security manager during an interview. Other flood incidences include those that took place on 22 March 2018, 4 February 2014 and 5 March 2014 as per the general manager's account and our confirmations from archival media reports. The floodwaters filled the entire basement, rising over 2 m. The Centurion Hotel general manager indicated that floodwaters swept three cars from the hotel's car park, including one rental car and a mini-bus, which ended up blocking the entrance/exit of the hotel basement.

By the time of our first fieldwork, front-end loaders could be seen clearing tons of mud and debris that covered the hotel basement and ground floor. The general manager indicated that guests had to be airlifted during the floods from the swimming pool area on the first floor. There was also severe damage to offices on the ground floor, with furniture damaged beyond repair. The scenario was the same at the reception and other areas. The perimeter walls were completely washed away, including one hotel skip that was deposited 500 m further into nearby Centurion Lake. The flood was devastating; in fact, many other properties surrounding the lake were flooded and empty office space could be observed adjacent to the hotel.

6.4.2 The Extent and Possible Causes of Flooding of The Centurion Hotel

The flooding of The Centurion Hotel in December 2019, as well as in the past, seems to be attributed

Fig. 6.4 Floodwater covering the entire basement and ground floor of The Centurion Hotel (right). (Source: Photograph supplied by Centurion Mall security)

Fig. 6.5 The Centurion Hotel after the 9 December flood disaster. (Source: Authors)

Fig. 6.6 Damage inside some offices on the ground floor. (Source: Authors)

Fig. 6.7 Perimeter brick walls knocked down completely. (Source: Authors)

to the hydro-meteorological and physical characteristics of the upper catchment of the Hennops River. It is also because of the bad management of the Centurion Lake, which was once part of a booming City of Tshwane recreational park. The main tributaries of the Hennops River near The Centurion Hotel include Sesmylspruit, Olifantspruit and Kaalspruit. There is also a medium-sized municipal dam on the upper catchment known as the Rietvlei Dam. The changes in the land cover and land use pattern of a catchment may induce hydrological responses that encourage surface run-off and at the same time reduce the lag time between the peak of rainfall and peak of river discharge. Significant changes observed to have occurred within the upper catchment of the Hennops River over the past 30 years include the conversion of forest and agricultural land into urban settlements, industrial parks and warehousing. These introduce impervious surfaces. There are also storm water drains that increase the rate at which precipitation is con-

Fig. 6.8 One of three cars washed into the lake area from the hotel parking area. (Source: Authors and Centurion Mall security (during the flood))

Fig. 6.9 Settlement encroachment into wetlands in Tembisa Township. (Source: Authors)

verted into run-off. Hence, there is more water reaching the streams in the catchment as surface run-off rather than base flow.

Field observations confirmed massive human activity and settlement encroachment on sensitive ecosystems such as wetlands on the upper parts of the Hennops River catchment (Fig. 6.9). Wetlands provide key ecosystem services such as flood attenuation. Like sponges, wetlands absorb rainwater and slowly release it to the surrounding river systems as base flow, which increases the lag time between the peak of rainfall and peak of discharge. The destruction of wetlands due to human activity increases the risk of flooding due to enhanced surface run-off especially during and soon after periods of heavy rainfall.

Land degradation is also occurring in the upper parts of the Hennops River catchment (Fig. 6.10). This is a direct result of sand poaching, urban informal agriculture and the expansion of informal settlements, well into river channels in the upstream area (Fig. 6.11). In most cases, these activities occur on riverbanks and lead to intensive soil erosion and sedimentation of river systems and dams. The siltation of rivers with sand reduces their capacity to handle huge volumes of discharge soon after rainfall events. This results in rivers flooding and sometimes bursting

Fig. 6.10 Illegal extraction of sand and other building material on Kaalspruit riverbanks. (Source: Authors)

Fig. 6.11 River channel narrowing due to informal settlement expansion. (Source: Authors)

their banks leading to the inundation of surrounding areas. The mentioned activities occurring in the upper part of the Hennops River catchment exacerbate the problem of flooding in the Centurion area, specifically the Centurion Lake, leading to the hotel flooding.

The expansion of informal settlements into the river channels is aided by the construction sector that dump rubble to create raised building surfaces. Field observations show that the construction vehicles regularly dump rubble in directed areas. Further investigations revealed that a truck load could cost as little as R100 (about US$6.50 as per the exchange rate on 1 December 2020). Also worrying is the seeming reluctance by officials from both the Ekurhuleni and City of Johannesburg metropolitan municipalities to enforce bylaws preventing the mushrooming of illegal urban agriculture, building-material extraction and informal settlements within wetlands and other undevelopable areas. The activities observed not only take place at night, but also during broad daylight. Raw sewage as well as open defecation were observed in the upper catchment area of the Hennops River and some of the sewage ended up in the Centurion Lake.

As indicated earlier, The Centurion Hotel is situated on the banks of Centurion Lake. Evidence from the satellite images show the progressive sedimentation of the lake over the past 20 years. The siltation of the lake has been a direct result of human activities alluded to on the upper parts of the catchment. As the amount of sediment grows in the lake, so too does the likelihood of flooding during storm events. This is because the islands created by the sediment hold back the flow of water, resulting in backflows that end up flooding the hotel and its surroundings. In 2002, the lake had minimal amounts of sediment as depicted in Fig. 6.12. The picture of Centurion Lake presented in the 2002 image corresponds with the signs observed on the ground during the 2019 fieldwork. It was clear that there were several activities permitted in the lake that included swimming, fishing and boating. Needless to say, as the lake's aesthetics deteriorated over the years due to a combination of factors, including upstream general environmental degradation and neglected maintenance by the City of Tshwane officials, all these activities are history.

By 2012, the lake had started to witness heavy sedimentation (Fig. 6.13), resulting in the formation of islands on the lake. The situation worsened by 2019 with the entire lake appearing silted (Fig. 6.14). During in-depth interviews soon after the 2019 flood event, hotel management indicated observing resistance to water flowing into the lake. They highlighted that the silting of the lake was responsible for exacerbating the flooding of the hotel. They reported that the lake was not supposed to be silted, but that the City of Tshwane had done nothing over the past 15–20 years to remove the silt. As the floodwaters were rising, the silt created a blockage, which pushed water back into the hotel. The lake is supposed to be 3 m deep, but it has islands that are up to a metre high. The situation is exacerbated by reed plants, which further increase the resistance to water, with the result that water does not flow fast enough to prevent backflows. Extensive rainfall in the upper area of the Hennops River catchment led to the overflow of the Rietvlei Dam, which also resulted in the dumping of excess water downstream, hence the occurrence of floods.

Cosgrove and Rijsberman (2000) argue that interconnected water-related hazards such as floods are aggravated by the fact that awareness of the problems is limited to only a few who deal with the water sector on a daily basis. The authors advocate for concerted efforts and extraordinary measures to face this challenge head on, and hence the need for the integration of water and environmental management under the concept of integrated water resources management. In

Fig. 6.12 Centurion Lake on 17 July 2002. (Source: Authors)

Fig. 6.13 Centurion Lake on 3 March 2012. (Source: Authors)

Fig. 6.14 Centurion Lake on 26 January 2020. (Source: Authors)

Colombo, Sri Lanka, Dammalage and Jayasinghe (2019) observed that increased land use change in the district's watershed, especially through urbanisation, contributed to the higher flood levels even though the recorded rainfall was moderately lower than that received during previous floods. On the other hand, Siregar (2018) sees land cover/land use as being key factors influencing hydro-ecological processes on watersheds because they influence the partitioning of rainfall. Hence, this enhances the vulnerability of areas to flooding, erosion of the soil and pollution of water.

6.4.3 Centurion Hotel DRR Protocols, City of Tshwane Remedies and Their Effectiveness

From in-depth interviews with key informants that included one divisional operations manager for The Legacy Hotels and Resorts Group and the general manager for The Centurion Hotel, it emerged that they are aware that floods are one of the biggest threats to the hotel facility. To this end, the hotel has a flood management plan and standard operating procedure for the risk should it arise. During the initial citing of the hotel, it was built on a 1-in-100-year flood line based on available data. This was meant to limit the flood incidences having an impact on the hotel, although this did not take cognisance of human-induced upper catchment developments. However, due to changes in the land cover/land use in the upper catchment of the Hennops River, particularly urban expansion, the 1-in-100-year flood line for the area has shifted. In addition, there were fewer bridges and less construction on the riverside in the past. By the time of writing this chapter, the hotel was still to determine the return period of the December 2019 flood for planning.

In 2014 when a 1-in-50-year flood hit the hotel, the management undertook both engineering and non-engineering interventions to counter the impact of future floods on the structure. Some of the interventions included the re-pegging of flood lines and their frequency of occurrence. This led to the creation of flood barriers to prevent the water from flowing into the hotel. These barriers were raised in the direction from which frequent floodwaters came, especially from the car park, the front gate and the hotel entrance. These flood barriers divert water into both the Hennops River and Centurion Lake. The basement of the hotel is the most vulnerable to flooding because of its low height above the channel base. The basement houses the hotel car park, workshop and also washing bays. To prevent it from flooding, overflow water is directed through drains in the basement to a holding pit and, from there, the water is pumped back into the river or lake.

In terms of electricity, which is one of the critical utilities affected during flooding, the electricity distribution boards (DBs) used at the hotel have a self-trip mechanism, which helps to prevent short circuits, possible fire outbreaks and electrocution. Once power from the national grid switches off during a flood, backup generators automatically switch on to provide power. These generators are elevated from the floor surface to reduce the risk of floodwater damage. In the event of the ground floor being flooded, the building is designed in such a way that every level has its independent generator to ensure that power cuts only occur on the submerged floors, not in the rest of the building. This enables normal functions of the hotel even during emergency periods.

As a result of past flooding, the hotel has a flood early warning mechanism and evacuation programme. Participants in these plans include the hotel maintenance and security staff, as well as the hotel patrons. When river flood levels reach a critical threshold, the management activates the evacuation plan. They start by removing all the cars and valuables in the hotel basement as this is more vulnerable to flooding. The cars are moved to the car park close to the Centurion Mall, which is more elevated and had not significantly flooded before the 2019 floods. To make it easy to contact car owners in case of an emergency, the hotel keeps a red book which every driver parking in the basement has to complete. Details include the drivers' names and their contact details. This helps to quickly reach out to the owners to move their vehicles in the event of a flood. If the flood levels eventually reach the hotel structure, the hotel triggers an evacuation by pressing the fire alarm panic button to wake up every guest and bringing them to a safe assembly point and alerting the emergency service authorities. At the emergency assembly point, hotel guests are briefed on the nature of the emergency, information available and procedures that need to be followed to evade the hazard.

Key offices in the running of the hotel are located on the ground floor of the hotel. These include reservations, accounts, payroll and hotel management. The ground floor is also the most vulnerable to damage during flood events. To avoid the loss of records, the different departments' servers are upstairs and not vulnerable to flooding. At the time of the December 2019 floods, the hotel was covered against floods by insurance, which helped it to recover faster.

To what extent were the DRR protocols effective during the December 2019 floods, one may ask? The straight answer is, to a large extent. For example, no data or records were lost during the flood event although the offices responsible for these records were flooded. This was mainly because all the servers for these offices are situated in an area safe from flooding. Data on accounts, payroll, reservations and human resources all survived and remained intact. However, there was little disturbance to the work of the administrative staff and minimal man-hours were lost. On the day of flood, the hotel had already established a temporary working space at the Group's headquarters operating at 100% capacity. Staff members who directly interact with patrons were granted leave during renovations since the facility was closed until early March 2020. Only core staff rotating in blocks continued reporting for duty. When the hotel reopened, most of the staff would then have close to zero leave days. This system of operation was therefore effective since the administration never had downtime due to the floods.

As soon as the emergency was triggered, cars from the basement were removed because the hotel staff were able to reach most of the car owners. The process of removing cars started the evening of Saturday, 7 December 2020, and by the morning of Sunday, 8 December, all the cars had been moved, except for two cars whose owners could not be reached. It is commendable that the hotel saved the cars from damage. The main reason why the two cars could not be moved was because the owners were foreign nationals not proficient in English who did not know how to respond to the emergency to remove their rental cars. To improve the effectiveness of the evacuation procedure in future, special attention must be paid to these vulnerable guests. Ritchie (2008) highlights the need for keeping up-to-date guest records to assist in timely warning and evacuation. The author further notes that some guests may not be familiar with the environment and sometimes do not understand the local language, and hence the need to give them special attention.

However, despite attempts by the hotel to prepare for future floods using their historical occurrences and given that the hotel is on a flood prone area, the 9 December 2019 flood overwhelmed all the structural measures put in place. For starters, the barriers put in place to counter the floods did not work, with the flood reaching levels never planned for. The barriers were an adaptation measure to stop floods from damaging the hotel. The pumps in the basement were overwhelmed and eventually ceased to operate as they were flooded. Even the open car park at the hotel flooded for the first time. As observed by the general manager, "Never in the past has that open car park been flooded, but this time it was. Three cars were also washed down into the lake". Cosgrove and Rijsberman (2000) argue that the failure of flood protection systems is a direct result of changes in environmental conditions in the catchment areas, especially the replacement of vegetation with urban surfaces and wetland destruction that reduce flood-retention capability.

As specified in their standard operating manual at 04:00 on Monday, 9 December 2020, the hotel activated the evacuation process of the guests. This was done by pressing the fire alarm panic button to wake up the entire hotel. The hotel staff managed to gather all 71 booked guests in the second-floor lobby, which was a flood-free area. The guests were informed about the impending flood and the procedures to be followed in the evacuation process. The guests were also told to pack their bags, bring them to the lobby and be ready for evacuation. The guests stayed together in the lobby and their questions were answered until breakfast was served to everyone. After breakfast the guests were allowed to return to their rooms to rest and wait for evacuation. The hotel team was in constant contact with the emergency service providers, including three firemen brought in from outside to assist with flood danger assessments. The firemen moved around the property with the fire chief and other rescue teams constantly providing telephonic advice. Eventually, all the guests were rescued individually using a military helicopter with neither injuries nor deaths recorded.

The group's divisional operations manager attributed the successful evacuation to the manner in which they regularly provided updated information to the affected guests. Although some guests were concerned with the evacuation method that used a boat, which was later changed to a helicopter. In addition, the hotel never ran out of power during the flood emergency. As soon as the main grid tripped due to the flood, backup power from the diesel-powered generators took over. Given that each floor had its own generator, only the one on the first floor cut out because it had been flooded. Therefore, the backup power systems at the hotel are very effective.

The hotel also took advantage of the flood to build back better and greener. During repeated visits by the authors, it emerged that the hotel was already investigating how it could counter future floods of similar magnitude. The hotel was insured against flood damage, which assisted in its quick rebound from the disaster. Two weeks after the flood, the hotel was reconnected to the main grid, and the pumps in the basement as well as the damaged electricity distribution boards had been replaced. Extensive clean-up and removal of all the damaged furniture was already in process as well as renovating the damaged car park. The hotel was temporarily closed in December 2019 and by 1 March 2020 it reopened its doors after extensive renovations. The hotel also adopted a policy of energy efficiency with some changes including the use of LED lights. The general manager highlighted the need to elevate the concept of building back better and building back greener. Lamanna et al. (2012) emphasise the need for effective disaster preparedness, quick response, as well as strategic recovery for tourism businesses affected by disasters. This would avoid long-term negative perceptions of the affected facilities. The hotel is also planning to build a stronger wall to prevent similar future flooding. However, by the time of completing this chapter, there was no tangible evidence to support this.

The City of Tshwane also came to the party to seek long-lasting remedial measures to The Centurion Hotel and other properties flooding. One of the solutions proposed and implemented was to remove the heavy silt and debris from the Centurion Lake. Work on this started in February 2020 (Fig. 6.15). The City of Tshwane planned to drain the lake permitting work to remove all the silt and other debris from the lake.

Given the amount of sewage that flows into the lake from upstream, the tenants and property owners surrounding the lake, including those from the Centurion Mall also, complained of bad odour. However, as of December 2020, the venture to remove silt and other debris seemed to have stalled, possibly due to Covid-19 as lockdowns persisted in the country. A further investigation with the manager of The Centurion Hotel following the heavy rains that flooded some areas in Centurion on 24 November 2020 indicated that the hotel had not been affected as yet. The general manager revealed that the City of Tshwane had cleared the debris from the sluice gates at the lake allowing excess floodwater to pass through without blockages. However, this needed to be done regularly if floods resulting from the flooding lake are to be managed effectively.

A follow-up conversation with one of the key informants from the Centurion Mall security management on 2 December 2020 confirmed that the City of Tshwane regularly sent tractor loader backhoes to clear the sluice gates at the lake. The key informant further highlighted that although the water once rose significantly in the lake, the river kept flowing. The informant further indicated that plastic waste from a nearby informal settlement upstream of The Centurion Hotel and mall was the main source of the waste. The informal settlement is less than 2 km away. This account confirms the author's observations (Figs. 6.16 and 6.17).

Figure 6.17 shows unwanted recyclables dumped on the banks of the Hennops River. As the rains come, the waste, especially polyethylene terephthalate (PET) bottles and other plastics, are swept away into the Centurion Lake less than 2 km downstream causing the sluice gates to clog, and sometimes flood The Centurion Hotel. This is what happened in December 2019. A conversation with one resident from Mushroom Ville (the name given to the informal settlement) indicated that approximately 50 waste pickers reside there.

The last two matters we wish to cover in this section concern insurance and the discontinued project to restore the Centurion Lake. Telephonic conversations with both the general manager and the divisional operations manager on 1 and 2 December 2020, respectively, revealed that The Centurion Hotel had indefinitely closed for business

Fig. 6.15 Opening of the clogged sluice gates at the lake to allow good drainage. (Source: Authors)

Fig. 6.16 Plastic and other waste choking Centurion Lake's sluice gates. (Source: Authors)

on 1 December 2020. The head office decision was taken as the hotel no longer had insurance for floods. The head office was exploring the possibility of constructing an overhead flood evacuation bridge linking the hotel and the Centurion Mall from its first floor and the flood barrier wall. All valuables including computers, printers and more were being migrated to flood-secure places within the hotel. Unfortunately, a quick solution to dredging the Centurion Lake seemed to wane as the City of Tshwane revealed in early December 2020 that the budget allocated to the project had been diverted to more pressing service delivery matters, such as electricity, water and road maintenance, since no tender had been approved and due to Covid-19 lockdown regulations (Ndlazi, 2020). A total of R28.6 million (about US$1.86 million as of 2 December 2020) had been allocated for the rehabilitation of the lake and the Hennops River to flow naturally. The City of Tshwane also hinted that The Centurion Lake Rehabilitation Plan was not yet ready and

Fig. 6.17 Informal settlement on the banks of Hennops River upstream from the Centurion Lake. (Source: Authors)

would be made available to the public once the yet-to-be-appointed engineering consultant was done with it (Ndlazi, 2020).

6.5 Conclusions

The repeated flooding of The Centurion Hotel over the past 6 years can be attributed to the shifting flood line. The 9 December 2019 floods were the most intense in the history of the hotel. Factors determining the magnitude of the flood at the hotel include the land cover and land use changes occurring on the upper parts of the Hennops River catchment, as well as the amount of rainfall received in the catchment. The catchment has witnessed increases in urbanisation, urban informal agriculture, wetland destruction and sand poaching along the tributary streams all tending to increase the amount of run-off in the streams, and reduce the lag time between the peak of rainfall and peak of discharge. The flood at the hotel was further amplified by the heavily silted Centurion Lake, which together with its blocked sluice gates prevented water from flowing fast enough to prevent its backflow. The already abnormally high discharge was also enhanced by the overflow of the Rietvlei Dam, which dumped more water into the Hennops River, bursting its banks.

Informed by previous flood events, The Centurion Hotel put in place measures to counter the effects of flooding on the property. These measures include creating flood barriers in the car park and front gate area, elevating the hotel entrance to prevent floodwaters from entering the hotel and putting in place water pumps in the basement to pump out the floodwaters. However, during the December 2019 floods, none of these worked, leading to the most severe floods to hit the hotel. The hotel had initially managed to identify the owners of vehicles parked in the hotel basement and moved them to the car park outside the hotel which in the past used to be a flood-free area. However, the parking area flooded for the first time damaging the cars previously removed from the basement. The evacuation procedures at the hotel went well with no recorded injuries. This was mainly due to a timely early warning system by the hotel and good communication between the hotel and guests as well as with other rescue teams.

It is recommended that an integrated water resource management system be implemented in the Hennops River catchment in order to deal with the problems of flooding. This is because problems that end up manifesting as floods are intertwined with changes in land use and/or land cover occurring in the catchment. All stakeholders need to be involved in the management of water resources and must be cognisant of the impact of their actions on the hydrological cycle. Discussion forums such as the Hennops River Forum, which include local governments, government environmental departments, members of academia, civic organisations and other experts, are a good starting point.

Lake Centurion needs to be managed and monitored in terms of the level of sediment, as well as the outlet valves that are in most cases blocked with sediment or various types of plastics. This will help to speed up the rate at which water will flow from the lake, thus preventing backflows that may end up flooding the hotel. There is also a need to give special consideration to foreign nationals, especially those from non-English-speaking countries during flooding or other emergencies because they have limited capacity to react swiftly to perceived threats as a result of the language barrier. Given the trend towards increased flood incidences, we further recommend that the hotel continues exploring new avenues to take insurance cover, which includes flood losses. This may be expensive but remains a necessity. Building back better and building back smarter is also recommended after flood damage in order to have structures that are not only flood smart but reduce the carbon footprint of the hotel. Without such measures, the hotel may end up closing doors permanently.

References

Ananda, G., Ophiyandri, T., & Hakam, A. (2018). Hotels contingency assessment in Padang city against coastal hazard. *MATEC Web of Conferences, 229*, 02015. https://doi.org/10.1051/matecconf/201822902015.

Arseculeratne, T., Ritch, A., & Wicklund, R. (2018). Tsunami 2004, India and International Impacts, International Disaster Management. *Great Plains Sociologist: The Official Publication of the Great Plains Sociological Association, 26*, 24–40.

Brown, N. A., Rovins, J. E., Feldmann-Jensen, S., Orchiston, C., & Johnston, D. (2017). Exploring disaster resilience within the hotel sector: A systematic review of literature. *International Journal of Disaster Risk Reduction, 22*, 362–370. https://doi.org/10.1016/j.ijdrr.2017.02.005.

Brown, N. A., Orchiston, C., Rovins, J. E., Feldmann-Jensen, S., & Johnston, D. (2018). An integrative framework for investigating disaster resilience within the hotel sector. *Journal of Hospitality and Tourism Management, 36*, 67–75. https://doi.org/10.1016/j.jhtm.2018.07.004.

Brown, N. A., Rovins, J. E., Feldmann-Jensen, S., Orchiston, C., & Johnston, D. (2019). Measuring disaster resilience within the hotel sector: An exploratory survey of Wellington and Hawke's Bay, New Zealand hotel staff and managers. *International Journal of Disaster Risk Reduction, 33*, 108–121. https://doi.org/10.1016/j.ijdrr.2018.09.014.

Chandrasekera, T., & Hebert, P. (2019). Emergency preparedness in the hospitality industry in Sri Lanka. *International Journal of Disaster Resilience in the Built Environment, 10*(2/3), 167–174. https://doi.org/10.1108/IJDRBE-01-2018-0001.

Choi, D. W., Lee, S., & Singal, M. (2019). An exploratory temporal analysis of the lodging market and state economic recovery from Hurricane Sandy (The lodging market recovery from Hurricane Sandy). *International Journal of Contemporary Hospitality Management, 31*(5), 2074–2094. https://doi.org/10.1108/IJCHM-06-2018-0513.

Cosgrove, W., & Rijsberman, F. (2000). *World water vision: Making water everybody's business*. London: Earthscan Publications.

Dammalage, T. L., & Jayasinghe, N. T. (2019). Land-use change and its impact on urban flooding: A case study on Colombo District Flood on May 2016. *Engineering, Technology & Applied Science Research, 9*(2), 3887–3891.

El Shaer, N. S. (2018). Managing Egyptian five-star hotels during times of crises. *International Journal of Heritage, Tourism and Hospitality, 12*(1/2), 210–232.

Eugene, J. (2018). Disasters' day out and Hotels' way out. *American International Journal of Research in Humanities, Arts and Social Sciences, 18*(310), 44–47.

Farquhar, J. D. (2013). *Case study research for business*. London: SAGE Publications Ltd..

Filimonau, V., & De Coteau, D. (2019). Tourism resilience in the context of integrated destination and disaster management (DM^2). *International Journal of Tourism Research., 22*(2), 202–222. https://doi.org/10.1002/jtr.2329.

Hall, S., Emmett, C., Cope, A., Harris, R., Setiadi, G. D., & Meservy, w. and Berrett, B. (2019). Tsunami knowledge, information sources, and evacuation intentions among tourists in Bali, Indonesia. *Journal of Coastal Conservation, 23*, 505–519. https://doi.org/10.1007/s11852-019-00679-x.

Ivkov, M., Blešić, I., Janićević, S., Kovačić, S., Miljković, D., Lukić, T., & Sakulski, D. (2019). Natural disasters vs hotel industry resilience: An exploratory study among hotel managers from Europe. *Open Geoscience, 11*, 378–390.

Karam, M. G. (2018). The impact of strategic planning on crisis management styles in the 5-star hotels. *Journal of Hotel & Business Management, 7*(1), 1–9. https://doi.org/10.4172/2169-0286.1000171.

Kim, M., Roehl, W., & Lee, S. K. (2019). Effect of hotels' price discounts on performance recovery after a crisis. *International Journal of Hospitality Management, 83*, 74–82. https://doi.org/10.1016/j.ijhm.2019.04.006.

Lamanna, Z., Williams, K. H., & Childers, C. (2012). An assessment of resilience: Disaster management and recovery for greater new Orleans' hotels. *Journal of Human Resources in Hospitality & Tourism, 11*(3), 210–224. https://doi.org/10.1080/15332845.2012.668653.

Lee, S. H., & Lee, J. (2019). Does price gouging happen in the lodging industry? Case of Hurricane Florence. *Journal of Revenue and Pricing Management, 19*, 151–161. https://doi.org/10.1057/s41272-019-00222-5.

Ndlazi, S. (2020). *Plan to restore Centurion Lake has been stopped*. Retrieved December 2, 2020, from https://www.iol.co.za/pretoria-news/news/plan-to-restore-centurion-lake-has-been-stopped-e9a14094-2678-46d6-a3db-1f2558b2187f

Nhep, T., Schott, C., & Sahli, M. (2018). Climate change adaptation in the coastal tourism hotel sector: The case of Sihanoukville, Cambodia. In T. Young, P. Stolk, & G. McGinnis (Eds.), *CAUTHE 2018: Get smart: Paradoxes and possibilities in tourism, hospitality and events education and research* (pp. 465–481). Newcastle, NSW: Newcastle Business School, The University of Newcastle.

Nguyen, D. N., Imamura, F., & Iuchi, K. (2017). Public-private collaboration for disaster risk management: A case study of hotels in Matsushima, Japan. *Tourism Management, 61*, 129–140. https://doi.org/10.1016/j.tourman.2017.02.003.

Nguyen, D. N., Imamura, F., & Iuchi, K. (2018). Barriers towards hotel disaster preparedness: Case studies of post 2011 Tsunami, Japan. *International Journal of Disaster Risk Reduction, 28*, 585–594. https://doi.org/10.1016/j.ijdrr.2018.01.008.

Rindrasih, E., Witte, P., Spit, T., & Zoomers, A. (2019). Tourism and disasters: Impact of disaster events on tourism development in Indonesia 1998-2016 and structural approach policy responses. *Journal of Service Science and Management, 12*, 93–115. https://doi.org/10.4236/jssm.2019.122006.

Ritchie, B. (2008). Tourism disaster planning and management: From response and recovery to reduction and readiness. *Current Issues in Tourism, 11*(4), 315–348. https://doi.org/10.1080/13683500802140372.

Rittichainuwat, B. N. (2013). Tourists' and tourism suppliers' perceptions toward crisis management on Tsunami. *Tourism Management, 34*, 112–121. https://doi.org/10.1016/j.tourman.2012.03.018.

Sawalha, I. H. S., Jraisat, L. E., & Al-Qudah, K. A. M. (2013). Crisis and disaster management in Jordanian hotels: Practices and cultural considerations. *Disaster Prevention and Management, 22*(3), 210–228. https://doi.org/10.1108/DPM-09-2012-0101.

Siregar, R. I. (2018). Land cover change impact on urban flood modeling (case study: Upper Citarum watershed). *IOP Conference Series: Earth and Environmental Science, 126*, 012027. https://doi.org/10.1088/1755-1315/126/1/012027.

Sydnor-Bousso, S., Stafford, K., Tews, M., & Adler, H. (2011). Toward a resilience model for the hospitality & tourism industry. *Journal of Human Resources in Hospitality & Tourism, 10*(2), 195–217. https://doi.org/10.1080/15332845.2011.536942.

Tang, J., Leelawat, N., Suppasri, A., & Imamura, F. (2019). An effect of tsunami to hotel occupancy: A case of Phuket, Thailand. *IOP Conference Series: Earth and Environmental Science, 273*, 012033. https://doi.org/10.1088/1755-1315/273/1/012033.

The Legacy Hotels & Resorts Group. (2019a). *Your Africa, Our Legacy*. Retrieved December 22, 2019, from https://www.legacyhotels.co.za/en

The Legacy Hotels & Resorts Group. (2019b). *The Centurion Hotel Factsheet*. Retrieved December 22, 2019, from https://www.legacy-hotels.co.za/storage/app/uploads/public/5d5/4f9/e5e/5d54f9e5eb46a029350344.pdf

The Legacy Hotels & Resorts Group. (2019c). *The Centurion Hotel closed following flooding: Expected to re-open March 2020*. Retrieved December 22, 2019, from https://www.legacyhotels.co.za/hotels/centurion-hotel

Tsai, C. H., & Chen, C. W. (2011). Development of a mechanism for typhoon- and flood-risk assessment and disaster management in the hotel industry—A case study of the Hualien Area. *Scandinavian Journal of Hospitality and Tourism, 11*(3), 324–341. https://doi.org/10.1080/15022250.2011.601929.

Tsai, C. H., Wu, T. C., Wall, G., & Linliu, S. C. (2016). Perceptions of tourism impacts and community resilience to natural disasters. *Tourism Geographies, 18*(2), 152–173. https://doi.org/10.1080/14616688.2016.1149875.

Vila, N. A., Toubes, D. R., & Brea, J. A. F. (2019). Tourism industry's vulnerability upon risk of flooding: The Aquis Querquennis Complex. *Environments, 6*, 122. https://doi.org/10.3390/environments6120122.

Wijayanto, D., Nainupu, N. A., Rahmawati, O. P., & Oktariyandari, S. (2019). Preparedness of tourist managers in Bali for facing disaster risk. In F. A. Kharis (Ed.), *Research articles disaster management study program* (pp. 79–84). Faculty of National Security Defence University.

Check for updates

7 Rethinking Harbours, Beaches and Urban Estuaries Waste Management Under Climate-Induced Floods in South Africa

David Chikodzi, Kaitano Dube, and Nonkululeko Ngcobo

Abstract

Globally there has been increased concerns over the state of ocean pollution which often become evident after extreme hydro-meteorological events such as floods. Chief among the concerns is the deposition of solid waste into harbours, beaches and estuaries by floods. A thought on harbours and beaches takes one to the blue (ocean) economy space, a subject that is embedded in the United Nations' Sustainable Development Goal (SDG) 14. The Durban Harbour is one of Africa's most prominent and busiest ports handling cargo for South Africa and several southern African countries. In addition, its beaches rank among the world's best. Using events study, document analysis, secondary data and in-depth interviews, the chapter aims to determine how the April 2019 floods revealed the challenges in the waste management systems in Durban, South Africa, and how the concerned stakeholders, among them government, the private sector and non-governmental organisations, responded to the challenge. The results showed that extreme floods caused a significant challenge at the Durban Harbour, along beaches and estuaries as plastic waste mainly comprising shopping bags heavily polluted these places. Heavy plastic pollution is a clear symptom of poor waste management in the upper catchment areas. In as much as various stakeholders have put forward some innovative solutions to the problem, these are largely inadequate because of the complexity of the problem. There is a need for the municipality and stakeholders to increase their efforts towards implementing environmental law enforcement to ensure the protection of the oceans and blue ocean economy. The problem requires a multi-dimension approach such as mass education, environmental law enforcement and capital resource investment into coastal cleaning among other such initiatives.

Keywords

Harbours · Floods · Beaches · Oceans · Waste management · Plastics · Urban estuaries

D. Chikodzi (✉)
Institute for Corporate Citizenship, University of South Africa, Pretoria, South Africa
e-mail: chikod@unisa.ac.za

K. Dube
Vaal University of Technology, Vanderbijlpark, South Africa
e-mail: kaitanod@vut.ac.za

N. Ngcobo
Department of Environmental Science, University of South Africa, Pretoria, South Africa
e-mail: 65942140@mylife.unisa.ac.za

G. Nhamo, L. Chapungu (eds.), *The Increasing Risk of Floods and Tornadoes in Southern Africa*, Sustainable Development Goals Series, https://doi.org/10.1007/978-3-030-74192-1_7

7.1 Introduction and Background

Oceans play a crucial role in human civilisation, supporting several ecosystem services. Oceans act as carbon sinks, deliver oxygen, food security and a socio-economic asset for the blue economy. The United Nations Agenda 2030's SDG 14 sees oceans as the centre of societal progress (United Nations, 2015). According to Agenda 2030, the oceans economy produces goods and services valued at over $3 trillion per annum, which is equivalent to over 5% of the global GDP. The significance of the ocean's economy is predominantly pronounced in the global south, where the ocean economy is central to coastal development particularly if one considers the centrality of coastal tourism to countries with a coastal line such as South Africa, Mozambique, Namibia and Tanzania to mention but a few (Dube et al., 2021).

Notwithstanding their value to human civilisations, oceans have faced numerous threats from several fronts, chief among them climate variability and change and pollution. These challenges threaten to undermine the socio-economic contribution of oceans in the development of societies. The Intergovernmental Panel on Climate Change (IPCC) special report on oceans and cryosphere laid bare the multiple threats facing oceans which must be dealt with to ensure sustainable development of inland and coastal communities globally (IPCC, 2019).

Andrady (2011) opines that the discovery of plastic made human life much easier. Plastic has key characteristics that make it preferable for use compared to other substances. These properties include its utility under different temperature regimes, relatively strong when related to its mass, limited thermal conductivity, low mass, and relatively cheap to produce (Naidoo et al., 2016). These qualities have made plastic appropriate to the fabrication of a diversity of products and provided a strong basis with its exponential increase in use. Plastic production rose from about 1.5–311 million tons per annum within the last 60 years (Schnurr et al., 2018). The properties that have led plastic to be popular also happen to be the same causing its serious environmental impacts. This has led to grave environmental apprehensions related to its use. Of particular concern is the widespread use of single-use plastics by modern societies which has accelerated the rate of degradation of aquatic ecosystems (Browne et al., 2011). Debris from plastic matter is now a common contaminant of the marine ecology and is potentially available for ingestion by marine life. Between 60 and 95% of marine litter is plastic, and many scientists have presented evidence of the occurrence of microplastics in the stomachs of aquatic organisms (Schnurr et al., 2018). Ryan (2015) posits that plastics are now omnipresent in every part of oceans worldwide, including remote oceanic gyres. Other places where plastic occurs include remote small islands, polar areas and deep oceanic ridges. However, the most polluted oceans are those occurring adjacent to highly industrialised urban settlements (Claessens et al., 2011; McCormick et al., 2014).

Flooding events and dredging are some of the pathways through which significant quantities of plastics are transported into the oceans (Bakir et al., 2014). Naidoo et al. (2016) observed estuaries to be one of the prime conduits through which plastics are moved from catchment areas to the seas, particularly in areas under coastal urban settings. This implies that nearshore ecosystems usually accumulate significant plastic materials in the aftermath of heavy rainfall or floods. These areas may include nearby beach resorts and port harbour facilities. SDG number 14 of the United Nations, 2030 Agenda for Sustainable Development, aims at achieving sustainable development through the protection and sustainable utilisation of ocean and maritime ecosystems (United Nations, 2015). Further Target 14.1 of SDG 14 aims at both preventing and significantly reducing any forms of maritime pollution especially those emanating from pollution on land by the year 2025. Target 14.2 emphasises the need for sustainable management and protection of maritime ecosystems in order to avoid their degradation and also enhance their resilience by 2020. Another target (14.7) aims at enhancing the economic benefits to countries in the Global South through the sustainable adoption of the blue economy activities like fisheries, aquatic culture and tourism (United Nations, 2015). Most of these targets are particularly at risk of failure to be fulfilled given the increased pace of contamination of seas and oceans by plastics and also increases in the occurrence and magnitude of flooding hazards that quicken the rate of transportation of the contaminant.

The deleterious impacts of plastic materials on the aquatic ecosystems have been flagged for some time, yet they continue to be underplayed (Derraik, 2002). Pollution by plastic is slowly becoming an emerging global problem, negatively impacting the blue economy, including aquatic wildlife, tourism and shipping (Dube et al., 2020). Despite its durability and being an ecological accumulator, the world has experienced an exponential increase in plastic production and its associated increase in marine ecosystems. The chapter's objective is to determine the impacts of the April 2019 extreme floods on plastic pollution at harbours, beaches, and urban estuaries in the KwaZulu-Natal province of South Africa. Further, the chapter reviews the challenges of waste management practices in South Africa and how the concerned stakeholders, among them government, the private sector and non-governmental organisations, have responded to the challenge.

The next section gives an overview of literature which addresses the impacts of plastic pollution on marine ecosystems, initiatives put in place at global and national levels to deal with plastic pollution.

7.2 Literature Survey

Most plastic materials are non-biodegradable and hence continue to accumulate in the environment either as litter or broken down into smaller sizes and will continue to exist in the natural environment for centuries (Schnurr et al., 2018). The oceans have been turned into huge warehouses of waste plastic generated on land, and of late, the build-up and potential effects of plastics on the ecology and integrity of oceans has been concerning (Gattringer, 2018). Generally, the seemingly abundant aquatic life combined with the immensity of seas has led many to dismiss plastic debris's proliferation as a potential hazard. However, many common forms of plastic can float on water and once dumped in oceans are being moved by waves to some of the most remote and relatively uninhabited places on earth such as the polar regions (Barnes et al., 2010). The accumulation of significant masses of plastic in specific areas of ocean has led to the coining of terms such as "garbage patch", and "ocean landfill" to refer to the phenomena. Most plastic waste pollution in the form of drink bottles, plastic bags and hygiene productions in oceans usually happens closer to densely populated coastal areas (Gattringer, 2018).

Van Rensburg et al. (2019) opine that the volume of plastic pollution in the seas is more rampant close to coastal cities of the Global South. According to Jambeck et al. (2015) in 2010, about 83% of plastics polluting the oceans came from just 20 coastal nations from Asia, Latin America as well as Africa. Thompson (2015) argues that improved waste management on land was the only way of stopping plastic materials from polluting the oceans. This was because consumer and industrial plastic materials ended up in the oceans due to the combined effects of improper waste management regimes accompanied by climate change and variability which have increased the intensity and frequency of storm events that lead to flooding and washing away of plastic into oceans. All this is further magnified by insufficient policies and regulation, poor enforcement, attitude, and individuals' behaviour. Derraik (2002) observes plastic litter in oceans as aesthetically distasteful and a threat to marine biota. Components of the blue economy such as tourism, shipping and fishing are very sensitive to the presence of plastics in the seas. Plastics impose huge economic costs when enter gill nets, damages equipment like propellers or get deposited on beaches (Jang et al., 2014).

Most of the costs incurred due to plastic pollution in the seas are incurred by the affected parties who are not necessarily the cause of the problem (Mouat et al., 2010). The most glaring effects are economic in nature, for example, reduced fishing time due to damaged equipment and the opportunity cost of time spend removing plastic pollution. In Europe, for example, over €630 million is incurred in an endeavour to clean plastic pollution from coastal areas and beaches every year (UNEP, 2018). In the United States, over US$520 million is used by communities of the west coast in fighting marine plastic litter and debris (Stickel et al., 2012). In Scotland, plastic pollution is responsible for over US$17 million (equivalent to 5% of total income) in losses incurred by the fishing industry per year (UNEP, 2011). Other considerations include the litter's intangible aesthetic costs which can influence visitor's perception of the poor state of environmental management. This perception can in the long term hurt economic progress especially for places dependent on tourist arrivals (Brink et al., 2009). Broken plastic also can injure or greatly inconvenience beach users (Hardesty et al., 2015). Some have observed the need to put more resources to stop plastic litter from reaching the oceans and even stop its manufacture altogether, but most initiatives have had limited success (Cole et al., 2011). Despite global efforts to combat ocean plastic litter, the problem has actually increased. This increase highlights the ineffectiveness of strategies put in place to address plastic pollution from sources which are either municipal or industrial. This also shows gaps in the enforcement of prevailing regulations and standards on limiting plastic pollution (Seltenrich, 2015).

Some countries have taken heed of the problems of plastic pollution and have put in place interventions to deal with the problem. Most of the interventions are in the form of policies, regulations or enforcement of international agreements on reducing plastic pollution. There are still, however, a sizable number of countries that either lack strategies to deal directly with plastic pollution or the adopted practices have been ineffective. Several global non-governmental organisations (NGOs) have been putting their focus on plastic litter at sea. Since 1997, the Algalita Marine Research Foundation has been active in carrying out research and surveys on plastic pollution in the Pacific, Atlantic and Indian Oceans (Algalita, 2011). The International Coastal Clean-up campaign led by the Ocean Conservancy is the biggest annual voluntary effort meant to remove marine litter and also gather data on the types of material accumulating as ocean debris. The clean-ups' objective is to harness the power of people to fight ocean trash and raise awareness on the extent of the problem. In 2018, more than a million volunteers from over 100 countries and locations collected ten million kg of mostly plastic from about 6000 coastal settlements (Ocean Conservancy, 2019). Cigarette filters are the most collected ocean litter with shopping plastic bags being number two. However, plastic bags have a greater potential for harming maritime ecosystems because they are ecological accumulators. Some NGOs like the Joint Group of Experts on the Scientific Aspects of Marine Environmental Protection and the Five Gyres Institute are also into the research of marine debris and carry out environmental education on the subject (Pettipas et al., 2016).

Governments of coastal countries have been fighting to reduce plastic litter reaching the oceans (Rochman et al., 2015). Through the International Convention for the Prevention of Pollution from Ships which was enacted in 1973, there was a ban imposed on the disposal of plastic in the oceans by 1988. More than 134 countries committed to stopping the disposal of plastics at sea. However, the marine litter problem has worsened since the signing of the convention (Xanthos & Walker, 2017). This regrettable situation is mostly attributed to the fact that plastic litter at sea is intricately linked with poor management of plastic waste on land and not so much at sea. The Honolulu Strategy outlines approaches to preventing and managing ocean litter (UNEP and NOAA, 2015). It has been adopted the world over to help different countries manage the problem of plastic litter at sea (Pettipas et al., 2016). Widely adapted from the Honolulu Strategy are two important approaches. The first uses market-based approaches such as fines and levies on new or single-use plastic to reduce waste. The other approach creates policies, laws and protocols that limit or ban the manufacture of certain forms of plastics considered to be polluting the environment (Xanthos & Walker, 2017). The Global Programme of Action for the Protection of the Marine Environment from Land-based Activities, which is housed at UNEP, acts as a worldwide plan of action meant to deal with ocean litter by addressing the inter-connections of river basins, coastal pollution and the open seas (UNEP/GPA, 2011). It suggests instruments for developing and implementing activities that deal with trans-boundary pollution of which plastic pollution is included (Xanthos & Walker, 2017).

According to Nhamo (2007) market-based environmental management policies meant to arrest plastic pollution on land can include taxes and levies on consumption. Regulatory measures can involve complete or partial bans on plastic production (Ibid.). The above policy options are now being used in many areas of the world albeit in different circumstances to deal with the problem of plastic pollution. Such policy instruments have become increasingly applied around the world in various forms. Taxes, for example, are meant to change user behaviour to become less wasteful, hence polluting less. Taxes and fines also raise funds for the authorities to help deal with the remedial actions after plastic pollution has occurred (Schnurr et al., 2018). Bans are recognised as the best approach to take if the ramifications of the pollution from plastics are very serious. Bans on plastic utilisation can sometimes only apply to plastics of certain thicknesses with thinner plastics usually banned because of the difficulty in recycling (Xanthos & Walker, 2017). In the context of single-use plastics, some governments have either completely or partially banned their use or employ market-based measures like levies or can employ a combination of these two measures (Ibid). The largest number of countries to have completely banned plastic are from Africa. Twenty-five countries have imposed bans and by 2018 close to 60% of these countries had moved to implement the ban (UNEP, 2018). In North America, the bans are usually effected at both the state and local government level. For example, in the United States, regulations on plastic utilisation have been implemented in states like Hawaii and California. At the local government level, municipalities such as Massachusetts, Oregon and New York City have effected bans in their areas of jurisdiction with over 150 local governments have banned or imposed levies on plastics (Schnurr et al., 2018).

On the African continent, success in dealing with plastic litter has mostly been achieved by countries that have instituted total bans. After the total ban on plastic bags by Eritrea in 2005, the country has significantly reduced costs associated with clogged drains, loss of livestock and spread of diseases (Xanthos & Walker, 2017). The ban was companied by very high fines that deter the would-be offenders and also effective monitoring. In the year 2008, Rwanda completely banned the utilisation, production and importation of plastic bags. The ban followed over 4 years of consultations accompanied by widespread sensitisation and education of the communities (Schnurr et al., 2018). The ban was effectively implemented with deterrent fines and even jail time for repeat offenders. Businesses and consumers quickly shifted to the use of paper and other alternatives which are relatively environmentally friendly (Danielsson, 2017). The ban on plastic has made Kigali one of Africa's cleanest cities and has brought a positive perception that is benefiting the tourism industry in the country (de Freytas-Tamura, 2017). Lessons learnt from the experiences of the above countries suggests that the most effective method of eliminating plastic litter is a complete ban on its use. For partial bans to be effective, they need to be complemented by widespread awareness and education programmes as well as making sure that affordable and environmentally friendly alternatives are made available (Ayalon et al., 2009).

Xanthos and Walker (2017) observed that education at schools is a very powerful tool in enforcing behaviour change especially in the youth; these could also cascade their newfound knowledge to the family level and the community at large. Since land-based activities are the major source of plastic pollution in oceans, educated communities are most likely to act upon the problem and can make a difference. Education is often seen as a more powerful catalyst for a change compared to stringent policies (Derraik, 2002). Nhamo (2007) observed participation to be central to waste management efforts and argues that in the South African context, special attention should be paid to women, the young, spiritual, conservational organisations, civil society, labour unions and service NGOs.

7.3 Methodology

The study is centred on the eThekwini Metropolitan Municipality, as shown in Fig. 7.1. eThekwini was created during the time the Municipal Demarcation Board drew-up new boundaries for local authorities in 2001. The city of Durban is the main urban area under the eThekwini municipality (Van Heerden, 2015). The municipality is part of the KwaZulu-Natal province and is the second biggest contributor to South Africa's GDP, with a share of about 16.6%. It has a population of about 3.5 million people, which translates to about a third of the population of the entire province (KZN Provincial Treasury, 2013). The chapter will focus mainly on the central beaches, the Durban Harbour and the uMgeni river mouth. The Durban port harbour in the study area is one of Africa's busiest cargo terminal catering for South Africa as well as other southern African nations like the Democratic Republic of Congo, Malawi, Botswana, Lesotho, Eswatini, Zambia and Zimbabwe (Van Heerden, 2015).

The chapter's main objective is to document the impacts of the April 2019 floods on plastic pollution at harbours, beaches, and urban estuaries in the KwaZulu-Natal province, specifically looking at what happened in Durban. The chapter also endeavours to unravel the challenges in the management of plastic waste in South Africa and how the concerned stakeholders have responded to the challenge. The chapter mainly relied on the "event study methodology" in which secondary data was the main source of information. The event studies

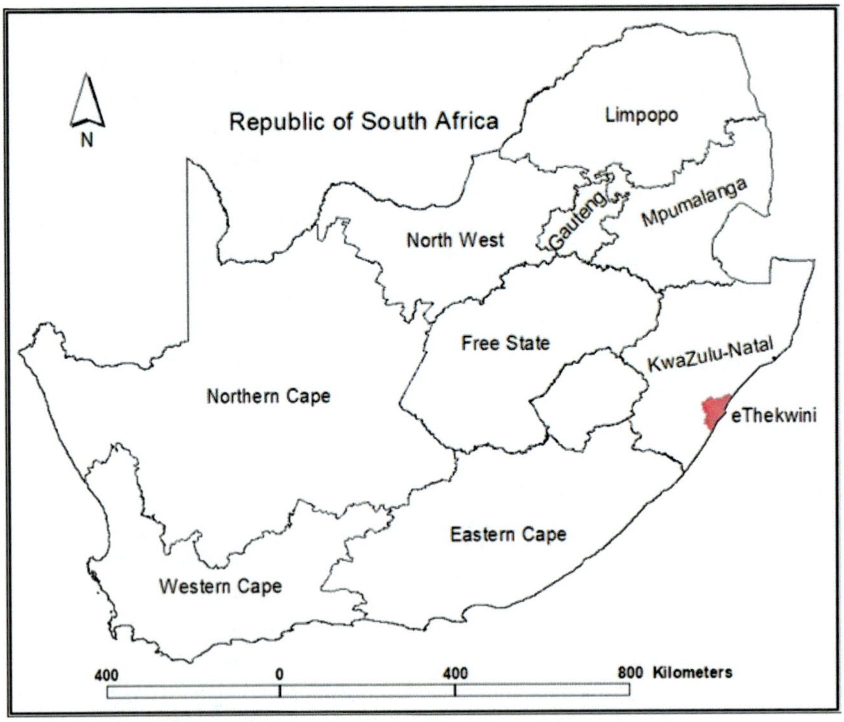

Fig. 7.1 Study area—eThekwini. (Source: Authors)

have their origins in finance, principally analysing the reaction of stock markets to and external events such as a pandemic or a natural hazard (Hayward, 2018). The method is designed to examine the outcome of an event on a specific dependant variable. In this paper, the event study method was utilised to analyse how plastic pollution (dependant variable), as a result of flooding (independent variable), impacted the tourism industry, Durban Harbour operations and urban estuaries. Further, the chapter examines how different stakeholders have responded to these pollution events.

The chapter also made use of a mixture of both quantitative and qualitative data obtained from secondary sources to fulfil the chapter objectives. Government agencies, NGOs, reports from representative bodies, and other scholarly sources provided the secondary data used in the chapter. The data derived was then grouped into themes using document analysis before the final analysis of the data was done using fine grain analysis. Field observation in the catchment area of the uMgeni and uMlazi rivers were also carried out. The Umgeni river stretches about 230 km from the Drakensberg to Durban and is the primary source of water for more than 3.5 million people in the riparian communities. The river feeds into various beaches within the city of Durban. Umlazi river, on the other hand, is a stream that originates south-west of Pietermaritzburg and flows right through to the Umlazi and Chatsworth townships and into the Durban sea. The uMgeni and Umlazi river basins are dammed at several places by a combination of large dams, weirs and small earth dams that store water for municipal supply as well as irrigating. The rivers are increasingly being put under severe strain as a consequent of competing uses and increasing populations that are deriving ecosystem services from them.

The next section presents findings of the impact of plastic pollution at the harbour, on beaches and on urban estuaries. The last part looks at how different stakeholders have tried to intervene to solve the problem.

7.4 Presentation of Data and Discussion of Findings

It was noted from the study that in April 2019, heavy rainfall and flash flooding affected most parts of KwaZulu-Natal and the eThekwini municipality was one of the hardest-hit areas. The area received 165 mm of rainfall in over 24 h breaking its previous record of 108 mm set in 2017 (Head, 2019). Simultaneously, there was also a spotlight on the deluge of plastic waste deposited by the floodwaters into rivers systems, beaches and the Durban Harbour. The huge amounts of plastic waste were observed to have come at a time when the City of Durban had been voted the greenest city in the world, beating 94 other cities (Baxter, 2019). The city won the title for its sustainability and green initiatives, including planting 80% indigenous trees and only 20% exotic plants (ibid). With such an award, naturally, the level of plastic pollution which became apparent during the April 2019 floods was deeply concerning and attracted worldwide attention.

The plastic pollution that ended up on beaches, Durban Harbour and urban river systems in particular estuaries resulted from intense flooding that flushed out all the plastic litter that had been improperly disposed of and built-up on riverbanks over the dry season. Observers noted with concern that the washing of huge volumes of litter into these places had, unfortunately, become the new norm after every episode of heavy rainfall in Durban. They highlighted that the problem was getting worse every year, and hence the need for drastic action to curb the pollution. Other key informants observed that the massive volume of plastic waste was indicative of a serious waste management problem upstream. Of note was the fact that the problem of plastic pollution in the rivers, ocean and harbours was worsening partly due to informal settlements expanding along rivers and poor service delivery in certain areas of the eThekwini municipality. Due to the non-collection of waste in some of these areas, there is the accumulation of waste at informal dumps or direct dumping into river systems. This waste material will then be washed and transported to the rivers by floodwaters during intense rainfall activity. Figure 7.2a–d shows uncollected plastic waste material in the Umlazi River catchment area. Most of this waste will eventually be washed into the rivers and transported to the ocean where it will accumulate.

The above illustrations (Fig. 7.2a–d) were taken on the month of March before flooding in Durban's city occurred and soon after the flood event. They illustrate heaps of uncollected garbage, comprising of plastic items in what seems to be illegal dumping spots in Umlazi. Plastic waste mismanagement has recently become a prevalent issue in some townships of the eThekwini municipality. This is despite some punctuated efforts such as recycling and re-use initiatives by the municipality to minimise plastic litter in its jurisdiction. Besides, repeated efforts by the municipality to stop illegal dumping have yielded limited success as offenders largely continue with the practice even after several rounds of clean-ups and erection of warning signs (Ogola, 2012). Weak institutional support for the shift to integrated waste management techniques and

Fig. 7.2 (**a**) Improperly disposed waste in Umlazi adjacent to Sibusiso Mdakane Road in a small stream leading to Umlazi river. (**b**) Improperly managed waste by the Umlazi river bank at Umlazi F section. (**c**) Heaps of uncollected litter in Umlazi F section that is carried down the road by rainwater and (**d**) Umlazi river adjacent to 107,343 street in Umlazi F section after flooding and plastic washed away. (Source: Authors)

urban population growth leading to the emergence of informal settlements have also proved to be additional contributors to persistent plastic waste problems within the Municipality (Menyuka et al., 2020). The increase of plastic pollution in water ways and rivers in Durban is directly related to the expansion of informal settlements in the area (Ibid.). Plastic pollution usually happens where waste is intentionally dumped or washed from the surface of the ground, and into the streams during the rainy season. Streams typically discharge their load into the nearest rivers before it is deposited into the sea. The illustration in Fig. 7.2d shows that the April 2019 floods washed away the entire waste that is shown in Fig. 7.2a–c into the ocean, specifically polluting Treasure and Brighton Beaches at the mouth of the Umlazi river.

The problem of plastic pollution in Durban also shows a bleak picture for one of the world's award-winning greenest cities. Illustrations in Fig. 7.2a–d show poor management of waste by the eThekwini municipality. Furthermore, this is only one road in a section of Umlazi; thus, one can only imagine the amount of improperly managed waste in the entire municipality. The waste is carried to the oceans specifically the Umlazi river that feeds into Treasure Beach and Brighton Beach. The waste had a considerable impact on the nature of the beaches, the harbours and other river systems. Other rivers such as the Umhlatuzana river which passes through Chatsworth Township on the north of Umlazi township and feeds into Umbilo river that feeds into Umgeni river also have high levels of plastic pollution. The Umgeni river feeds into the coastlines of Durban, the North beach specifically. Thus, the improperly managed waste from inner towns reaches the beach and essential river systems such as the mangroves swamps. This affects the sustainability pillars of the Durban beaches. This reflects the importance of tackling the improperly managed waste from the origin; the organisation that does clean the beaches only clean the surface "the tip of the iceberg" as one would say. In tackling the inner towns improper waste management, there is a need to take the integrated environmental management approach by the eThekwini municipality and other relevant affected and interested parties that would also include each community's locals. This will also educate them on adequately managing waste and not to throw waste at the river.

The problem of plastic litter pollution in the Durban harbour, urban estuaries such as the uMgeni and Umlazi river mouth and beaches in the area was highlighted to likely continue because these areas were located at the outlets of urban catchments that were poorly handling plastic waste. The Durban Harbour, for example, lies at the outlet of a combined catchment area of the rivers, canals and storm water drainage systems that were over 200 km^2 (Ardé, 2019). Several pathways of plastic pollution were noted in the study area during the research. These mainly included rivers that transverse polluted areas, through canals and storm water drains and trash which flies into the ocean from open bins on the coast. For the Durban Harbour, three rivers and 57 storm water drain feed into it. Open bins were observed to release high amounts of plastic when subjected to the wind, which will lift and carry waste items. Plastic items noted to be easily dislodged by the wind included lightweight plastics, such as straws, food wrappers and plastic bags (Lishivha, 2018). At the Port of Durban, which is one of the busiest ports in Africa, more than 300 tons of mainly plastic dirt has been removed during the clean-up operation after plastic pollution during the floods (Goba, 2019). The clean-up was led by Transnet National Ports Authority, together with the eThekwini Municipality and other interested organisations as well as volunteers. However, although the port remained fully operational, progress in cleaning up plastic pollution due to the April 2019 flooding was slowed down due to the massive volume of material that continued to be washed into the port through the previously identified pathways. Some of the plastic debris that washed into the port posed a threat to the navigation equipment in the Ocean and impacted the movement of vessels.

On Durban beaches, plastic pollution is a big problem, especially after flooding events such as those witnessed in April 2019. This has the potential to damage the good destination image that the city has got. The KZN Department of Economic Development, the national Tourism and Environmental Affairs Department and the eThekwini Municipality are cognisant of the significance of tourism in the area especially the impact of the Golden Mile beaches as a major tourism asset (Van Heerden, 2015). According to Durban Tourism (2013), tourism in the metro is worth over R5.7 billion which is about 3% of its annual economy. This then highlights the need for improved management of the plastic waste management environment to avoid the shocking sights of plastic litter, which affected some of the popular beaches during the 2019 floods. The most identified plastic items during the flood-induced plastic pollution in Durban included plastic bags, cool drink bottles, food wrappers, disposable plastic cutlery, asthma pumps, single-use baby diapers and disposable syringes (Lishivha, 2018). A study done by Leonard (2004) in the KZN province reviews that over to 45% of medical waste produced in the area was unaccounted for. This suggests that some of the material was being illegally disposed of at unofficial sites. Ardé (2019) opines that in Durban, the plastic pollution that ends up in the harbour and on the beaches represents about 20% of what makes it into the seas as plastic bottles with no lids and other plastic pollutants with no air in them quickly sink in the Ocean and become ecological accumulators.

Coastal litter has always been a problem for South Africa, especially in Durban even without the floodwaters flushing out the plastic to the sea (Ocean Conservancy, 2019). Figure 7.3 shows the amount of waste picked up during the annual coastal clean-up days of the years in South Africa. As reflected from Fig. 7.2a–d, significant amounts of waste contaminate the coastline, which is not suitable for both the development of tourism and marine ecosystems. Studies by Naidoo et al. (2016) and Van Rensburg et al. (2019) have also observed the problem of plastic litter pollution on Durban beaches and estuaries.

Durban's prevailing subtropical climate results in its beaches being open to tourists throughout the year. The warm summer, however, signals the peak bathing season. This may create a risk of potential plastic pollution on the beach during the peak season since the area is characterised by fairly high rainfall compared to other parts of South Africa. The high amounts of rainfall are normally associated with higher amounts of run-off that periodically enter the beaches through storm drains that convey water from the inner city and also carry huge volumes of plastic litter. Increased rainfall activity with a tendency towards high intensity results in abnormally high discharge in the uMgeni River, which can also bring massive amounts of plastic into the sea and eventually carried to the beaches by wave action. Due to a combination of climate and inefficient waste management system, Van Heerden (2015) highlights that the eThekwini municipality started to find it increasingly difficult if not impossible in some cases to comply with the Blue Flag Beaches requirement on their beaches especially those to do with the quality of water, upkeep of amenities and plastic litter.

Durban's main river mouth on the Umgeni River is also the biggest contributor to plastic pollution into the ocean in South Africa (Caboz, 2019). The river mouth is located less than 6 km from coastlines of Durban popular beaches, namely, North Beach, Blue lagoon beach, South Beach, Country Club beach and the Laguna beach where thousands of holiday-makers swim. Key informants observe the amounts of plastic pollution in this river to be deleterious to coastal marine ecosystems and affect recreational activities like fishing. Studies by Naidoo et al. (2016) review high levels of plastic contamination on Durban's urban estuaries, with the highest peak being at the Durban Harbour and significantly reducing as one moves towards estuaries situated further from the urban centre. This implies that the nearshore environment receives more plastic pollution soon after flood events, which then accumulates in these ecological systems.

Several initiatives have been put in a plan by different government levels, the private sector, and the civic society to deal with plastic pollution problems in South Africa and in the KZN province in particular. The Litter boom project is one of the most prominent. The project works with volunteers to curb the movement of plastic in the

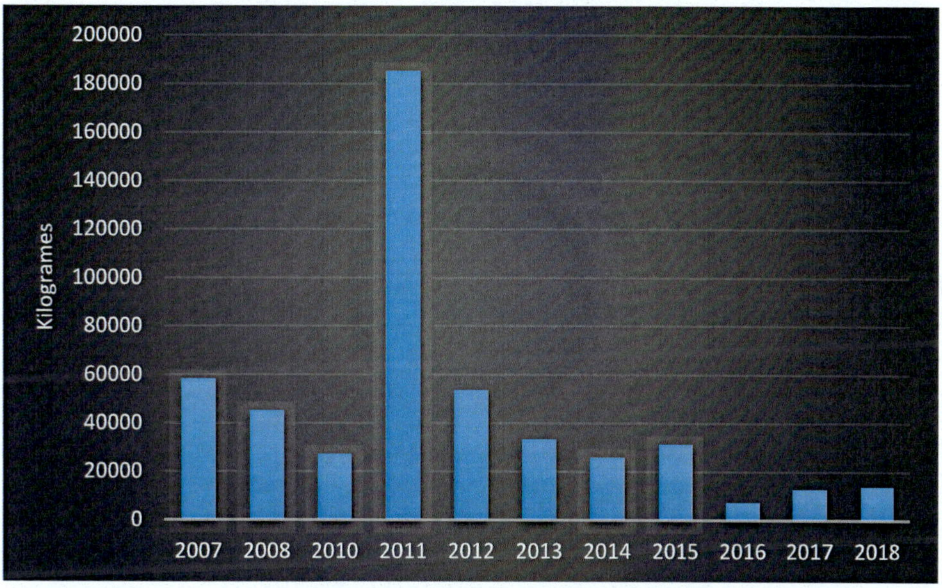

Fig. 7.3 Amount of plastic waste picked up during clean-up campaigns. (Source: Authors, data from Ocean Conservancy, 2019)

river systems and local water sources around Durban city and beyond. This is done through the use of physical barriers, community initiatives and education programmes. The project has targeted five of Durban's most polluted river systems. At each site, 2–3 L booms are installed. The booms are made of sealed, robust pipes installed in a series to act as fail-safe mechanisms when there is high discharge in rivers. The booms are easy to maintain for teams in the communities and very effective when cleared daily. Each day, the amount of collected waste is weighed and documented to measure effectiveness in the collection over time and identify the biggest polluting materials.

Other NGOs that have come to the fore to deal with plastic pollution in Durban include Plastics SA, which through teaming up with other organisations such as the KZN Marine Waste Network helped remove over 300 tons of litter from the Durban Harbour and beaches after the 2019 flood induced plastic pollution. Plastic Producer responsibility organisations such as the Polystyrene Association and individual plastic companies stepped up to the challenge to clean up and remove the litter from some of the most affected areas. The South Africa government also enacted instruments that both banned certain forms of plastics and also imposed levies on the other. The ban stopped the fabrication of plastics below 30 microns, and levies were imposed on those wishing to use thicker plastics. Levies led to an initial reduction in plastic bag consumption by users, but as time went by, the deterioration in public support accompanied by limited public education has seen the utilisation of plastic bags raising to above the pre-levy levels (Nhamo, 2008; Dikgang et al., 2012). The levy has also had a limited impact on plastic bag consumption in South Africa because less than half of the proceeds from it have been officially allocated to the recycling sector. Ryan (2015) observes that levies put in place with limited enforcement mechanisms like environmental education are likely to be effective only in the short-term. Nhamo (2007) highlights that participation remains central to waste management because it creates ownership and gives room for working together in the creation and implementation of viable solutions to the problem of waste management. In a study done by Van Rensburg et al. (2019) on ways to solve the plastic problem in Durban, 43% of the participants were of the view that a total embargo on plastic bag use was likely to be more effective in solving the problem when compared to a partial ban, voluntary actions or increasing the cost of plastic bags. Reusing plastic bags has been recommended as a possible measure of reducing one's ecological footprint. However, it does not imply that the reused bags will not end up polluting the environment after use. Van Rensburg et al. (2019) observe that repeated use of plastic bags by the South African people will go a long way in temporarily limiting the problem of plastic pollution but was not an effective long-term remedy. They highlight that users usually forget to bring with them reusable bags when they go shopping and hence end up buying single-use plastic bags that are more damaging to the environment.

7.5 Conclusions and Recommendations

The chapter's objective was to determine the impacts of the April 2019 extreme floods on plastic pollution at harbours, beaches and urban estuaries in the KwaZulu-Natal province of South Africa. It also reviewed how different stakeholders have tried to deal with the problem. The chapter revealed that the beaches, harbours and urban estuaries were hit hard by plastic litter after the floods. The serious problem of plastic in these areas was shown to be a symptom of poor waste management in the upper catchments as these places were at the catchment outlets. Plastic pollution posed a serious threat to marine ecosystems that provide life support services and threatened the tourism industry and the movement of ships at one of Africa's busiest port. Initiatives have been put in place to deal with plastic pollution from the government, private and civic society. This has seen very noble initiatives, but they remain inadequate as the problem has worsened.

We argue the problem of plastic pollution in Durban is complex, and no one institution can deal with it. There is a need for a paradigm shift in dealing with the problem which must be tackled from multiple fronts including tourism, retail, rivers, communities, industry and government. The solutions must be suited to regions and communities with limited resources, and societal challenges hence can only be arrived at through intensive consultation processes. There is a need for more investment in waste management and the environment, including both the hardware and software components. Awareness raising and mass environmental education relating to the deleterious effects of the use and improper disposal of plastics need to be promoted at schools, on social media, radios and in newspapers. These and many more relevant media need to be harnessed and widely used to educate and draw attention to the public on the dangers of plastic pollution in the oceans.

References

Algalita. (2011). *Algalita Marine Research Foundation*. Retrieved May 20, 2020, from http://www.algalita.org

Andrady, A. L. (2011). Microplastics in the marine environment. *Marine Pollution Bulletin, 62*, 1596–1605.

Ardé, G. (2019). *Durban floods: Profiting from plastic*. New Frame. Retrieved April 20, 2020, from https://www.newframe.com/durban-floods-profiting-from-plastic/

Ayalon, O., Goldrath, T., Rosenthal, G., & Gossman, M. (2009). Reduction of plastic carrier bag use: An analysis of alternatives in Israel. *Waste Management, 29*, 2025–2032.

Baxter, J. (2019). Durban's Plastic Pollution Makes World Headlines (Whilst Voted Greenest City in the World). *SA People*. Retrieved April 20, 2020, from https://www.sapeople.com/2019/12/19/watch-durbans-plastic-pollution-makes-world-headlines-whilst-voted-greenest-city-in-the-world/

Bakir, A., Rowland, S., & Thompson, R. (2014). Enhanced desorption of persistent organic pollutants from microplastics under simulated physiological conditions. *Environmental Pollution, 185*, 16–23.

Barnes, D. K., Walters, A., & Gonçalves, L. (2010). Macroplastics at sea around Antarctica. *Marine environmental research, 70*(2), 250–252.

Brink, P., Lutchman, I., Bassi, S., Speck, S., Sheavly, S., Register, K., & Woolaway, C. (2009). *Guidelines on the use of market-based instruments to address the problem of marine litter*. Brussels, Belgium/Virginia Beach: Institute for European Environmental Policy (IEEP)/Sheavely Consultants.

Browne, M. A., Crump, P., Niven, S. J., Teuten, E., Tonkin, A., Galloway, T., & Thompson, R. (2011). Accumulation of microplastic on shorelines worldwide: Sources and sinks. *Environmental Science and Technology, 45*, 9175–9179.

Caboz, J. (2019). *Shocking footage caught tons of plastic crashing in the waves off Durban following heavy rains, Business Insider SA*. Retrieved April 20, 2020, from https://www.businessinsider.co.za/tags/topics/durbantons-of-plastic-churns-through-waves-on-south-african-beach/

Claessens, M., Meester, S. D., Landuyt, L. V., Clerck, K. D., & Janssen, C. R. (2011). Occurrence and distribution of microplastics inmarine sediments along the Belgian coast. *Marine Pollution Bulletin, 62*, 2199–2204.

Cole, M., Lindeque, P., Halsband, C., & Galloway, T. S. (2011). Microplastics as contaminants in the marine environment: A review. *Marine Pollution Bulletin, 62*(12), 2588–2597. https://doi.org/10.1016/j.marpolbul.2011.09.025.

Danielsson, M. (2017). *The plastic bag ban in Rwanda: Local procedures and successful outcomes. A case study on how Rwanda implemented a nation-wide ban on plastic bags*. MSc Thesis, Uppsala University.

de Freytas-Tamura, K. (2017). *Public shaming and even prison for plastic bag use in Rwanda*. Retrieved November, 2019, from https://www.nytimes.com/2017/10/28/world/africa/rwanda-plastic-bags-banned.html

Derraik, J. G. B. (2002). The pollution of the marine environment by plastic debris: A review. *Marine Pollution Bulletin, 44*(2002), 842–852. https://doi.org/10.1016/S0025-326X(02)00220-5.

Dikgang, J., Leiman, A., & Visser, M. (2012). Analysis of the plastic-bag levy in South Africa. *Resources, Conservation and Recycling, 66*, 59–65.

Dube, K., Nhamo, G., & Chikodzi, D. (2021). Rising sea level and its implications on coastal tourism development in Cape Town, South Africa. *Journal of Outdoor Recreation and Tourism, 33*, 100346.

Dube, K., Nhamo, G., & Mearns, K. (2020). Beyond's response to the twin challenges of pollution and climate change in the context of SDGs. In G. Nhamo, G. Odularu, & V. Mjimba (Eds.), *Scaling up SDGs implementation* (Sustainable Development Goals Series). Cham: Springer. https://doi.org/10.1007/978-3-030-33216-7_6.

Durban Tourism. (2013). *Taking Durban to the world: Visitor marketing strategy 2013–2020*. Durban: eThekwini Municipality.

Gattringer, C. W. (2018). A revisited conceptualisation of plastic pollution accumulation in marine environments: Insights from a social ecological economics perspective. *Marine Policy, 96*, 221–226. https://doi.org/10.1016/j.marpol.2017.11.036.

Goba, T. (2019). 300 tons of dirt removed from harbour after #DurbanFloods. *Daily News*. Retrieved April 20, 2020, from https://www.iol.co.za/dailynews/news/kwazulu-natal/300-tons-of-dirt-removed-from-harbour-after-durban-floods

Hardesty, B. D., Good, T. P., & Wilcox, C. (2015). Novel methods, new results and science-based solutions to tackle marine debris impacts on wildlife. *Ocean and Coastal Management, 115*, 4–9. https://doi.org/10.1016/j.ocecoaman.2015.04.004.

Hayward, R. (2018). Foreign exchange speculation: An event study. *International Journal of Financial Studies, 6*, 22. https://doi.org/10.3390/ijfs6010022.

Head, T. (2019). *Durban floods: Here's how much rain has fallen since Monday*. Retrieved June 17, 2020, from www.thesouthafrican.com/news/weather/durban-floods-how-much-rainfall-kzn-23-april/

Intergovernmental Panel on Climate Change (IPCC). (2019). *Special report on the ocean and cryosphere in a changing climate—SROCC*. Geneva, Switzerland: Intergovernmental Panel on Climate Change.

Jambeck, J., Geyer, R., Wilcox, C., Siegler, T., Perryman, M., & Andrady, A. (2015). Plastic waste inputs from land into the ocean. *Science, 347*(6223), 768–771.

Jang, Y. C., Hong, S., Lee, J., Lee, M. J., & Shim, W. J. (2014). Estimation of lost tourism revenue in Geoje Island from the 2011 marine debris pollution event in South Korea. *Mar.Pollut.Bull, 81*(1), 49–54.

KwaZulu-Natal (KZN) Treasury. (2013). *Municipal finance 4th quarter review, Close-out report 2012/2013*. KwaZulu-Natal Provincial Government: Pietermaritzburg.

Leonard, L. (2004). *Health care waste in Southern Africa: A civil society perspective*. In: Proceedings of the International Health Care Waste Management Conference and Exhibition, Johannesburg, South Africa.

Lishivha, W. (2018). Help save our seas on International Coastal Cleanup Day 2018. *Getaway News*. Retrieved April 20, 2020, from https://www.getaway.co.za/travel-news/save-the-seas-international-coastal-cleanup-day-2018/

McCormick, A., Hoellein, T. J., Mason, S. A., Schluep, J., & Kelly, J. J. (2014). Microplastic is an abundant and distinct microbial habitat in an urban river. *Environmental Science & Technology, 48*, 11863–11871.

Menyuka, N. N., Sibanda, M., & Bob, U. (2020). Perceptions of the challenges and opportunities of utilising organic waste through urban agriculture in the durban south basin. *International Journal of Environmental Research and Public Health, 17*(4), 1158. https://doi.org/10.3390/ijerph17041158.

Mouat, T., Lopez-Lozano, R., & Bateson, H. (2010). *Economic impacts of marine litter*. KIMO (Kommunenes Internasjonale Miljøorganisasjon/Local Authorities International Environmental Organisation.

Naidoo, T., Smit, A. J., & Glassom, D. (2016). Plastic ingestion by estuarine mullet Mugilcephalus (Mugilidae) in an urban harbour, KwaZulu-Natal, South Africa. *African Journal of Marine Science, 38*(1), 145–149. https://doi.org/10.2989/1814232X.2016.1159616.

Nhamo, G. (2007). Environmental law and policy reform surrounding packaging waste management in South Africa. *Packaging Waste Management in South Africa, SAJELP, 14*, 136–157.

Nhamo, G. (2008). Regulating plastics waste, stakeholder engagement and sustainability challenges in South Africa. *Urban Forum, 19*, 83–101. https://doi.org/10.1007/s12132-008-9022-0.

Ocean Conservancy. (2019). *Internarial Coastal Cleaning 2019 report*. Retrieved June 5, 2020, from www.oceanconcervancy.org/annualreport2019

Ogola, J. S. (2012). *Management of solid wastes: A case study in Limpopo Province, South Africa*. Retrieved from http://www.Intechopen.com/InTech

Pettipas, S., Bernier, M., & Walker, T. R. (2016). A Canadian policy framework to mitigate plastic marine pollution. *Marine Policy, 68*, 117–122.

Rochman, C. M., Kross, S. M., Armstrong, J. B., Bogan, M. T., Darling, E. S., Green, S. J., Smyth, A. R., & Verı́ssimo, D. (2015). Scientific evidence supports a ban on microbeads. *Environmental Science & Technology, 49*, 10759–10761.

Ryan, P. G. (2015). A brief history of marine litter research. In M. Bergmann, L. Gutow, & M. Klages (Eds.), *Marine anthropogenic litter* (pp. 1–25). Berlin: Springer.

Schnurr, R. E. J., Alboiu, V., Chaudhary, M., Corbett, R. A., Quanz, M. E., Sankar, K., Srain, H. S., Thavarajah, V., Xanthos, D., & Walker, T. R. (2018). Reducing marine pollution from single-use plastics (SUPs): A review. *Marine Pollution Bulletin., 137*, 157–171. https://doi.org/10.1016/j.marpolbul.2018.10.001.

Seltenrich, N. (2015). New link in the food chain? Marine plastic pollution and seafood safety. *Environmental Health Perspectives, 123*, 34–41.

Stickel, B. H., Jahn, A., & Kier, W. (2012). *The cost to west coast communities of dealing with trash, reducing marine debris* (p. 21). Prepared by Kier Associates for U.S. Environmental Protection Agency.

Thompson, R. C. (2015). Mircoplastics in the marine environment: Sources, consequences and solutions. In M. Bergmann, L. Gutow, & M. Klages (Eds.), *Marine anthropogenic litter* (pp. 185–200). Heidelberg: Springer.

United Nations Environment Programme (UNEP). (2011). *Plastic debris in the Ocean, UNEP year book 2011*. Nairobi: UNEP.

UNEP and NOAA. (2015). *The Honolulu Strategy: A global framework for prevention and management of marine debris*. Retrieved November 15, 2019, from http://unep.org/gpa/documents/publications/honolulustrategy.pdf

UNEP/GPA. (2011). *Global programme of action for the protection of the marine environment from land-based activities*. Retrieved from http://www.gpa.unep.org/

United Nations Environment Programme (UNEP). (2018). *Single-use plastics: A roadmap for sustainability*. Nairobi, Kenya: United Nations Environment Programme.

United Nations. (2015). *Transforming our world: The 2030 agenda for sustainable development*. New York: United Nations.

Van Heerden, P. (2015). *The politics of knowledge: Case study of the management of beach water quality in Durban, South Africa*. Submitted in fulfilment of the academic requirements for the Master of Social Science Degree in the School of Built Environment and Development Studies, University of KwaZulu-Natal, Durban.

Van Rensburg, M. L., Nkomo, S. L., & Dube, T. (2019). The 'plastic waste era'; Social perceptions towards single-use plastic consumption and impacts on the marine environment in Durban, South Africa. *Applied Geography, 114*, 102132. doi.org/10.1016/j.apgeog.2019.102132.

Xanthos, D., & Walker, T. (2017). International policies to reduce plastic marine pollution from single-use plastics (plastic bags and microbeads): A review. *Marine Pollution Bulletin, 118*(1–2), 17–26.

Check for updates

The Need for Effective Storm Water Management to Build Flood Resilient Communities: A Case of Port St Johns, South Africa

8

Hlengiwe Precious Kunene,
Nompe Gladys Ntombela, and Asanda Madikizela

Abstract

Globally, urbanisation attracts informal settlements. Such settlements often do not comply with official spatial planning guidelines and are hardly provided with basic services, including drainage infrastructure. Storm water management is particularly important to build resilience in cities, especially with the increasing impact of climate change and the efforts to achieve the Sustainable Development Goals (SDGs) by 2030. It is against this background that this chapter explores the factors influencing the poor provision and maintenance of storm water management systems in different areas located in Port St Johns, South Africa. Using a household survey, key informant interviews and observations as the main research instruments to collect data, the study points to the importance of storm water management as a key infrastructure component that requires sufficient funding to reduce flood risks. In addition, the study recommends community-led initiatives to enable storm water intervention strategies. It further proposes that the local government develops prevention measures against the building of houses in flood-prone areas.

H. P. Kunene (✉) · N. G. Ntombela · A. Madikizela
Institute for Corporate Citizenship, University of South Africa, Pretoria, South Africa
e-mail: Kunenhp@unisa.ac.za; ntombgn@unisa.ac.za; Madika@unisa.ac.za

Keywords

Drainage systems · Sustainable · Port St Johns · Storm water · Flooding

8.1 Introduction

Storm water drainage systems are considered vital infrastructures that carry storm water runoff away from human settlements (Armitage, 2011; Gupta, 2016; Malulu, 2016). Despite their importance, it has often proven difficult to provide effectively functioning storm water management structures for informal settlements in South Africa (Williams & Sutherland, 2018). A combination of factors has influenced the formation of informal settlements in South Africa and this includes the rapid urban population growth, and the invasion of unused land, which is usually unsuitable for construction owing to its physical or environmental features (Parkinson, 2003; Parkinson et al., 2007; Ramkissoon et al., 2015). These factors are often complicated by a situation where storm water management is a neglected aspect in overburdened and under-capacitated municipalities. Additionally, housing structures within informal settlements are habitually built without consent from official authorities and

G. Nhamo, L. Chapungu (eds.), *The Increasing Risk of Floods and Tornadoes in Southern Africa*, Sustainable Development Goals Series, https://doi.org/10.1007/978-3-030-74192-1_8

rarely conform to building regulations and proper construction standards (Butler & Davies, 2011; Hetz & Bruns, 2014). To aggravate these problems, the majority of informal settlements are established in the precarious areas of the urban periphery that are susceptible to flooding. These locations often include low-lying areas near streams or rivers, in floodplains, disturbed wetland areas, and steep slopes that are prone to mud- and rockslides (Hallegatte et al., 2015).

By law, informal settlements in South Africa are illegal and therefore are often excluded from receiving municipal basic services. Drainage infrastructure is among the amenities that are often not provided for informal dwellers (Williams & Sutherland, 2018). This is because storm water management is generally funded through municipal rates, which informal settlers do not pay (Fisher-Jeffes & Armitage, 2013). In cases where infrastructure funding is provided, insufficient budgets are set aside for the provision and maintenance of drainage infrastructure (Fisher-Jeffes & Armitage, 2013). This misallocation could be linked to the inadequate understanding by some municipal officials of the importance of storm water management infrastructure provision and maintenance. Insufficient technical capacity and experience to maintain storm water structures by municipalities is also regarded as another hindrance to reliable service delivery. A combination of these challenges often subjects populations located in informal settlements to high risks of flooding. This is a threat that will exponentially increase as the climate changes and weather patterns become more erratic and intense (Delgado & Guibrunet, 2017). This is of great concern, especially considering that the world is pursuing its ambitious 2030 Agenda for Sustainable Development and the Sustainable Development Goals (SDGs) (UN, 2018).

At a micro level, these goals serve as a roadmap for the local government towards more equitable urban development in a world that is increasingly becoming urbanised. The implicit mention of storm water management across several SDGs relating to sustainable water management, resilient infrastructure, inclusive, safe and resilient cities, healthy lives and well-being highlight a crucial element in enhancing the resilience of communities. This study aims to unveil the challenges associated with the inadequate provision and maintenance of storm water management structures in the context of flooding in Port St Johns (Eastern Cape, South Africa). The study calls for storm water management structures to be part of discussions and policies focused on the basic service provision for informal settlements.

8.2 Literature Review

Storm water management in urban informal settlements is a critical but complex issue. This section presents background literature that provides insights on storm water management and drainage in the context of informal settlements. It outlines key concepts in storm water management and reviews the consequences of poor drainage systems in the informal settlements. Lastly, it examines the interplay between climate change, storm water management and informal settlements.

8.2.1 Definition of Key Concepts in Storm Water Management

Storm run-off, greywater and wastewater, at their source as well as at the receiving water bodies, are potentially detrimental to the environment. Storm water usually remains above the ground in the event of precipitation. Greywater, on the other hand, is derived from domestic activities such as dishwashing, bathing and laundry, while wastewater contains human body waste (Watershed et al., 2017). These three water types pose various risks to humans, thus necessitating proper drainage and management, especially in informal settlements where there is limited infrastructure and poor service delivery.

Storm water management typically serves two primary objectives that are aimed at waste management and water volume control (Zhou et al., 2013). Storm water drainages are intended to provide benefits associated with improved water run-off velocity, which can eventually translate to the minimised risk of flooding and the delivery of additional water resources to supplement existing water supplies for different reasons, such as irrigation (Jia et al., 2017). However, concerns have been raised regarding conventional drainage systems and their long-term sustainability on the environment (Randall et al., 2019). One of the concerns is that traditional systems may not meet the general criteria of sustainability, and hence the call to adopt sustainable drainage systems that can potentially minimise the risk of flooding, in line with the SDGs agenda.

In efforts to potentially minimise storm water run-off, both structural and non-structural measures are used. Structural storm water management strategies are typified by physical structures that are designed to improve the draining of water run-off (Ruhnke, 2011). Structural drainage infrastructures intended to reduce run-off velocities, collect and transport storm water while releasing it to relevant destinations far from habitable areas. However, the costs and time needed for the installation of drainage networks are tremendous, which often renders them undesirable (Keeley et al., 2013). Non-structural measures involve human practices to prevent and minimise pollutants from entering drains to maximise the volume of storm water penetrating through them (Sakigele et al., 2012). In South Africa, for example, local government is the overseer of such issues, and given its proximity to the people, it is considered a sphere where most of the operational activities relating to disaster management occur. Among other interventions, the National Department of Public Works, in partnership with local municipalities through its Expanded Public Works Programme (EPWP), employs local community members to inspect and tidy local drains. These initiatives are effective in communities with inadequate waste collection systems.

8.2.2 The Impact of Urbanisation on Informal Settlements

Urbanisation is regarded as one of the factors influencing drainage challenges in cities. The continuous flock of unplanned population to cities overloads drainage infrastructures, which frequently block and overflow. In addition, high densities and the congested nature of informal developments tamper with the natural terrain which is often replaced by compact soil that decreases run-off infiltration. For instance, Du et al. (2016) found that the rapid increase of populations in Chinese urban areas resulted in impervious surfaces. They further discovered that with an increase in urban development, cities experienced high run-off even in places where drainage systems exist. This is particularly detrimental to the urban poor, who often poorly construct housing on marginal lands such as riverbanks and floodplains with little to no consideration of storm water management systems (Taylor et al., 2015). Besides, poor housing conditions and inappropriate zoning makes drainage provision difficult due to lack of space. Furthermore, some of these settlements are established in coastal towns and river estuaries (Ngie, 2012). These areas typically have high average rainfall and flat terrain which make the draining of water more difficult, making them susceptible to flooding (Olorunfemi, 2009). However, Thorne (2014) found that for informal dwellers, the benefits of living close to sources of employment and urban services generally outweigh the disadvantages associated with flooding. Informal dwellers generally perceive flooding as a natural and seasonal event. Another study by Harries (2013) conducted in England and Wales had the same outcome where communities in informal settings perceived flooding to be low risk and a part of life compared to the benefits provided by living close to the city.

8.2.3 The Nexus Between Climate Change, Storm Water Drainage Systems and Informal Settlements

Climate change is recognised as a global issue and its widespread impacts are noted for storm water management infrastructure (Madsen et al., 2018). According to Jury (2018), there is an expected increase in general infrastructural damage due to climate change. It is highly likely that heavy precipitation would increase in many regions of the world, including South Africa (Intergovernmental Panel on Climate Change (IPCC), 2012). In essence, current traditional drainage systems in South Africa are not spared in this regard. This is because traditional systems are designed with limited consideration for climate-induced stresses (Madsen et al., 2018). As a result, they are often overwhelmed by severe capacity problems in events of intense rainfall. Additionally, current drainage infrastructures are built without the regard for frequent rainfall (Löwe et al., 2018). Local governments often add to existing problems by not maintaining existing drainage infrastructure (Ngie, 2012). Armitage (2011) identifies poor solid waste management as one of the maintenance issues, where uncollected solid waste often enters drains, causing blockages and reduced flow capacity.

While climate change impacts are anticipated to be devastating for vulnerable communities in various ways, the absence of and poorly maintained storm water management systems are likely to worsen the impact of flooding (Williams & Sutherland, 2018). Generally, people living in informal settlements have fewer resources available to them and receive limited to no support from government. The damage caused by flooding to their livelihoods places extra demand on the limited resources, exposing them more to other segments of vulnerability (Hetz & Bruns, 2014). Among others, women, children and people with disabilities usually experience difficulties in dealing with the aftermath of flooding and are particularly more vulnerable to suffer health effects from floods (Okaka et al., 2017). Women, in particular, tend to be disproportionately affected as they deal not only with the economic devastation and disruption of livelihood systems but are also often left to cope with the psychological turmoil that comes from dealing with death, diseases and food shortages during post-flooding events (Azad et al., 2013).

8.2.4 Implications of Inadequate Storm Water Management Systems on the SDGs

The implementation of sustainable storm water management systems can contribute immensely to the attainment of several Sustainable Development Goals (SDGs). It is worth considering the extent to which the SDGs stress the importance of storm water management in the context of flooding. Urban infrastructure and services in general influence and are influenced by several SDGs and targets. For instance, SDG 6 (clean water and sanitation) makes mention of sustainable water use and management of water resources concerning water supply to urban areas. The goal addresses sustainable storm water drainage systems which have proven to be a cost-efficient solution to manage flood risks while also delivering wider benefits.

Given a range of economic activities in cities (SDG 8—decent work and economic growth), urban infrastructure needs to be able to withstand the impacts of extreme weather events associated with climate change. Hence, investments or funding for such systems remain necessary (SDG 9—industry innovation and infrastructure, SDG 11—sustainable cities and communities and SDG 13—climate action). SDG 13 also places emphasis on the investments of resilient infrastructure for the protection of communities living in disaster-prone areas. In the event of deficits in such systems, the extent of damage is likely to be enormous. Moreover, these infrastructures must be carefully planned, in order to be delivered with a minimal carbon footprint.

Urbanisation is associated with the generation of a considerable quantity of waste. This is mainly the case in informal settlements where solid waste management services are not provided. People living in informal settlements end up dumping waste illegally, which has implications for other services and flooding. Waste is a huge concern for storm water management and its reduction is fundamental for the achievement of SDG 12 (sustainable consumption and production); otherwise waste can have a direct impact on ecosystems and biodiversity (SDGs 14 and 15). These linkages show that proper storm water management plays a central role in facilitating all the dimensions of sustainability, and hence its significance to the achievement of the 2030 Agenda on Sustainable Development and SDGs.

8.3 Materials and Methods

8.3.1 Study Area, Population and Sampling

Data was collected in Port St Johns, a local municipality under the OR Tambo District Municipality, South Africa, as shown in Fig. 8.1. Port St Johns is a town that stretches about 300 km along the scenic Wild Coast of the Eastern Cape in South Africa. It is characterised by a steep coastal escarpment comprising perennial rivers and uneven topography in the form of hills and valleys. Similar to other towns worldwide and in South Africa, urbanisation is a prominent phenomenon in Port St Johns IDP (2019a). Urbanisation is regarded as a contributory factor towards an increase in urban settlements, which are characterised by informality and inadequate social and infrastructural facilities. Port St Johns municipality comprises informal settlements (usually located in close proximity to town), low-income housing and rural villages. Villages are isolated with poor infrastructure and limited resources and are relatively ethnically homogenous and socially cohesive. However, there are salient differences between villages as some have bigger yards for building and subsistence farming. Houses are built with concrete while others are mud-plastered. Informal settlements, on the other hand, are nucleated linear settlements with limited or no spaces in-between for gardens. The narrow streets and walkways between these informal settlements are particularly prone to damage when flooding occurs. In addition, the land on which these settlements are built is generally on high elevation, encompassing steep hills and inclines along which storm water easily flows through the valleys and gorges.

According to the Port St Johns flood incident report (2019b), 882 people from different wards and villages were directly affected by the 2019 floods in Port St Johns. However, this research study was unable to survey all the affected wards, due to time constraints, resource limitations and

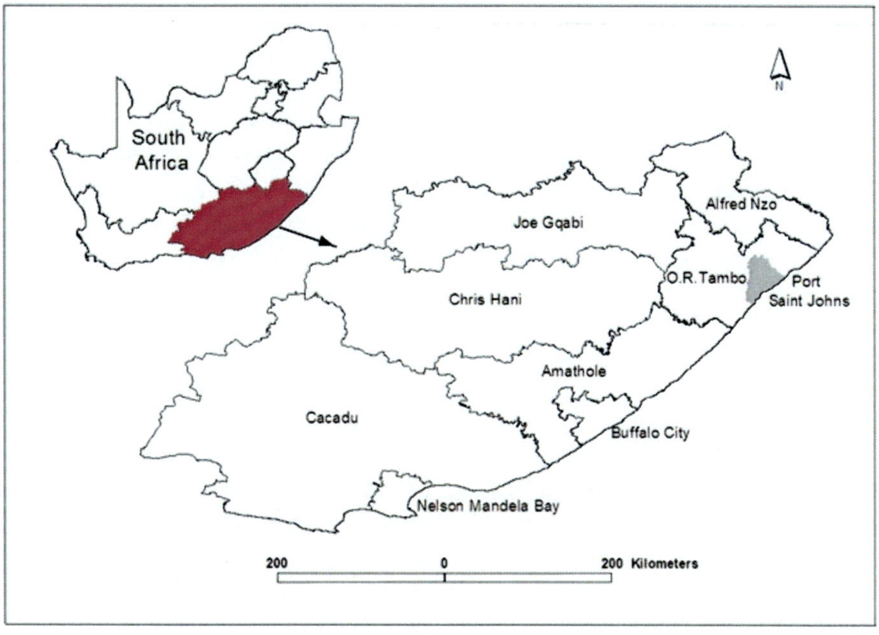

Fig. 8.1 The study area, Port St Johns, Eastern Cape, South Africa. (Source: Authors, 2020)

Table 8.1 Respondents from Port St Johns

Village/township	Ward number	No. of respondents	%
Green Farm, Town, Mpantu, Mthumbane	6	102	52%
Cremone	10	2	1%
Elithengela, Qhoboshendlini, Luhengele	8	18	9%
Emdlakala, Isandlube, Luqhoqhweni, Mdlankala	15	24	21%
Gogogo, Mswakadzi	1	13	7%
Skhilikhi Village, Esalwani, Mbabalane, Polar Park, Qhubushwayo, Selwane, Tyityane—rural	16	36	18%
Lower Ntafufu, Mahalo	12	2	1%

Source: Authors, 2020

accessibility challenges. The study therefore employed purposive sampling, by means of surveying respondents who were directly affected by flooding, as recorded in the municipality's 2019 Flood Incident Report. Out of the total of 227 surveys administered, 197 households were willing to participate in the study. Questionnaires were administered in English, using the QuestionPro online survey software. In cases where respondents could not understand English, questions were translated into isiXhosa which is the local language in the area.

The 197 respondents who participated in the survey were from different wards with the majority (52%) from ward 6, while 21% were from ward 15, 18% from ward 16 and the remaining respondents from the other wards, as per Table 8.1.

8.3.2 Research Design

This study adopted a mixed method research approach for data collection. The approach is described as a procedure that uses both quantitative and qualitative research methods in a single study to understand a research problem (Creswell & Plano Clark, 2011). Informal settlements are at the heart of environment problems. However, both are affected by the lack of basic drainage systems, which results in natural disasters such as flooding. While this impacts on the community in terms of well-being disruption, it also puts a strain on government resources. Hence, the study investigates the impacts of these challenges on all parties (community members and government representatives), serving as the motivation for adopting the mixed methods research approach.

The mixed methods approach can be deployed in different ways, primarily to suit the research problem at hand. For this study, the quantitative research was carried out first, to obtain the demographic characteristics of the respondents. This method was closely followed by the qualitative research approach, which was used to explore the closed-ended questions that were posed in the survey questionnaire. For the quantitative approach, the study used a community household questionnaire survey. This approach was necessary as the aim of this study was to understand the links between the absence of drainage systems, widespread flooding and the resulting impacts. For the qualitative approach, the study employed key informant interviews which were undertaken in face-to-face settings with council members, municipality contractors and community representatives. This approach is attributed to the presumption that council members, municipality contractors and community representatives often possess first-hand information about the communities they serve. This was done to understand if there were plans to minimise the impacts of the recurrent flooding in the area as well as current or anticipated challenges in executing such plans. Observations were also used as a data collection instrument in the form of pictures to give a visual representation of the situation in the area.

A household questionnaire survey and interviews were used as primary sources. The study used secondary sources, which included the Port St Johns local municipality integrated development plan (IDP), the Port St Johns 2019 floods incident report, observations, press releases, census data, online journal articles, online and printed books, reports, web pages and grey literature related to the research. Six key informants from the Port St Johns municipality were interviewed in this study, namely the Port St Johns municipal manager, municipality strategic manager, two municipal contractors and two local councillors. Furthermore, lodge managers who had been impacted by the flooding event were identified from the municipality's flood incidents report and were also interviewed. Findings from key informants were audio recorded and these recordings were later transcribed by the researchers.

8.4 Presentation and Discussion of Findings

This section analyses survey results from the local community and key informants on whether proper drainage systems could have played a role in mitigating the impacts of the 2019 floods. Also, respondents were asked about their perceptions on the challenges that the municipality has with regard to maintaining existing and installing new Storm water drainage infrastructure. As a result, three questions in the household questionnaire survey aimed at understanding the consequences of poorly maintained Storm water drainage systems, as experienced by the people of Port St Johns. Community responses to the three questions are presented in Subsects. 8.4.1–8.4.3 while key informant responses are covered in Subsect. 8.4.4.

8.4.1 Possible Reasons That Exacerbated the Risk of Flooding

Respondents were asked to provide reasons perceived to be contributing to flooding. Respondents were given four options to choose from, namely, lack of storm water drainage systems, absence of early warning systems, location of household and other reasons. The results are presented in Fig. 8.2.

The majority of respondents (40%) believed that the lack of storm water drainage systems is the ultimate reason for worsened risk of flooding in the area, followed by the location of household at 36% and absence of early warning systems, while 1% of respondents suggested other reasons. When considering the topography of the worst affected area as per the incident report, one would agree with the above statistics. Respondents were further informally interviewed about the location of their households. Those located in disturbed wetlands indicated that when it rains heavily for long periods, rain often doesn't drain, causing flooding in the area. Other respondents were from Green Farm informal settlement, which is visibly located at the foot of the valley that receives water from the surrounding high areas. One key informant from the municipality indicated that they are well aware of the lack of storm water drainage systems. However, he stated that storm water management lies outside their jurisdiction and responsibility as a local municipality, given that the budget allocation for such lies with the OR Tambo metropolitan municipality.

Furthermore, these findings concur with a study conducted by Armitage (2011) who states that because of the illegal status of informal settlements, the local government responsible for the provision of urban drainage infrastructure is not officially allowed to provide services to these areas. This is because funds allocated to local governments for basic services on official households by the National Treasury are usually not enough (StatsSA, 2020). This usually puts local governments in a predicament where they either have to evict informal dwellers or provide them with the basic services (Taing, 2017).

Nevertheless, irrespective of the legitimacy of these settlements, most local municipalities frequently lack resources necessary to provide adequate drainage systems to growing urban populations. Even in the better-resourced municipalities, the provision of services to informal settlements is often frustrated by the fragmentation of those responsible for service delivery into a number of departments that do not always work together or even communicate well. It was also revealed in this study that despite poor drainage systems and the lack of maintenance of the avail-

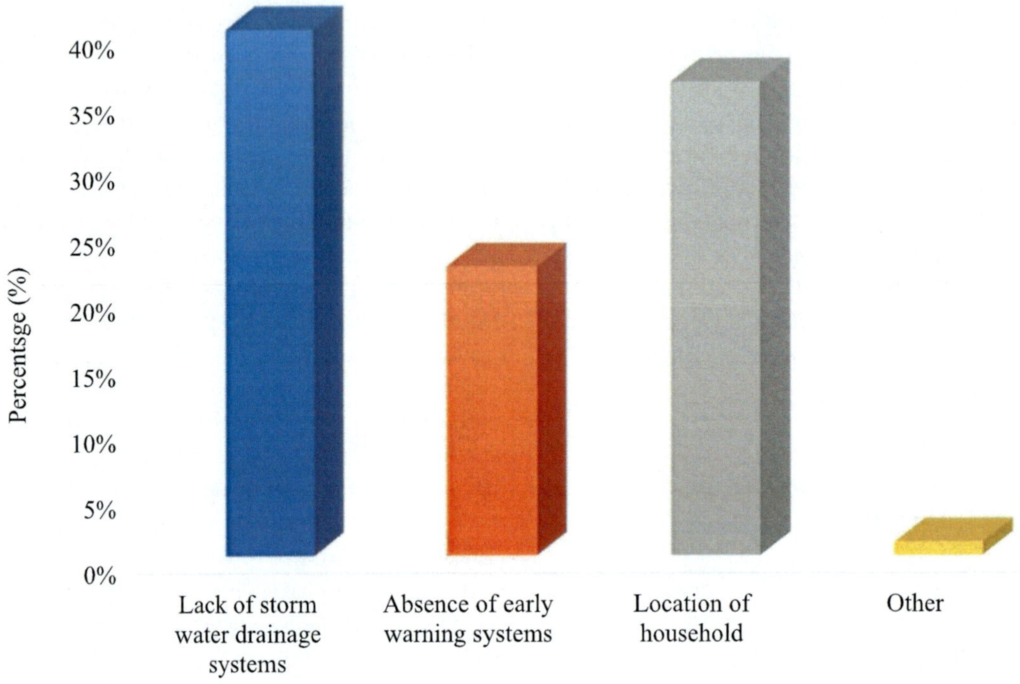

Fig. 8.2 Reasons perceived to have worsened the risk of flooding in Port St Johns. (Source: 2019 Eastern Cape fieldwork)

able drainages, residents displayed their resilience by employing several practices and alternatives that sought to manage storm water in order to prevent flooding while they also devised coping strategies to flooding events. As such, raised houses and self-constructed water channels were witnessed at the Green Farm informal settlement. Some of these informal structures appeared very unsafe given the residents' financial status and limited construction skills and capabilities of those who constructed them, as shown in Fig. 8.3.

8.4.2 Flood Risk Exposure in the Absence of Proper Storm Water Management Infrastructure

Respondents were asked if they believed that their flood risk exposure would have been minimised by a proper drainage system. Respondents were given two options, yes or no. Pursuantly, 87% indicated their agreement, while only 13% disagreed. These results indicate that poor storm water drainage run-off is a significant problem when flooding occurs in the area. The lack of proper drainage systems increases residents' exposure to flood risk. In addition to flooding, the lack of maintenance (Tas et al., 2013) results in informal drainage systems increasingly becoming the source of all kinds of waste including solid waste that eventually affects the health and well-being of the residents, as shown in Fig. 8.4. The state of a storm water management system has a ripple effect on other public services such as basic sanitation and decent transportation. Therefore, a good drainage system would require proper waste management, which in turn needs comprehensive land use planning and management. This study further found a high level of household health vulnerability to flooding, associated with the occupation of low-lying marginal areas. Local communities in Port St Johns are exposed to flooding every time they experience heavy rains. Additionally, due to their poor economic standing and the lack of support from the local government, their adaptive capacity is low.

Fig. 8.3 A raised house made of bamboo sticks after the 2019 floods at Green Farm, with a self-constructed water channel. (Source: 2019 Eastern Cape fieldwork)

Fig. 8.4 Soil build-up in a storm water drain witnessed in Green Farm informal settlement. (Source: 2019 Eastern Cape fieldwork)

This study also revealed that the Port St Johns local municipality has not adequately invested in risk reduction to enhance disaster preparedness of people living in informal settlements, thus increasing their vulnerability to flooding. Furthermore, the study reveals that the issue of run-off as a result of inadequate drainage systems is a regular occurrence, virtually every time it rains, resulting in the disruption of transport networks, making it difficult to get to nearby towns.

Interestingly, despite this situation, some respondents were uninterested when asked about relocating to higher ground areas with better drainage systems. Several reasons for this unwillingness included concerns about moving further away from job opportunities with some just wanting to stay in town. This finding resonates with a study by Harries (2013) where study respondents ranked flooding as very low risk compared to other problems, such as lack of job opportunities. According to Wachinger et al. (2013), previous flood experience can decrease risk perception, especially if it did not do significant damage. Engagements with the respondents also confirmed this point, with some asserting that they had become used to flooding and know that it comes and goes.

8.4.3 Hindrances in the Provision of Proper Storm Water Management Structures in Informal Settlements

Respondents were further asked about their perceptions regarding hindrances in the provision of proper storm water drainage systems. They were presented with six options: lack of capacity, lack of funding, lack of maintenance, frequent rainfall, all the above or other factors. The results are presented in Fig. 8.5.

The majority of respondents (30%) identified a lack of funding allocated to drainage systems as the main factor resulting in poor systems, 19% felt that the lack of maintenance was the problem, followed by all of the above at 17%, frequent rainfall at 15%, other reasons at 10% and only 9% believed that the lack of capacity was hindering the provision and maintenance of proper drainage systems as shown in Fig. 9.5. According to these results, the main reason for the failure of the Port St Johns local municipality to provide sustainable urban drainage to its citizens was due to funding constraints. The funding issue relates to a city's revenue which ordinarily comes from rates payments. However, informal settlement residents are not usually required to make these payments.

Rate-payers, who are usually members of the middle class, understandably want to see their money invested in their own sections of the city. Therefore, very little funds are available to appropriately address these conditions in informal settlements. According to Le Jallé et al. (2013), in countries such as England and Northern Ireland, residents are required to pay for the provision and management of storm water infrastructure, despite living in informal settlements. This suggests that any money raised from this revenue should be spent on those who pay for the service (Okoth & Beneah, 2019). In South Africa, however, this is problematic as many cannot afford to pay these rates. Lack of maintenance together with institutional incapacity and insufficient resources on the part of the Port St Johns municipality were considered as further constraints that could also be associated with institutional fragmentation. At the Port St Johns local municipality, the water and sanitation, and the waste management departments all operate independently with minimal or no collaboration. There was also a disjuncture between the expectations of the local community and the plans and goals of the municipal officials. Local community members felt that the government is failing to fulfil its moral obligation of making opportunities available for the poor and most vulnerable. Most of the information gleaned from the interviews seemed to suggest that the residents of Port St Johns were faced with three crucial problems: unemployment, proper road drainage systems (Fig. 8.6) and the lack of proper housing.

8.4.4 Key Informants' Findings

During the engagements with municipal officials, there was an indication of a plan to move the town to higher ground to a piece of land owned by tribal authorities. However, an issue that was identified as a concern by the fieldworkers was seeing new building constructions in the same town that was supposedly to be moved. In response to this concern, municipal officials stated that they could not prohibit developers from building. That was also the same reason

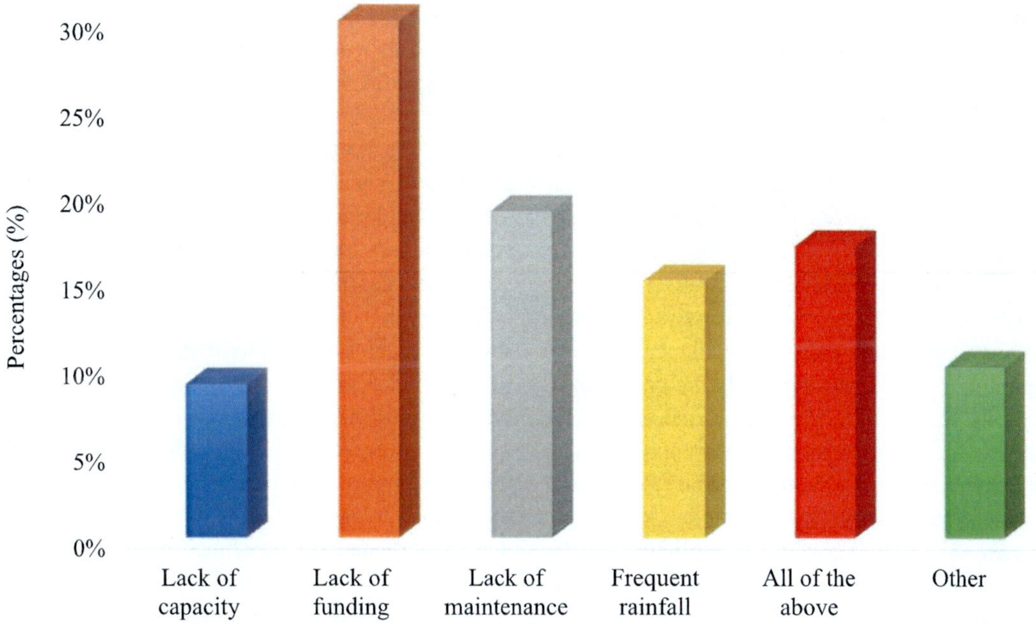

Fig. 8.5 Respondents' perceptions—hindrances in the provision of proper storm water drainage systems. (Source: 2019 Eastern Cape fieldwork)

Fig. 8.6 Drainage infrastructure in Port St Johns abandoned due to lack of funding. (Source: 2019 Eastern Cape fieldwork)

they provided when questioned about why they were allowing the informal settlements to mushroom everywhere.

There was an uncomfortable admission by one municipal official that floods of that nature could have been foreseen based on historical occurrence but despite that acknowledgement, there were no clear guidelines for action. Interestingly, there appears to be a reactionary stance by the municipality as it seemed to wait to see what happens before sanctioning a response. When asked about early warning systems, one official narrated a story of how they still in some cases use the indigenous early warning methods such as seeing 'foam' on the riverbanks days prior to the actual floods. This is congruent with Mwaura's (2008) suggestion that the resilience of local communities to environmental change is linked to how they utilise their sophisticated indigenous knowledge systems for environmental management and coping mechanisms. Hence, a blend of science, technology and traditional knowledge systems presents opportunities for better disaster management, prevention, response and mitigation.

Another municipal official was at pains, lamenting the limited funding of the local municipality, in dealing with disaster issues as well as the provision of basic services to informal settlements. In some instances, municipal budgets only allowed for maintenance and not for the expansion of infrastructure to cope with the influx of people into the city. As noted by the official, Green Farms informal settlement (Ward 6), situated in town, did not have an adequate storm water drainage system and as a result suffered severe damage in comparison to other wards. Also, as a result of budget constraints, it was difficult for the municipality to attract experienced technicians.

8.5 Conclusion

Without a doubt, flooding has compounded the negative impacts on the livelihoods and lives of those living in informal settlements. The underfunding of proper storm water drainage systems coupled with the lack of basic service delivery, and proper urban planning further worsen the consequences on the poor and the most vulnerable. From this deduction, this study offers the recommendations outlined below.

Proper storm water systems largely depend on the financial resources and human capital with the requisite skills. It is therefore critical for local government to have a structure that is capable of dealing with storm water management. As things stand, storm water management resides within the roads and transport department. Instead, there should be an integrated institutional structure that comprises the roads and transport department and other departments within local government such as planning and environment. There is a need for the proactive management of storm water, rather than treating it as an issue that is only dealt in times of floods; that should minimise the impact of flooding. The lack of investment in drainage also results in the rapid deterioration and ageing of the infrastructure.

This study recommends that partnerships between the local government and community members be established to respond to the flooding events. This can be through the formation of community-led initiatives to reduce flood risk vulnerabilities within the community. An example of this initiative could include constructing storm water channels that will direct water during rainy seasons. Another example could be the establishment of a local team of volunteers to collect waste in the area on a weekly basis and dispose of it at a designated spot. This can prevent waste from getting into and blocking storm water drains. These interventions certainly require some sort of incentives for the work done from the local government's side. A good example of the partnership between the locals and local government in flood risk prevention can be observed in two informal settlements in Johannesburg, South Africa (Olumuyiwa, 2015). The findings of the study reveal a possibility of synergy between government and community-led initiatives towards storm water management, provided that everyone works together. Furthermore, the partnership could enable residents to present inputs and provide localised solutions to their challenges.

The study calls also for storm water management infrastructures to be sufficiently funded, given its implications on the broader SDGs framework and socio-economic dimensions of the poor. The lack of funding coupled with the lack of maintenance for storm drains are highlighted as a serious challenge that worsens the impact of flooding, especially in areas such as Green Farm settlement, which witnessed more damage. Given the findings of this case study, the local municipality needs to reprioritise and invest in storm water management approaches that are in line with international best practice. One of the approaches that municipalities could explore is the sustainable drainage systems (SuDS), where storm water is treated close to the source. As it is, municipalities are struggling to maintain their piped systems which in turn contribute to the ongoing degradation of the environment.

Needless to say, poverty is a pressing issue in this local municipality, but adequate storm water management ought to bring about the desired improvements in these communities. For example, when a flooding event strikes where there is pre-existing poverty, the poor often suffer the most as they tend to lose more, which results in multidimensional poverty. However, if storm water management is prioritised, the damage from flooding could be drastically reduced in relation to the already impoverished households.

This study further proposes that the local government devises measures to prevent the construction of houses/settlements in flood-prone areas. As much as the supervision of people flocking into the area can be difficult, the local government needs to establish a team to monitor the growth of informal settlements, especially around town. As mentioned before, the current informal settlement which is Green Farm is a ticking time-bomb in terms of its location, in the path of the waterways. Hence, the current households need to be relocated to a more habitable piece of land. Although this might be costly, it is the only feasible way of saving lives from the clear signs of danger.

References

Armitage, N. (2011). *The challenges of sustainable urban drainage in developing countries.* In Proceeding: SWITCH Paris Conference, Paris, pp. 132–144. Retrieved from http://www.switchurbanwater.eu/outputs/pdfs/W22_GEN_PAP_Challenges_of_sustainable_urban_drainage_in_developing_countries.pdf

Azad, A. K., Hossain, K. M., & Nasreen, M. (2013). Flood-induced vulnerabilities and problems encountered by women in northern Bangladesh. *International Journal of Disaster Risk Science, 4*, 190–199. https://doi.org/10.1007/s13753-013-0020-z.

Butler, D., & Davies, J. W. (2011). *Urban drainage* (3rd ed.). New York: Spon Press.

Creswell, J. W., & Plano Clark, V. L. (2011). *Designing and conducting mixed methods research* (2nd ed.). London: Sage Publications Ltd..

Delgado, G. C., & Guibrunet, L. (2017). Assessing the ecological dimension of urban resilience and sustainability. *International Journal of Urban Sustainable Development, 9*(2), 151–169.

Du, J., Kong, F., Du, S., Li, N., Li, Y., & Shi, P. (2016). Floods in China. In P. Shi (Ed.), *Natural disasters in China* (IHDP/Future Earth-Integrated Risk Governance Project Series). Berlin, Heidelberg: Springer. https://doi.org/10.1007/978-3-662-50270-9_5.

Fisher-Jeffes, L., & Armitage, N. P. (2013). Charging for storm water in South Africa. *Water SA, 39*(3), 429–436. https://doi.org/10.4314/wsa.v39i3.13.

Gupta, A. (2016). Storm water drainage and urban flood mitigation: Lateral Solutions. *Water Digest*, pp. 64–82. Retrieved from https://www.researchgate.net/publication/309770187_Stormwater_Drainage_and_Urban_Flood_Mitigation_Lateral_Solutions

Hallegatte, S., Fay, M., Bangalore, M., Kane, T., & Bonzanigo, L. (2015). *Shock waves: Managing the impacts of climate change on poverty.* Washington, DC: World Bank.

Harries, T. (2013). Responding to flood risk in the UK. In H. Joffe, T. Rossetto, & J. Adam (Eds.), *Cities at risk: Living with perils in the 21st century* (pp. 45–72). Germany: Springer. https://www.researchgate.net/publication/259442018_Responding_to_Flood_Risk_in_the_UK.

Hetz, K., & Bruns, A. (2014). Urban planning lock-in: Implications for the realization of adaptive options towards climate change risks. *Water International, 3*(6), 884–900.

Intergovernmental Panel on Climate Change (IPCC). (2012). *Managing the risks of extreme events and disasters to advance climate change adaptation.* Special Report of the Intergovernmental Panel on Climate Change. Edinburg: Cambridge University Press. Retrieved from https://www.ipcc.ch/report/managing-the-risks-of-extreme-events-and-disasters-to-advance-climate-change-adaptation/

Jia, H., Wang, Z., Zhen, X., Clar, M., & Yu, S. L. (2017). China's sponge city construction: A discussion on technical approaches. *Frontiers of Environmental Science and Engineering, 11*(18). https://doi.org/10.1007/s11783-017-0984-9.

Jury, M. R. (2018). Climate trends across South Africa since 1980. *Water SA, 44*(2), 297–307. https://doi.org/10.4314/wsa.v44i2.15.

Keeley, M., Koburger, A., Dolowitz, D. P., Medearis, D., Nickel, D., & Shuster, W. (2013). Perspectives on the use of green infrastructure for storm water management in Cleveland and Milwaukee. *Environmental Management, 51*, 1093–1108.

Le Jallé, C., Désille, D., & Burkhardt, G. (2013). *Urban Storm water management in developing countries. La gestion des eaux pluviales urbaines dans les pays en développement—Step three: Reuse the Storm water, 1–10. Novatech 2013*, City Scale Strategy. Retrieved from http://documents.irevues.inist.fr/handle/2042/51221

Löwe, R., Urich, C., Kulahci, M., Radhakrishnan, M., Deletic, A., & Arnbjerg-Nielsen, K. (2018). Simulating flood risk under nonstationary climate and urban development conditions-experimental setup for multiple hazards and a variety of scenarios. *Environmental Modelling Software, 102*, 155–171. https://doi.org/10.1016/j.envsoft.2018.01.008.

Madsen, H. M., Andersen, M. M., Rygaard, M., & Mikkelsen, P. S. (2018). Definitions of event magnitudes, spatial scales, and goals for climate change adaptation and their importance for innovation and implementation. *Water Research, 144*, 192–203. https://doi.org/10.1016/j.watres.2018.07.026.

Malulu, I. C. (2016). *Opportunities for integrating Sustainable urban Drainage Systems (SuDS) in informal settlements as part of Storm water management.* MPhil dissertation, Stellenbosch University.

Mwaura, P. (2008). *Indigenous knowledge systems in disaster management in Africa.* Nairobi, Kenya: UNEP.

Ngie, A. (2012). *A GIS approach to flood vulnerability and adaptation analysis in Diepsloot, Johannesburg.* Master of Science Dissertation, University of Johannesburg.

Okaka, F. O., Omondi, P., & Odhiambo, B. D. (2017). Urban residents' perception of health risks of climate change and its impact on behavior change in Mombasa City, Kenya. *International Journal of Human Social Studies, 5*(2), 248–257.

Okoth, F., & Beneah, O. (2019). Health vulnerability to flood—Induced risks of households in flood-prone informal settlements in the Coastal City. *Natural Hazards, 99*(2), 1007–1029. https://doi.org/10.1007/s11069-019-03792-0.

Olorunfemi, F. (2009). *Climate change and flood risk in the informal settlements of Cape Town: Understanding vulnerability and adaptation options.* ACCFWP Fellow Final Technical Report.

Olumuyiwa, B. A. (2015). State-led versus community initiate: Storm water drainage and informal settlement intervention in Johannesburg, South Africa. *Environment and Urbanization, 27*(2), 407–420.

Parkinson, J. (2003). Drainage and Storm water management strategies for low-income urban communities. *Environment and Urbanization, 15*(2), 115–126. https://doi.org/10.1177/095624780301500203.

Parkinson, J., Tayler, K., & Mark, O. (2007). Planning and design of urban drainage systems in informal settlements in developing countries. *Urban Water Journal, 4*(3), 137–149. https://doi.org/10.1080/15730620701464224.

Port St Johns Local Municipality. (2019a). *Draft Integrated Development Plan.* 2019/20. https://www.psjmuni.co.za/council-idp-2019-2020/

Port St Johns Local Municipality. (2019b). *Flood Incident Report 2019.* Unpublished.

Ramkissoon, H., Smith, L., & Kneebone, S. (2015). *Accelerating transition to water sensitive cities (Behaviour Assessment Database Report).* Melbourne: Cooperative Research Centre for Water Sensitive Cities.

Randall, M., Sun, F., Zhang, Y., & Jensen, M. B. (2019). Evaluating Sponge City volume capture ratio at the catchment scale using SWMM. *Journal of Environmental Management, 246*, 745–757. https://doi.org/10.1016/j.jenvman.2019.05.134.

Ruhnke, A. (2011). Storm water management: Designing urban hydrological systems as infrascapes. *Journal of Bio Urbanism, 1*(1), 59–73. https://journalofbiourbanism.files.wordpress.com/2013/01/jbu1_2001_amiyo-r-ruhnke.pdf.

Sakigele, T., Lupala, J., & Shueya, S. (2012). Flooding, flood risks and coping strategies in urban informal residential areas: The case of Keko Machungwa, Dar es Salaam, Tanzania. *Jamba: Journal of Disaster Risk Studies, 4*(1), 46–56.

Statistics South Africa (StatsSA). (2020). *How unequal is South Africa?* Retrieved from http://www.statssa.gov.za/?p=12930

Taing, L. (2017). Informal settlement janitorial services: Implementation of a municipal job creation initiative in Cape Town, South Africa. *Environment and Urbanization, 29*(1), 299–314. https://doi.org/10.1177/0956247816684420.

Tas, M., Tas, N., Durak, S., & Atanur, G. (2013). Flood disaster vulnerability in informal settlements in Bursa, Turkey. *Environment and Urbanization, 25*(2), 443–463. https://doi.org/10.1177/0956247813501134.

Taylor, A. L., Dessai, S., & Bruine de Bruin, W. (2015). Communicating uncertainty in seasonal and interannual climate forecasts in Europe. *Philosophical Transactions of the Royal Society A, 373*(2055), 1–16.

Thorne, C. (2014). Geographies of UK flooding in 2013/4. *Geography Journal, 180*(4), 297–309.

United Nations (UN). (2018). *Global indicator framework adopted by the General Assembly (A/RES/71/313) and annual refinements contained in E/CN.3/2018/2 (Annex II) 1–21.* Retrieved from https://unstats.un.org/sdgs/indicators/indicators-list/

Wachinger, G., Renn, O., & Begg, C. (2013). The risk perception paradox—Implications for governance and communication of natural hazards. *Risk Analysis, 33*(6), 1049–1065.

Watershed, S., Cho, J. H., & Lee, J. H. (2017). Storm water runoff characteristics and effective management of nonpoint source pollutants from a highland agricultural region in the lake. *Water, 9*(10), 784. https://doi.org/10.3390/w9100784.

Williams, D. S., & Sutherland, C. (2018). Water, informal settlements and flooding: Identifying strengths and weaknesses in local governance for water management. *Water, 10*(7), 871. https://doi.org/10.3390/w10070871.

Zhou, Q., Panduro, T. E., Thorsen, B. J., & Arnbjerg-Nielsen, K. (2013). Adaption to extreme rainfall with open urban drainage system: An integrated hydrological cost-benefit analysis. *Environmental Management, 51*, 586–601. https://doi.org/10.1007/s00267-012-0010-8.

Check for updates

9 Impacts of Cyclones Idai and Kenneth and the 2019 Floods on the Insurance Sector in South Africa and Mozambique

Shelton Siwedza, Simbarashe Mukonzo, Chalo Ngambi, and Soul Shava

Abstract

The insurance industry plays an important role in the management of risks, including risks from extreme weather events such as tropical cyclones and floods. This chapter explores the impacts of the aftermath of the recent cyclones Idai and Kenneth and their associated floods on the insurance sector in Mozambique and the impacts of the 2019 floods in South Africa. Data was collected through an online questionnaire survey in Mozambique and a desktop-based analysis in South Africa. Results show that the insurance sector was susceptible to the cyclone and flood events in both countries. The study concludes that the insurance sector was ill-prepared for such risks, with substantial losses to the sector, and recommends improvements by the sector in response to the predicted increase in frequency and intensity of cyclone and flood events in southern Africa.

Keywords

Insurance · Cyclones · Floods · Climate change · Catastrophic environmental risks · Adaptation · Mitigation

9.1 Introduction and Background

The insurance industry plays an important socio-economic role in the management of economic risks through the provision of relevant insurance products to society. It does this by providing a mechanism of financial contracts that transfer risks from individuals and corporations to the financial markets, thereby allowing risk reduction using risk diversification and spreading.

Cumulative effects of interactions between humankind and the biophysical environment have resulted in increasing frequency, intensity and extent of "natural" disasters such as droughts, floods, hailstorms, veld fires, cyclones, biodiversity loss and extinction (Niang et al. 2014). In southern Africa these extreme weather events include droughts, floods and storm surges (Davis-

S. Siwedza
African Reinsurance Corporation, Johannesburg, South Africa
e-mail: siwedza.shelton@africa-re.com

S. Mukonzo
Mukoma Financial Services, Johannesburg, South Africa
e-mail: simbamukonzo@mukfin.co.za

C. Ngambi
Emeritus Resseguros, S.A., Sandton, South Africa
e-mail: chalong@emeritusre.com

S. Shava (✉)
Department of Science and Technology Education, University of South Africa, Pretoria, South Africa
e-mail: shavas@unisa.ac.za; https://orcid.org/0000-0003-4103-0559

G. Nhamo, L. Chapungu (eds.), *The Increasing Risk of Floods and Tornadoes in Southern Africa*, Sustainable Development Goals Series, https://doi.org/10.1007/978-3-030-74192-1_9

Reddy et al. 2017). At the household level, these events destroy homes and the asset base for income generation, whilst at regional and national levels the destruction of infrastructure and communication services causes severe disruption to economic activities (Franzke 2017). On one hand, adequate insurance cover and timely insurance advice on impeding extreme weather events helps to mitigate the catastrophic impact of natural disasters on affected communities. However, on the other hand these extreme weather events can also translate to insurance risks that have negative implications on the insurance industry, which has borne the biggest brunt of the cost of these "natural" disasters. These insurance risks are exacerbated by the increasing frequency and intensity of extreme weather events, especially if they cover large areas.

The focus of this chapter is on insurance challenges and responses to recent cyclone and flood risks in southern Africa, with particular reference to Mozambique and South Africa.

9.2 Literature Review

9.2.1 Global Cyclone and Flood Risks

Globally, the insurance sector has been confronted with increasing frequency and intensity of large-scale climate change-related catastrophes, including cyclone and flood-related damages (Crichton 2002; Hawker 2007; Kron 2009; Siwedza and Shava 2020). Such catastrophes present both challenges and opportunities to the insurance industry. The insurance industry's financial interest is very much dependent on the climate and weather in that they increase demand for insurance on one hand, but at the same time climate and weather, if not properly managed, may adversely affect the sustainability of the industry due to overexposure in high-risk areas as a result of an inability to make informed estimates of the extent of the economic consequences of future events resulting therefrom. To properly underwrite risks, insurers need to understand the risks that are insurable and how they can price and provide cover for such. For risks to be insurable, they have to comply with certain underwriting acceptance criteria (Valsamakis et al. 2010; Karten 1997; Njore 2016), being namely:

There must be a large number of homogenous risks or risks of a similar type.

The losses arising from such risks have to be fortuitous or accidental (i.e. out of control of the insured).

Such losses also have to be determinable and measurable both in terms of time and quantum.

The chance or probability of such losses also has to be calculable.

Lastly, the price or premium charged has to be reasonable or economically feasible vis-à-vis the potential loss.

It is therefore essential for the insurance sector to be able to (i) identify the risk, (ii) quantify the risk and (iii) estimate the chance of occurrence of the event and the extent of losses likely to be incurred. This enables insurers to set the right premiums for the different classes of potential clients (Kunreuther and Michel-Kerjan 2007). However, predicting and quantifying risks against a background of increasing frequency and intensity of climate change-induced weather events such as cyclones and flood make this a difficult task.

Worldwide, economic losses from natural catastrophes increased from \$528 billion in the decade 1981–1990 to \$1197 billion during 1991–2000 and \$1213 billion during 2001–2010 (Munich Re 2013). Considering the ten most expensive catastrophes between 1950 and 2005 from a global perspective, the majority were the result of extreme weather events in the form of hurricanes, typhoons, winter storms, with most of them (six) happening between 2004 and 2005 (Munich Re 2006). In 2011 alone, economic losses amounted to over \$400 billion, in large part due to the March 2011 Japan earthquake and resulting tsunami; 2012 brought another \$170 billion in economic losses (Munich Re 2013). Insured losses have dramatically increased as well. Between 1970 and the mid-1980s, annual insured losses from natural disasters worldwide were only in the

$3 billion to $4 billion range (Kunreuther and Michel-Kerjan 2007). However, this has increased drastically since then. Hurricane Hugo's impact in South Carolina, United States of America in 1989, resulted in insured losses of $4.2 billion (Kunreuther and Erwann 2014). Global reinsurance giant, Munich Re reported in 2000 that the costliest weather-related insurance losses globally were being caused by tropical cyclones and winter storms, coupled with hailstorms, bush/wildfires as well as floods (Munich Re 2000). According to the global reinsurer, Swiss Re, during the period 2001–2010, insured losses from weather-related disasters alone averaged $30 billion annually (Swiss Re 2011). Overall, weather-related catastrophic events have large-scale impacts which result in extremely large economic losses that threaten the solvency of the insurance industry. Against the background of an ever-changing environment due to climate change impacts, the potential losses from extreme weather events are not only rising but becoming more volatile and therefore riskier.

It is therefore imperative for the insurance industry to understand the long- and short-term risks resulting from cyclones and floods as this has implications on the sustainability of the insurance business (Hawker 2007). According to Coleman (2003, p. 6), "Having a good understanding of changes to weather-related risk is central to the insurance sector's business and bottom line". However, when knowledge of our past weather events becomes an unreliable guide to the behavior of future weather events in an ever changing environmen, assessing risk in a rapidly changing climate becomes increasingly arduous. This uncertainty in assessing risks is because minute changes in atmospheric conditions can result in very significant changes in weather patterns, leading to large-scale disasters. The biggest question we then need to address is how much the changing climate will affect southern Africa socio-economically, and our biggest challenge lies in minimizing the additional risk associated with human-induced climate change.

Table 9.1 Cyclone categories

Category	Wind speed
4	209–251 km/h
3	178–208 km/h
2	154–177 km/h
1	119–153 km/h
Tropical storm	63–118 km/h
Tropical depression	<63 km/h

Source: Hanlon (2019)

9.2.2 Cyclone and Flood Risks in Mozambique and South Africa

The African continent is vulnerable to disproportionate exposure to climate change-related impacts such as cyclones and floods compared to the more industrialized areas of the globe. Cyclones and floods are an emerging climate change-related catastrophic risk in which a drastic increase in frequency and intensity has been realized in southern Africa (Fauchereu et al. 2003), with major economic implications for the insurance sector. Extreme weather events resulted in damages estimated to be USD 10 billion between 1980 and 2015 in southern Africa (Davis-Reddy et al. 2017). More recent cases in point are the Durban and Johannesburg (South Africa) flood storms at the end of 2017 as well as in April 2019 and Cyclones Idai and Kenneth in Mozambique, Malawi and Zimbabwe.

According to Hanlon (2019), cyclones are categorized according to the Saffir-Simpson scale, with maximum sustained winds per minute as shown in Table 9.1.

9.3 Materials and Methods

Data was collected from the short-term insurance industry in Mozambique and South Africa. The choice of the two countries was determined by the location of the authors in these countries and their ability to access information from colleagues in the insurance sector. This study used both quantitative and qualitative data collection techniques to gather data for estimating the aggregated economic effects of the cyclones and floods derived from both direct and indirect losses (Hesse-Biber 2010). A desktop-based document analysis was employed for South Africa, and a questionnaire survey (Axinn and Pierce 2006) was used for Mozambique. The document analysis was mainly done to explore the effects of floods on the insurance sector in South Africa, while the questionnaire survey was done to examine the effects of cyclones and floods on the insurance sector in Mozambique. An online survey questionnaire (on surveymonkey.com) was administered to 20 short-term insurance companies/underwriters and short-term insurance broking firms in Mozambique which focused on the coverage of cyclones and floods by the insurance industry, economic impacts of cyclone damage to the insurance industry and insurance industry preparedness for cyclone and flood events. Secondary data was gathered from a desktop analysis of national-level reports, statistical databases, news articles, and the respective insurance industry records on floods and cyclone-related damage in South Africa. The collected data was coded into emerging themes that are discussed in the findings. The framework helped to identify the main aspects of economic impacts of natural disasters, identify the underlying dynamics, provide estimates of the magnitude and importance of their impacts, and overall, to determine the ability of the insurance industry to predict their effects.

9.4 Results and Discussions

9.4.1 Impacts of Cyclones in Mozambique

Mozambique, due to its location in a coastal area, is prone to storms and cyclones (see Neumann et al. 2013). However, over the last 30 or so years there has been a dramatic increase in the number of cyclones that have made landfall in Mozambique and affected mainly the east coast of Mozambique. Since 1984, due to the increased weather-related natural catastrophes in the world, the cyclones in Mozambique have also been increasing in frequency and intensity.

Notable cyclones with devastating results in Mozambique were Cyclone Eline in 2000, Cyclone Favio in 2007, Cyclone Idai in March 2019 and Cyclone Kenneth in April 2019. With the advent of global warming, the incidence of weather-related events is on the rise, with six category 2 and above cyclones recorded as having made landfall in Mozambique in the last 12 years. The most recent cyclones (Mozambique News Reports and Clippings 2019) that have made landfall are listed in Table 9.2. These cyclone events have resulted in economic losses to the insurance industry in Mozambique.

Cyclones Idai and Kenneth are to date the most powerful cyclones ever recorded in the southern hemisphere, and Cyclone Idai was by far the worst tropical cyclone on record to affect the African sub-continent and the southern hemisphere. The long-lived storm caused catastrophic damage in Mozambique as well as in Zimbabwe and Malawi. The impact was further exacerbated by institutional weaknesses, such as a lack of an advance evacuation plan, resources, poorly constructed and defective buildings that flouted building regulations and overall low insurance coverage for properties. The city of Beira was severely affected due to its positioning as a port city and being below sea level. Being a critical port city with many warehouses for incoming and outgoing goods from Mozambique and other hinterland countries such as Zimbabwe, Zambia, Malawi and the Democratic Republic of the Congo (DRC), there was significant water damage to goods in the various warehouses, with the resultant ripple effect to economies of these hinterland countries.

According to the World Bank (Bloomberg 2019), Tropical Cyclone Idai, which caused very

Table 9.2 Most recent cyclones in Mozambique

Cyclone Name	Category	Occurrence
Favio	3	February 2007
Jokwe	3	March 2008
Funso	2	January 2012
Dineo	2	February 2017
Idai	3	14th March 2019
Kenneth	4	25th April 2019

Source: Hanlon (2019)

severe and deadly cyclone-provoked flooding in Mozambique, as well as Zimbabwe and Malawi, is expected to have cost the Southern African nations more than USD2 billion to repair the damage. The storm is said to have affected more than 3 million people across the region in its trail (Bloomberg 2019). In Mozambique alone, where the cyclone made landfall and caused the most destruction, as much as USD773 million in assets is anticipated to have been destroyed (Bloomberg 2019).

The ten short-term insurance companies operating in the insurance industry in Mozambique received over 5000 claims relating to Cyclones Idai and Kenneth, which cost the industry an estimated MZN 4.5 billion or USD 71.4 million (Emeritus Resseguros SA 2020).

An online survey questionnaire was sent to 20 short-term insurance companies/underwriters and short-term insurance broking firms in Mozambique focusing on insurance coverage of cyclones and the economic impacts they had on the insurance industry (https://www.surveymonkey.com/results/SM-GJXQNGVN7/). A total of 13 respondents to the questionnaire (7 short-term insurance companies and 6 short-term insurance broking firms) were received out of the 20, giving a response rate of 65%. The respondents were executive managers within these short-term insurance companies and broking firms. The majority of the respondents (83%) agreed that the economic impact of cyclones to the local population was significant. This is represented in Fig. 9.1.

Most of the insurance industry respondents (76%) concurred that insured losses from cyclone damage were considerably high (Fig. 9.2). Such losses have implications on insurance products in cyclone-prone areas.

From the responses, 85% stated that they had policy wording covering cyclones (Fig. 9.3). The lack of cyclones in the policy wording of some insurers might indicate that they do not cover cyclone damage despite cyclones being a frequent phenomenon in Mozambique. Of the above 13 short-term insurance companies and brokers, 46% stated that they charged an additional premium for cyclones, while the rest (54%) did not (Fig. 9.4). The additional premium charge is indicative of the rising insured losses from cyclone damage.

The majority (80%) of the respondents claimed that the local insurance industry did not have relevant solutions to cover cyclone damages (Fig. 9.5). The lack of products to cover cyclone damage indicates a gap that needs to be addressed by the insurance sector, given the increasing frequency and intensity of cyclones in Mozambique. Regarding the Mozambique insurance sector's preparedness to deal with cyclone damage, most of the respondents (76%) felt the local insurance industry was poorly or ill-prepared due to limited catastrophic reinsurance (Fig. 9.6). This relates to the lack of insurance solutions to cover the cyclone damage presented in Fig. 9.5.

Some of the participants indicated the need for improved Geographical Information Systems and other related remote sensing techniques data to enhance cyclone and flood risk awareness, prediction and mapping in the insurance sector. They suggested that this could be achieved through closer cooperation between the insurance industry and the national metrological departments.

Overall, the above findings reveal limited disaster risk assessment and management and inadequate cyclone and flood risk insurance financing and transfer despite Mozambique being highly prone to the cyclone and associated flood damage.

9.4.2 Impacts of Floods in South Africa

While South Africa has not experienced the effects of cyclones, it has not been spared the effects of extreme weather events such as severe weather storms and floods (see Els 2011), which have been concentrated mainly in KwaZulu-Natal (Mather and Stretch 2001), Gauteng, Western Cape and the Eastern Cape provinces. It was also noted that most of the disasters in South Africa are weather-related.

During the period December 2004 to January 2005, the Garden Route region of the Southern Cape of South Africa, a prime holiday destina-

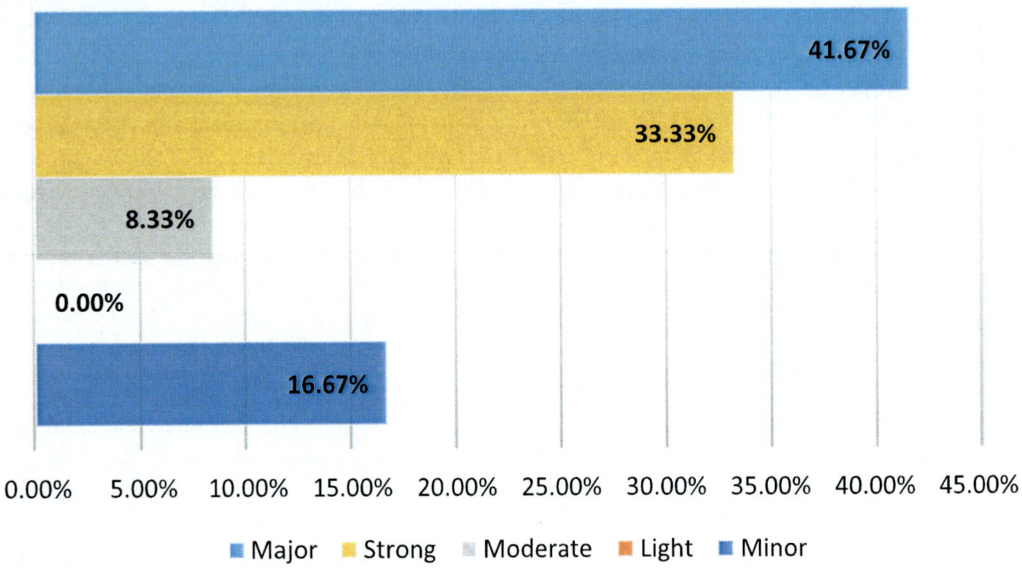

Fig. 9.1 Insurers' assessment of the economic impacts of cyclones

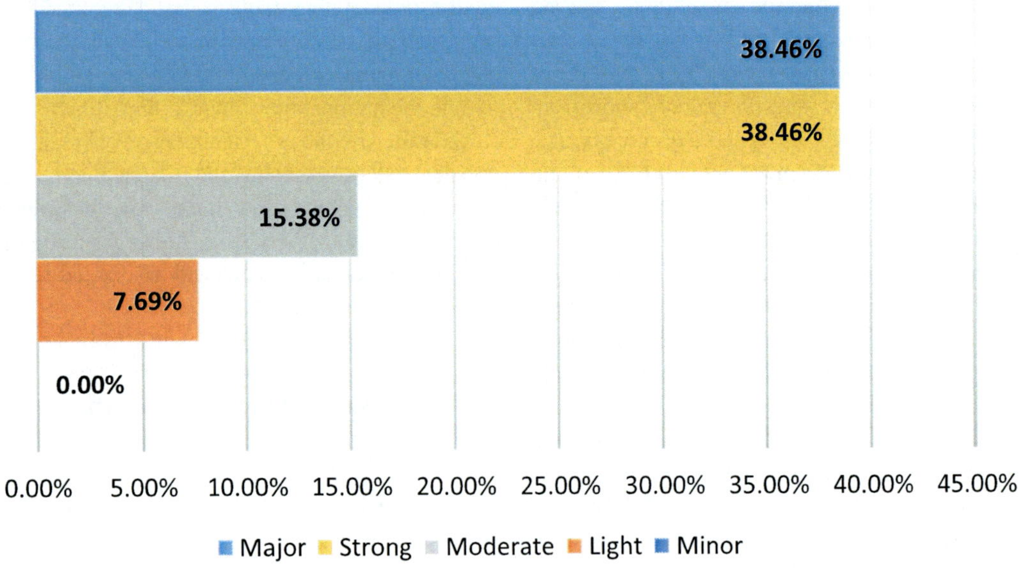

Fig. 9.2 Insurance losses caused by cyclones

tion, was hit by intense floods. The floods had a significant impact on local economic activities, including tourism and local development (Tempelhoff et al. 2009). Heavy rainfall and consequent floods in December 2004 resulted in significant damage of infrastructure being recorded in many areas with an estimated cost of R24 million (Tempelhoff et al. 2009). However, these were much higher as one insurer paid out approximately R20.4 million to claims totaling 1224 (Tempelhoff et al. 2009). All members of the tourism sector recorded direct losses.

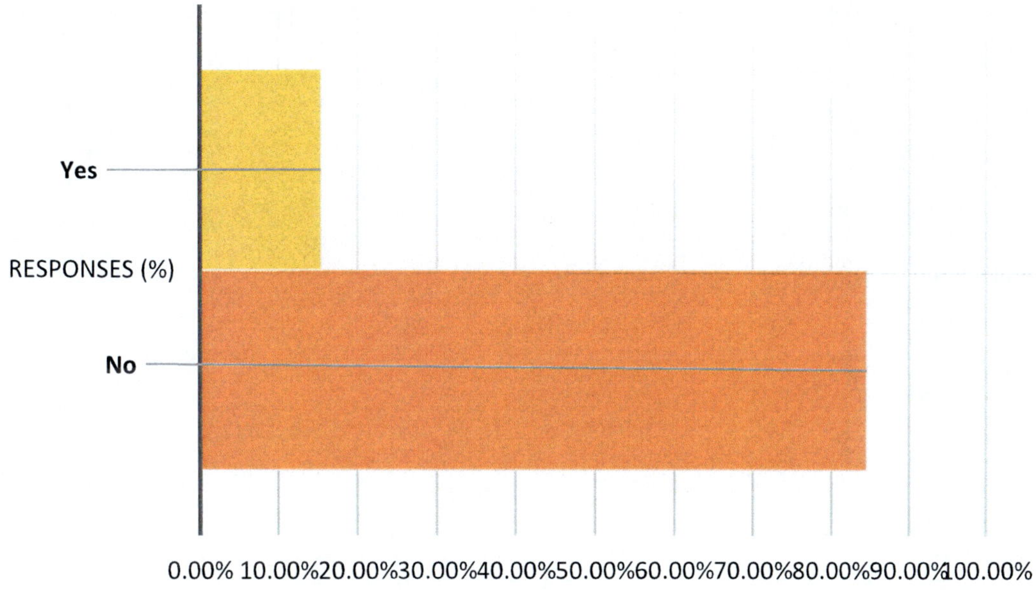

Fig. 9.3 Availability of insurance policy wording on cyclones

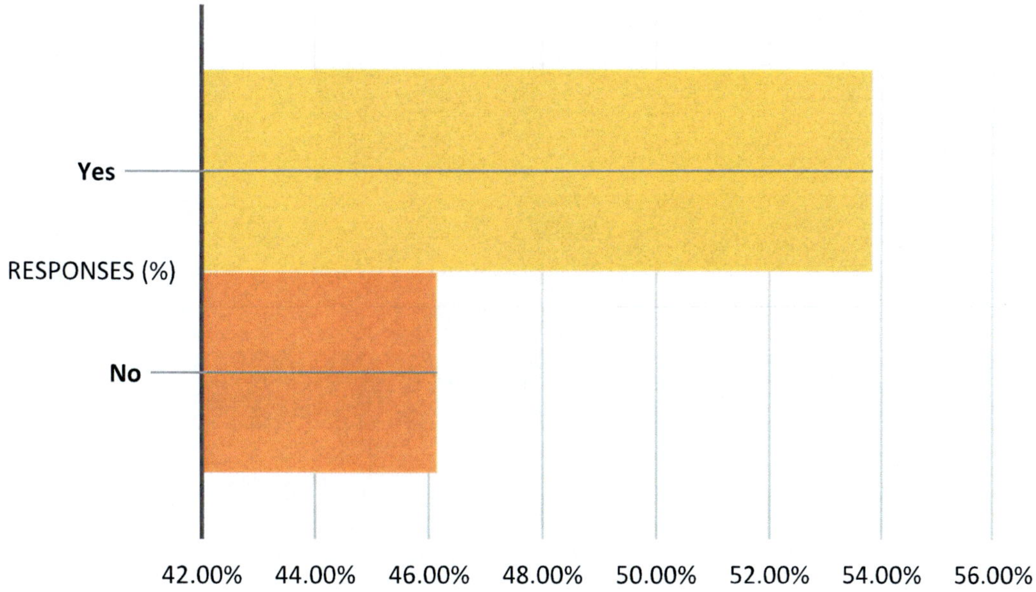

Fig. 9.4 Additional insurers' policy premium charges for cyclone cover

Flood disasters were recorded in Limpopo and Mpumalanga provinces in 2014 (National Disaster Management Centre 2015). Also, in August 2014, floods hit the Western Cape, with Khayelitsha Township built on the Kuils River flood plains being the worst hit (Cape Argus 2014). Khayelitsha is highly vulnerable to flood risks, and most of the low-income homes, particularly in the informal settlements, are not insured. The after-effects of the floods included property losses or damage to homes, infrastructure, vehicles and loss of human lives.

Over the past few years, there was an increase in the frequency of extreme weather events

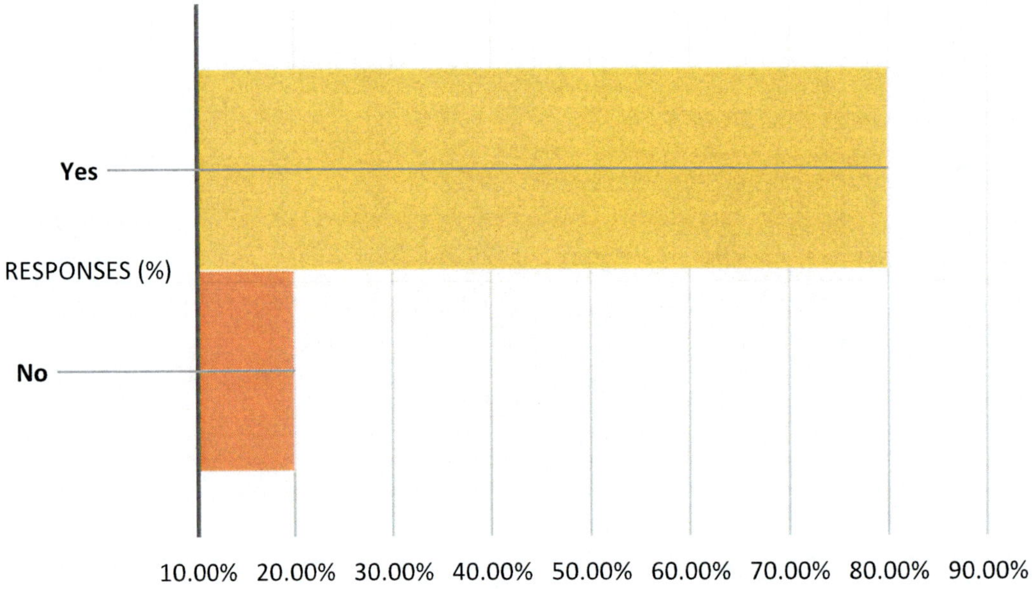

Fig. 9.5 Assessment of the availability of relevant insurance solutions to cover cyclone damage

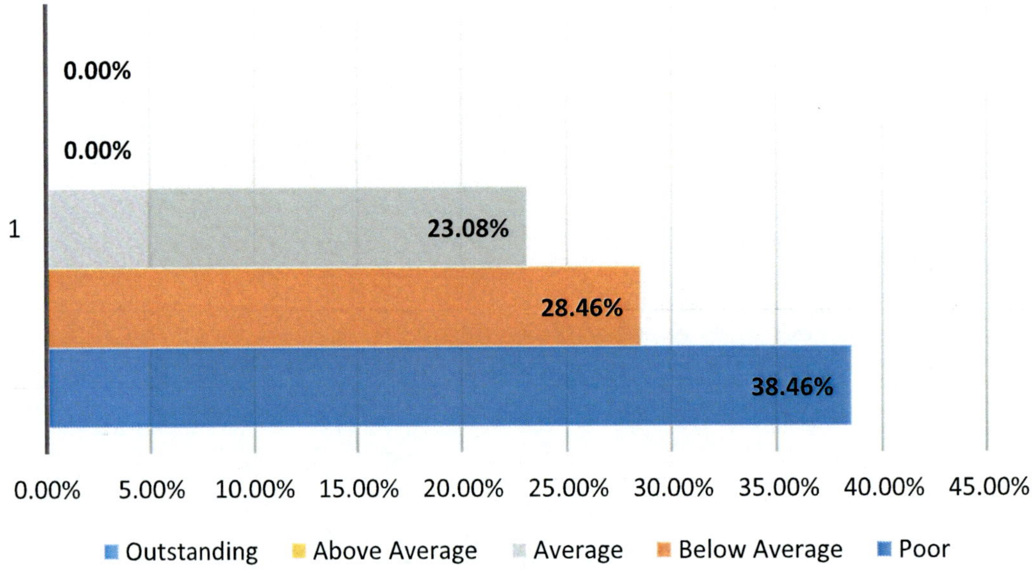

Fig. 9.6 Rating of insurance preparedness to deal with cyclones

according to Aon South Africa (Sunday Times 2017), with significant weather-related catastrophes for the South African insurance industry. Flood events in 2016 alone racked-up R700 million worth of insured losses, whilst the Knysna wildfires and Cape Town storms in June 2017 cost over R4 billion worth of damages (Sunday Times 2017).

According to the South African National Space Agency (SANSA 2019), in April 2019 the Durban region of South Africa received heavy rainfall, which resulted in the flooding of various parts of the eThekwini Metropolitan Municipality, with mudslides, destruction of road infrastructure and housing. The heavy rainfall also spread to and affected the Eastern Cape, with areas such as

Port St Johns, Lusikisiki and Mbizana being heavily hit. More than a thousand people were affected, with over 70 people reported dead. The estimated cost of damage was R1.1 billion (News24 2019). eThekwini Metropolitan Municipality's estimated financial loss in terms of required intervention measures to assist displaced residents and the cost of repairs to damaged infrastructure was over R650 million (SANSA 2019).

KwaZulu-Natal was again hit by four tornadoes and heavy storms in November 2019, within a space of fewer than 2 weeks apart, with an estimated cost of damage of R560m. More than 15,500 people and 3072 households were affected (eNCA 2019).

The flood events discussed above are evidence of increased frequency and intensity of extreme weather events in South Africa, with severe damage to property and infrastructure. This has negative implications on the insurance industry and points to the dire need for broader disaster risk assessment (Els 2011) and management strategies, early warning systems, public awareness, investment in disaster-resistant infrastructure, and insurance risk financing and transfer.

9.5 Proposed Recommendations for Insurers in Southern Africa

There is now a dire need for the insurance industry in Southern Africa to urgently respond to the threat that cyclones and floods are posing to the industry's profitability and overall sustainability.

According to Leggett (1993), the options available for the insurance industry are as follows:

- Continue with the current business status quo.
- Drastically review and overhaul the way business has been done by the industry players (i.e. brokers, insurers and reinsurers).
- Collaborate with and lobby stakeholders (i.e. consumers, industry, governments) to pursue climate-friendly social and commercial activities that reduce the threat on the environment.

Following the research on the impact of cyclones, insurers, governments, and the private sector in Southern Africa are advised to consider the following recommendations:

9.5.1 Increase Insurance Penetration by Providing Innovative Micro-Insurance Solutions

Increasing urbanization, industrial development and rising property values contribute to higher economic losses from extreme weather events in areas vulnerable to cyclones and their associated floods. The insurance industry can play a key role in persuading their clients in the corporate, domestic and government sector to consider climate change-induced cyclones and floods as a core business issue and the mitigation of and adaptation to cyclone and flood damage in their operations through catastrophe preparedness. However, the insurance markets are largely concentrated in the industrialized areas, meaning most African countries, including the southern African region, are under-insured. This means that the ratio of economic losses to insured losses (L/I) is very high. It is also notable that the socio-economic impacts of cyclones and floods largely affect the poor, who are most vulnerable and usually uninsured or poorly insured for such catastrophic events. There is extremely low insurance penetration in Africa, with most of the insurable assets either uninsured or under-insured (Olayungbo and Akinlo 2016). Due to this, Africa cannot effectively fund its recovery in the aftermath of cyclone and floods-induced damage and the subsequent threat to lives. This places considerable strain on limited government resources and negatively impacts economic growth.

The southern African insurance industry can therefore play an important role toward driving insurance awareness and uptake in previously uninsured communities through civic education on insurance products and provision of micro-insurance solutions in, that ease recovery from the impacts of cyclone and flood disaster, as well as impressing on the importance cyclone and flood risk sustainability, best practice, corporate governance and risk management practices among its business clients (Statistica 2017).

The insurance industry has a critical role to play to conscientize the public and the government about how insurance works and the various solutions available in the market. One of the major obstacles is that most of the clients do not buy insurance due to lack of knowledge on the availability and necessity of the insurance products. The other dimension is that those clients who have policies usually assume that they are covered for cyclones and flood losses, yet this risk is excluded from their policy. The insurance industry needs to adopt a policy of treating customers fairly by first providing them relevant advice about the risks they are exposed to and making sure that they agree or decline the cover. The premium charged for the cyclone and flood risk and the excesses applicable should also be stated clearly in the policy. This will enable the client to determine how much insurance cover to purchase based on a cost–benefit analysis.

9.5.2 Establish Disaster Compensation Schemes

Financial vulnerability is the risk of a local or national government not having adequate funds from either domestic or foreign sources to meet the financing of reconstruction and provision relief after a disaster. As a result of the financial gap, governments are unable to fully rebuild infrastructure, assist the poor in the wake of a major disaster such as cyclone and flood damage and to progress with social and economic programs. This can severely impact on the populace and the country's economy (i.e. vital economic infrastructure such as roads, railways, electricity provision and water supply). Since infrastructure is essential for economic growth and development, there is a direct link between poor infrastructure and poverty. Post-disaster financial vulnerability is quite common problem in developing countries (OECD 2005).

Insurance companies in southern Africa can assist in reducing the financing gap, being the difference between a government's loss potential and the available internal financing resources at the governments' disposal to meet their post-disaster requirements for infrastructure reconstruction and investment relief. The insurance industry can help to alleviate this financial gap by generating the necessary funds through large-scale catastrophe compensations schemes.

9.5.3 Provide Sustainable Risk Financing Solutions

Environmental degradation from human activities (which result in biodiversity loss, land and atmospheric pollution, deforestation and desertification) contributes to climate change, whose impacts include increased frequency and intensity of cyclone and flood events. The insurance industry can assist through providing risk financing solutions and cover against environmental risks such as pollution from the industrial and commercial activities that can cause health problems and property damage, natural resource contamination, habitat loss and biodiversity loss. In collaboration with governments, insurers can assist governments to enforce a compulsory environmental liability and/or pollution insurance for environmentally high hazard economic activities (OECD 2003).

Through relevant insurance products, the insurance industry in southern Africa can play an active toward mitigation and adaptation to climate change and thereby be an effective part of contextually relevant adaptation strategies to address climate change threats., By responding to the corresponding environmental risks through providing insurance solutions, the insurance sector can contribute toward achieving the Sustainable Development Goals (SDGs) (United Nations 2015). In line with this, some insurers have developed index-based insurance cover for farmers to protect them against climate change-related damage to their crops. Another emerging innovative insurance product is insurance-linked securities (ILS), specifically for low-frequency but high-severity natural catastrophe events, which is provided in the form of catastrophe bonds and weather derivatives (Fisher et al. 2018; Hessu and Syronka 2004).

There is still need for the insurance industry in southern Africa to explore initiatives to provide insurance products cushion governments from low national food production resulting from climate change-induced catastrophes. This will help to ensure food security (SDGs 2 and 3) (United Nations 2015). According to Wesselink (2013, p. 28), insurers can help reduce flood risk by:

- Helping to identify areas of risk, such as improvements in early warning and evacuation systems (for cyclone and flood events).
- Modeling of flood catastrophe bonds.
- Enforcement of building codes and discouraging development in flood plains by refusing to insure properties in such areas.
- Collecting data on flood damage costs for flood management schemes.
- Promoting post-flooding temporary defense solutions by providing incentives to policyholders that undertake such activities.
- Capacitating the insurance industry for it to take on the majority of catastrophe (CAT) risks.
- Reviewing reinsurance programs in order to improve treaty balances, rate-on-lines and pay-back periods.
- Improving risk management within insurance companies in order for them to effectively monitor and manage risk accumulation.

9.5.4 Funding for Climate Change Research Initiatives

Insurance companies need to invest in socioeconomic resilience to natural hazards by offering pro-active risk management and providing funding for research and development into adaptation to extreme weather-related events and risks.

There is an urgent need for increased private–public partnerships in catastrophic insurance and for exploring alternative approaches to global climate risk financing, building on the competences and resources of the insurance industry to address catastrophe risk management, the volatility and severity of impacts of natural disasters, as well as their overall impact on the SDGs (Dlugolecki and Hoekstra 2006). In this regard, the insurance industry can collaborate with governments to provide government-backed insurance for risks that would not otherwise be insurable, provide government-backed compulsory insurance schemes to communities, as well as provision of risk statistics and information on climate change impacts, threats and impending natural disasters.

9.5.5 Improvement of Regulation and Policies on Climate Change-Related Cyclones and Floods

The insurance industry is well organized through various industry bodies and associations. The industry organization can interact with government at all levels. The insurance industry can therefore play an important role in influencing policy formulation, regulation and legislation by government on cyclone and flood risk and other climate change-related environmental disasters through the application of statistics collected from losses and predictions provided by advanced risk modeling tools. The industry should influence governments to provide a regulatory framework for industries that ensures that insurance becomes a compliance item in their business operations, especially those in cyclone and flood-prone areas. The insurance industry has an important role in helping governments to include climate change and natural disaster funding and preparedness in their fiscal budgets, as well as directing them to access global reinsurers and risk managers for capacity support on natural disasters (Gurenko 2006).

9.5.6 Invest in Sustainable Solutions

Cyclones and floods in Southern Africa are increasing in frequency and intensity due to the long-term effects of climate change. To promote the SDGs within the financial services industry, the United Nations Global Compact in partnership with KPMG International produced the "SDG Industry Matrix", which outlines opportunities for financial services companies to "create value for their business whilst creating a more sustainable and inclusive path to economic growth, prosperity and well-being" (United Nations Global Compact and KPMG International 2015). The matrix identified four sets of opportunities, namely increasing financial inclusion, investing in, financing and insuring renewable energy and infrastructure projects, leveraging risk expertise to influence customer behavior and positively influencing environmental, social and governance practices of corporate clients and investment companies (Jones et al. 2017).

The insurance industry should actively promote the adoption of renewable energy solutions (SDGs 7 and 13) by availing relevant and affordable products. The long-term goal is to enable business practices that preserve the environment and hence reducing the negative impact of climate change. Governments also need to complement the effort of the private sector by providing access to patient capital for renewable energy projects.

The growing urbanization and industrialization in most parts of Africa are leading to an increase in the usage of fossil fuels to meet the energy requirements for households and industry. In most cases, these fossil fuels are cheaper compared to clean energy sources. The insurance industry can actively contribute toward reducing global carbon emissions and encouraging climate-friendly corporate behavior (SDGs 7, 9, 11, 12 and 13) (United Nations 2015) by supporting the gradual transition to low-carbon economies through provision of risk transfer solutions to renewable energy initiatives and refusing to ensure fossil-fueled power plants. Global insurance companies such as AXA, Zurich, Allianz, Munich Re and Swiss Re are taking the lead in such initiatives through de-risking on fossil-fueled power plants in favor of renewable energy (Bosshard 2018). The transition to a low-carbon economy is a mitigation strategy leads to a reduction of greenhouse gas emissions, thereby contributing toward reducing climate change impacts such as cyclones and floods.

The insurance industry should also promote the construction of climate and flood-resistant infrastructure (SDG 9), which ensures long-term adaptation to climate change-induced extreme weather events.

9.5.7 Collaboration between Governments and the Insurance Industry

Despite the southern African insurance not being a polluter or climate policy developer (Golnaraghi 2018), it can still play an important role being part of the solution by collaborating with governments and other interested institutions as it is well-positioned to provide in-depth information on the climate change-associated risks and can. In this regard, the insurance industry can help in filling the data deficiency void through sharing with industry, NGOs, government authorities and the general public the scientific data that it has collected over the years on the nature and frequency of losses and understanding of multiple stressors of environmental risks (Robyn et al. 1998).

Insurers are in a unique position which enables them to measure insured losses at source, that directly to the specific insured by location and monetary value. The insurance industry can leverage on this information to create risk models which can be adopted by the government in developing sustainable mitigation solutions to deal with climate change losses. It is evident from this research that the uninsured losses are mostly associated with the poor and disadvantaged sections of society. The government would benefit immensely by forging cooperation with the insurance industry in developing long-term solutions to cushion the vulnerable groups in society.

One of the common recommendations from the respondents to a questionnaire survey carried out in Mozambique is the need for insurers and reinsurers to collaborate with the government to reduce the costs of catastrophe insurance. This can be done through the establishment of an extreme weather pool or one specifically for cyclones. By so doing, the country can access cheap catastrophe insurance and reinsurance on the international market, including catastrophe bonds on the financial markets. The insurance industry in turn will benefit from the collaboration through minimizing the risk of insolvency by broadly spreading the transfer of very large risks.

9.6 Conclusion

The future sustainable development of southern Africa is very much dependent on the region's ability and capacity to pro-actively mitigate and adapt to the impacts of climate change such as cyclones and floods. The impacts of cyclones and associate flood events can result in significant socio-economic damage in affected areas in southern Africa. However, catastrophes such as cyclones and floods do not just strike; there are early warnings that tell of the impending disaster before it strikes. Insurers, by virtue of them being risk-takers, handlers and managers, can play a key advisory role (Njore 2016). Research data-driven insurance models for predicting and mapping cyclone and flood risks can assist in sending early warnings to clientele and in estimating exposure as well as pricing insurance products. This can help increase people's understanding of localized cyclone and flood activities through raising the necessary awareness, encouraging adaptive avoidance practices (risk-reducing behavior) and adoption of insurance products, thereby developing a culture that promotes loss prevention, reduction and risk management. Similarly, innovative insurance products that incentivise the establishment of the cyclone and flood-resistant infrastructure through providing economic incentives (lower premiums) can contribute significantly to socio-economic and ecological adaptation, thereby helping to develop relevant cyclone and flood resilience practices in areas prone to such disasters. The insurance sector also plays an important role in cushioning clients from economic shocks resulting from cyclone and flood damage through assessing damage and providing timely finance in the form of claim pay-outs (risk transfer). The insurance industry also contributes to climate change mitigation through promoting carbon emissions reduction, such as refusing to insurer carbon-heavy industrial practices, for example, coal mining and coal-powered electricity generation. Insurance companies are therefore in a powerful position to play an important role in promoting the transition to a sustainable global future. Southern Africa has limited cyclone and flood risk insurance coverage, which provides an opportunity for the development of catastrophic risk insurance market in this area.

References

Axinn, W. G., & Pierce, L. D. (2006). *Mixed method data collection strategies*. New York: Cambridge University Press.

Bloomberg. (2019). *Southern African cyclone damage cost seen topping $2 Billion*. Retrieved from April, 12, 2019 https://www.bloomberg.com/news/articles/2019-04-12/cyclone-damage-cost-southern-africa-2-billion-world-bank-says.

Bosshard, P. (2018). *Insuring coal no more: The 2018 scorecard on insurance, coal and climate change*. Unfriend Coal Report. Retrieved December, 2018, from https://unfriendcoal.com/wp-content/uploads/2018/11/Scorecard-2018-report-final-web-version.pdf

Cape Argus. (2014). *The worst flooding in 10 years*. Retrieved from https://www.iol.co.za/news/south-africa/western-cape/the-worst-flooding-in-10-years-1743122

Coleman T. (2003). *The impact of climate change on insurance against catastrophes*. The Institute of Actuaries of Australia 2003 Biennial Convention Shaping the Future: In a World of Uncertainty, Retrieved May 18–21, 2003, from https://actuaries.asn.au/Library/Events/Conventions/2003/1050%20coleman7a.pdf

Crichton, D. (2002). *The implications of climate change for the insurance industry. Wilton Park Conference, "Climate change, what can be done?"*, 15th May, 2002.

Davis-Reddy, C., Vincent, K., & Mambo, J. (2017). Socio-economic impacts of extreme weather events in southern Africa. In C. Davis-Reddy & K. Vincent (Eds.), *Climate risk and vulnerability: a handbook for southern Africa* (2nd ed., pp. 30–46). Pretoria: CSIR. https://www.csir.co.za/sites/default/files/Documents/SADC%20Handbook_Second%20Edition_full%20report.pdf.

Dlugolecki, A., & Hoekstra, E. (2006). The role of the private market in catastrophe insurance. *Climate Policy, 6*(6), 648–657. https://doi.org/10.1080/14693062.2006.968563.

Els, Z. (2011). *Data availability and requirements for flood hazard mapping in South Africa*. Unpublished master of natural sciences thesis, Stellenbosch University, Stellenbosch, South Africa.

Emeritus Resseguros SA. (2020). Statistics provided 5 March 2020 (Personal communication).

eNCA. (2019). KZN Flood damage estimated at R1.1bln. Retrieved April 29, 2019, from https://www.enca.com/news/kzn-flood-damage-estimated-r11bn.

Fauchereu, N., Trzarska, S., Roualt, M., & Richard, Y. (2003). Rainfall variability and changes in southern Africa during the 20th century in global warming. *National Hazards, 29*(2), 139–154. https://doi.org/10.1023/A:1023630924100.

Fisher, E., Hellin, J., & Jensen, N. (2018). Index insurance and climate risk management addressing social equity. *Development Policy Review, 37*(5), 581–602.

Franzke, C. L. E. (2017). Impacts of a changing climate on economic damages and insurance. *EconDisCliCha, 1*, 95–110.

Golnaraghi, M. (2018). *Climate change and the insurance industry: Taking action as risk managers and investors*. Zurich: The Geneva Association. Retrieved from https://www.genevaassociation.org/sites/default/files/research-topics-document-type/pdf_public/climate_change_and_the_insurance_industry_-taking_action_as_risk_managers_and_investors.pdf

Gurenko, E. N. (Ed.). (2006). *Climate change and insurance: Disaster risk financing in developing countries*. London: Earthscan.

Hanlon, J. (2019). Unprecedented 2nd intense cyclone hits Mozambique. *Mozambique News Reports & Clippings, 454*, 1–3.

Hawker, M. (2007). Climate change and the global insurance industry. *The Geneva Papers, 32*, 22–28. https://www.jstor.org/stable/41952918.

Hesse-Biber, S. N. (2010). *Mixed method research: Merging theory with practice*. New York, London: The Guilford Press.

Hessu U. & Syronka J. (2004). *Weather-based Insurance in Southern Africa. The case of Malawi*. Agricultural & Rural Development Discussion Paper 13. The World Bank.

Jones, P., Hillier, D., & Comfort, D. (2017). The sustainable development goals and the financial services industry. *Athens Journal of Business and Economics, 3*(1), 37–50. https://doi.org/10.30958/ajbe.3.1.3.

Karten, W. T. (1997). How to expand the limits of insurability. *The Geneva Papers on Risk and Insurance, 22*(85), 515–522. Retrieved from https://www.jstor.org/stable/41952850.

Kron, W. (2009). Flood insurance: From clients to global financial markets. *Journal of Flood Risk Management, 2*(1), 68–75. https://doi.org/10.1111/j.1753-318X.2008.01015.x.

Kunreuther, H., & Erwann, M. (2014). Economics of natural catastrophe risk insurance. In M. K. Machina & W. K. Viscusi (Eds.), *Handbook of the economics of risk & uncertainty* (Vol. 1, pp. 651–699). New York: Elsevier.

Kunreuther, H. C., & Michel-Kerjan, E. O. (2007). Climate change, insurability of large-scale disasters, and the emerging liability challenge. *University of Pennsylvania Law Review, 155*(6), 1795–1842.

Leggett, J. (1993). *Climate change and the insurance industry: Solidarity among the risk community?* (2nd Ed.), Greenpeace.

Mather, A. A., & Stretch, D. D. (2001). A perspective on sea level rise and coastal storm surge from southern and eastern Africa: A case study near Durban, South Africa. *Water, 4*(1), 237–259. https://doi.org/10.3390/w4010237.

Mozambique News Reports & Clippings. (2019). Unprecedented 2nd intense cyclone hits Mozambique. Mozambique News Reports & Clippings 454 26 April 2019. http://www.open.ac.uk/technology/mozambique/sites/www.open.ac.uk.technology.mozambique/files/files/Mozambique_454-26Apr19_2ndseverecyclone.pdf

Munich Re. (2000). *Review of natural disasters 1999*. Munich Re, Report No. 2946-M-e.

Munich Re. (2006). *Hurricanes: more intense, more frequent, more expensive*. Munich: Munich Re, 44 pp.

Munich Re (2013). *Economic consequences of natural catastrophes: Emerging and developing economies particularly affected—Insurance cover is essential*. Munich Re Economic Research Position Paper.

National Disaster Management Centre. (2015). *Annual Ieport 2014–2015*. Republic of South Africa: Department of Cooperative Governance.

Neumann, J. E., Emanuel, K. A., Ravela, S., Ludwig, L. C., & Verly, C. (2013). *Assessing the risk of cyclone-induced storm surge and sea level rise in Mozambique. WIDER Working Paper, No. 2013/036*. Helsinki: The United Nations University World Institute for Development Economics Research (UNU-WIDER).

News24. (2019). Tropical cyclone Idai: The storm that knew no boundaries. News24, 24 March 2019. https://www.news24.com/news24/Analysis/tropical-cycloneidai-the-storm-that-knew-no-boundaries-20190324

Niang, I., Ruppel, O. C., Abdrabo, M. A., Essel, A., Lennard, C., Padgham, J., & Urquhart, P. (2014). Part B: Regional aspects. Contribution of Working Group II to the Fifth Assessment Report of the Intergovernmental Panel on Climate Change. In V. R. Barros, C. B. Field, D. J. Dokken, M. D. Mastrandrea, K. J. Mach, T. E. Bilir, M. Chatterjee, K. L. Ebi, Y. O. Estrada, R. C. Genova, B. Girma, E. S. Kissel, A. N. Levy, S. MacCracken, P. R. Mastrandrea, & L. L. White (Eds.), *Climate change 2014: Impacts, adaptation, and vulnerability* (pp. 1199–1265). Cambridge, United Kingdom and New York: Cambridge University Press.

Njore, L. (2016). *Reinvention of insurance to support sustainable development goals*. LinkedIn. Retrieved from https://www.linkedin.com/pulse/reinvention-insurance-support-sdgs-sustainable-goals-lawrence-njore/

Olayungbo, D. O., & Akinlo, A. E. (2016). Insurance penetration and economic growth in Africa: Dynamic effects analysis using Bayesian TVP-VAR approach. *Cogent Economics & Finance, 4*(1), 1150390. https://doi.org/10.1080/23322039.2016.1150390.

Organisation for Economic Co-operation and Development (OECD). (2003). Environmental Risks & Insurance: a comparative analysis of the role of Insurance in the management of environment-related risks Policy Issues in Insurance No. 6. OECD.

Organisation for Economic Co-operation and Development (OECD). (2005). Catastrophic Risks & Insurance. Policy Issues in Insurance No. 8. OECD.

Robyn, F., Mead, C. D., & Williams, W. P. (1998). *Environmental risk assessment: Approaches, experiences and information sources*. Copenhagen: European Environmental Agency.

Siwedza, S., & Shava, S. (2020). Insurance, increasing natural disaster risks and the SDGs: A focus on southern Africa. In G. Nhamo, G. O. A. Odularu, & V. Mjimba (Eds.), *Scaling up SDGs implementation: Emerging cases from state, development and private sectors* (pp. 129–137). Cham: Springer.

South African National Space Agency (SANSA). (2019). *Update on recent flooding in KZN and Eastern Cape using satellite imagery*. Retrieved April 29, 2019, from https://www.sansa.org.za/2019/04/29/update-on-recent-floodingin-kzn-and-eastern-cape-using-satellite-imagery/.

Statistica. (2017). *Rate of insurance penetration in Sub-Saharan Africa in 2017, by country*. Retrieved October 23, 2019, from https://www.statista.com/statistics/727403/insurance-penetration-in-sub-saharan-africa-by-country/.

Sunday Times. (2017). *More extreme weather coming in the next few years*. Retrieved October 12, 2017, from https://www.timeslive.co.za/news/sci-tech/2017-10-12-more-extreme-weather-coming-in-the-next-few--years/.

Swiss Re. 2011. *Natural catastrophes and man-made disasters in 2010: A year of devastating and costly events*. Sigma 01/2011. Retrieved from https://www.swissre.com/dam/jcr:d58dea24-f966-4614-8b90-3acd9b0a361f/sigma1_2011_en.pdf

Tempelhoff, J., van Niekerk, D., van Eeden, E., Gouws, I., Botha, K., & Wurige, R. (2009). The December 2004–January 2005 floods in the garden route region of the southern cape, South Africa. *JÀMBÁ: Journal of Disaster Risk Studies, 2*(2), 93–112.

United Nations. (2015) Transforming our world: The 2030 agenda for sustainable development. Retrieved from https://sustainabledevelopment.un.org/content/documents/21252030%20Agenda%20for%20Sustainable%20Development%20web.pdf

United Nations Global Compact and KPMG International. (2015). SDG industry matrix. UN Global Gompaxt LEAD and KPMG International.

Valsamakis, A. C., Vivian, R. W., & Du Toit, G. S. (2010). *Risk management* (4th ed.). Cape Town: Pearson.

Wesselink, A. H. (2013). *A comparative analysis of the role of insurance companies in flood risk management*. Master's thesis, European spatial and environmental planning; Nijmegen School of Management, Radboud University Nijmegen.

Part IV

Challenges from Rural Flooding

10 Exploring the Migration Effects of Cyclones and Floods in Southern Africa: A Focus on 2019 Floods and Cyclones Idai and Kenneth

Nthivhiseni Mashula, Nompe Gladys Ntombela, Hlengiwe Precious Kunene, Desiree Sehlapelo, and Nedson Pophiwa

Abstract

The cyclones Idai and Kenneth as well as the 2019 floods were declared the worst to have hit the southern African region in decades. Strong winds and heavy rains placed the region in a state of crisis, claiming lives, flattening buildings and triggering massive floods that damaged critical infrastructure and farmlands and submerged entire communities. Their aftermath left affected people without shelter, food, safe drinking water, sanitation and hygiene, a situation which prompted migration. Using the case study research design, this chapter investigates the impacts of these meteorological events on migration patterns of people in Zimbabwe and South Africa focusing on Chimanimani, Zimbabwe, and Port St Johns, South Africa. The chapter further explores whether cyclones, floods and related environmental catastrophes will be the next drivers of intra- and international migration within the southern African region. The findings of the study were based on a survey comprising household questionnaires. The results showed that even though migration decisions are influenced by economic, social and political drivers, people generally prefer to move internally despite the reoccurrence of disasters. Instead, migration to other countries is typically triggered by household economic constraints and tends to be voluntary. The findings resonate with other studies that explore the relationship between environmental catastrophes and internal migration.

Keywords

Migration · Floods · Cyclone Idai · Cyclone Kenneth · Displacement · Southern Africa

N. Mashula (✉) · N. G. Ntombela · H. P. Kunene
Institute for Corporate Citizenship, University of South Africa (UNISA), Pretoria, South Africa
e-mail: mashun@unisa.ac.za; ntombng@unisa.ac.za; kunenhp@unisa.ac.za

D. Sehlapelo
Department of Environmental Science, University of South Africa (UNISA), Pretoria, South Africa

N. Pophiwa
The National Consumer Commission, Pretoria, South Africa
e-mail: N.Pophiwa@thencc.org.za

G. Nhamo, L. Chapungu (eds.), *The Increasing Risk of Floods and Tornadoes in Southern Africa*, Sustainable Development Goals Series, https://doi.org/10.1007/978-3-030-74192-1_10

10.1 Introduction

This chapter contributes to scholarship that documents the effects of cyclones Idai and Kenneth, which devastated parts of the southern African region in 2019, by exploring the impact that these disasters had on the displacement and the migration of affected communities. The chapter takes cognisance of the fact that the region has historically experienced forced migration as a result of civil wars (e.g. in Angola and Mozambique), economic collapse and political crisis. In the case of Zimbabwe, adverse weather conditions have also driven temporary displacement among affected communities. Cyclone Eline, which struck the region in 2000, was previously the most devastating sudden-onset disaster in the region. It affected 5 million people and displaced as many as 1.25 million people across southern Africa, of which 250,000 were displaced in Mozambique alone. The aftermath of such disasters often lasts a lifetime once the emergency relief programmes exit the recipient communities and the people have to rebuild their lives. Almost two decades later, after Eline, cyclones Idai and Kenneth were declared the worst to have hit the region. The strong winds and heavy rains put the region in a state of crisis, claiming lives, flattening buildings and triggering massive floods that damaged critical infrastructure and farmlands and submerged entire communities, thus leaving affected people without shelter, food, safe drinking water, and sanitation and hygiene. The chapter is embedded in scholarly debates on the nexus between extreme weather or climate variability and human mobility or displacement. A 1985 report by the United Nations Environment Programme (UNEP) (cited by Biermann and Boas 2010, p. 62) is one of the early studies of the concept of environmental refugees. The report defined them as 'people who have been forced to leave their traditional habitat, temporarily or permanently, because of a marked environmental disruption (natural and/or triggered by people) that jeopardized their existence and/or seriously affected the quality of their life'. Since then, many more terms have been coined to describe refugees or displaced persons who have fled as a result of climate-induced disasters. In more recent years, there has been consensus that it is undisputable that there are links between climate change disasters and other causes of displacement (UNHCR 2017, p. 5). At a global level, since the 2004 Indian Ocean earthquake and tsunamis, there has been growing international concern that adverse weather and disasters related to climate change have an impact on population movement.[1] The focus has also been on seeking solutions to managing migration in the context of these disasters. Biermann and Boas (2010, p. 75) suggested the following five principles to address what they referred to as 'climate refugee' problems: (1) planned relocation and resettlement; (2) resettlement instead of temporary asylum; (3) collective rights for local populations; (4) international assistance for domestic measures; and (5) international burden sharing. The situation in the most impoverished parts of the world shows that more needs to be done to avert the vulnerability of local communities who live in areas that are prone to climate-induced disasters.

For decades, climate change and disastrous weather-related events have been proven to be among the forces behind human migration (Marino and Lazrus 2015). Several studies have concluded that although climate change and its consequences have an impact on human migration, sometimes there is inconclusive evidence that environmental impacts are among the pressing issues that influence human migration (Piguet et al. 2011). Usually, migration is influenced by pre-existing factors, such as economic constraints and opportunities, socio-cultural networks and the political context, that are prevalent on the continent. That, of course, does not mean that high-frequency disasters (cyclones, floods,

[1]While it may be beyond the scope of this chapter, it should be noted that there have been numerous initiatives and frameworks to address migration caused by climate-induced disasters, such as the Sustainable Development Goals, the Sendai Framework for Disaster Risk Reduction, the World Humanitarian Summit, the New York Declaration for Refugees and Migrants, the Nansen Initiative on Disaster-induced Cross-border Displacement and its successor, the Platform on Disaster Displacement. Refer to https://disasterdisplacement.org/the-platform/the-context (accessed on 22 October 2020).

drought, typhoons, hurricanes, heatwaves, tornadoes, storms, etc.) do not also play a major role in migratory patterns (Piguet et al. 2011). Piguet et al. (2011) continue to point out that extreme weather events inform individuals' decisions to relocate from their home countries to the nearest accessible safer countries. Extreme weather events, such as floods and cyclones, have in many cases proven to pose a great threat to humankind worldwide. Approximately 99 million (yearly) and 39 million people have been affected by floods and cyclones, respectively, between 2000 and 2008 globally (Piguet et al. 2011). According to Raleigh et al. (2017), there have been cases where people relocated from their places of residence due to extreme weather events like cyclones and extreme floods. Although research shows that there is a relationship between environmental factors and migration, there is room for more evidence-based studies to be conducted.

It is upon this basis that this study was conducted a few months after the cyclones in two areas, namely the Eastern Highlands of Zimbabwe in Chimanimani and the Wild Coast of South Africa in Port St Johns. The study sought to establish, among other things, the probability that migration (local or international) would be an intended option for the affected communities to cope with the destruction of their livelihoods. According to a report by USAID (2019), about 144,751 people were affected, 10,097 homes were destroyed and 37,483 homes were partially destroyed by cyclone Idai in two Zimbabwean provinces (Manicaland and Masvingo), and the entire population of Port St Johns was affected by floods in 2019. This chapter seeks to unpack how extreme weather events influence the migratory patterns of humans, with a view to answering the following research question: To what extent can cyclones and flood-related environmental catastrophes become the next drivers of intra- and international migration within the southern African region?

The chapter proceeds with an extensive review of literature on historical developments of migration, policies and instruments related to displacement and migration, and characteristics of migration and displacement. It goes on to describe the methodology used to collect and analyse data as well as present and discuss the results of the study and finally provide a conclusion and recommendations.

10.2 Literature review

Historical antecedents of displacement and migration and policies.

According to Ionesco et al. (2017) in *the atlas of environmental migration* people have been migrating in search of better climatic conditions as far back as 45,000 years, to a great extent as a result of land degradation, ecosystem disruption and resource depletion as climates change (Ionesco et al. 2017). An early example of displacement associated with climatic conditions occurred in 1755 when the city of Lisbon experienced an earthquake and tsunami, which displaced many people into camps throughout Europe (ibid). Another example was the Great Irish Famine of 1845–1852, where warmer and damper weather conditions led to crop destruction, resulting in famine in the country. This epidemic caused many people to flee from their country and to seek refuge in other countries (ibid). Even though it is difficult to establish global migratory figures, it is widely accepted that international migrants have constituted 3.5% of the world's population (272 million) as of September 2019 (UN DESA 2019). In 2000, the global population of international migration stood at 2.8% (51 million).[2] Migration is an adaptation strategy for extreme weather events, as people relocate to avoid the impacts and the stress that come with extreme weather events (Marino and Lazrus 2015). Depending on the circumstances, migration can be temporary (from a few days to months) or permanent, within and across borders (Hassani-Mahmooei and Parris 2012).

[2]UN Department of Economic and Social Affairs (DESA) (2019). The number of international migrants reaches 272 million, continuing an upward trend in all world regions, says UN. Available at: https://www.un.org/development/desa/en/news/population/international-migrant-stock-2019.html (accessed 2 June 2020).

The migratory pattern of a population will be determined by the affordability of migration or the availability of resources to migrate, considering that while some victims may wish to migrate, a lack of resources might prevent them from doing so. They then return to the disaster zone and reconstruct their homes there or move within a short distance of the area (Piguet et al. 2011). For the most part, displaced populations tend to migrate within their own countries as opposed to migrating to neighbouring countries, unless they live near a border. In Bangladesh, for example, it has become the norm for people to relocate from their cities within their country during seasons of flooding (Black et al. 2011). They also migrate to neighbouring India. In 2000, a record number of approximately 20 million Bangladeshis relocated to India, largely fleeing floods and searching for better opportunities (Quencez 2012). The literature is awash with examples of other parts of the world where similar patterns of movement have occurred.

The fact that migration features prominently among the Sustainable Development Goals (SDGs) shows the seriousness with which the world seeks to address both migration and the impacts of climate change. Take, for instance, SDG 10.7.2, which deals with migrants and international migration by calling upon countries to facilitate orderly, safe, regular and responsible migration as well as the mobility of people through the implementation of policies. The migration target contributes to other 2030 targets, particularly those related to poverty, health, education and gender equality. However, the SDGs do not entirely link climate change and migration. SDG target 10.7 calls for signatories to 'facilitate orderly, safe, and responsible migration of people, including through implementation of planned and well-managed policies' (Podesta 2019). Even so, there is no clear indication and projections on where migrants will go in the future, therefore it is hard to precisely point out the impacts of human mobility in relation to SDG13. Nevertheless, migration tends to be one of the vital features of the SDGs. Therefore, a better understanding of the relationships between migration and other key development issues dealt with in the SDGs is essential (Wilkinson 2016).

Historically, intraregional migration has been a common phenomenon in Africa. With the drawing up of African borders at the infamous Berlin Conference in 1885, colonial governments began to regulate movement across these borders. History shows that the cross-border migration of people to South Africa was influenced by three streams of movement: contract mine migration (migrants who came to South Africa to work in the mines), other categories of voluntary migration and refuge migration (Wentzel 2003). In the case of the mining sector, the discovery of diamonds in Kimberley in 1867 and, later, gold in 1886 on the Witwatersrand meant that South Africa's need for cheap African labour increased. This led to the sourcing of labour from the then colony of Nyasaland (Malawi), northern Rhodesia (Zambia), in some cases under strict regulation from southern Rhodesia (Zimbabwe)[3] and the neighbouring protectorates of Bechuanaland (Botswana), Basutoland (Lesotho) and Swaziland to work on the mines. Labour migration historians have documented in detail the dynamics associated with migration during the colonial period (Van Onselen 1976; Rwamatwara 2005). Most of these writings did not pay attention to the influence of environmental change on human mobility in the region. They focused on migration due to labour as a result of growing urbanisation, which was a magnet for rural to urban migration (De Haan 2002). Besides migration for work in the region as of the late 1970s, there was also forced movement of people across borders as a result of civil wars in countries such as Angola and later Mozambique.

The literature also documents changes in the migration landscape that occurred with the advent of democracy in South Africa from 1994. New patterns of migration occurred because migrants no longer sought work on mines and

[3] Southern Rhodesia also required labour for its mines and agricultural enterprises, therefore there were agreements to regulate labour migration to South Africa. Of course, such regulation was not easy to implement and clandestine labour migration continued to occur.

farms but in urban areas of the country in various sectors and in entrepreneurship since old apartheid restrictions on the freedom of movement of black people had been lifted. This ushered in a new phase of policy responses on the part of the post-1994 government, which had to regulate international migration. The most recent legislation at the time had been passed in 1991 as the Alien Control Act, which served the interests of the apartheid regime. In 1995 the Aliens Control Amendment Act was adopted, turning the official migration policy of the country to focus on skilled labour migration. The Act also provisions the formation of an immigration selection board to ensure selection based on academic qualifications.

In the same spirit of formulating migration policies that served the democratic dispensation, the Mandela administration enacted the Refugees Act of 1998, which applied a right-based approach towards asylum seekers concerning education and work. The Refugees Act was driven by a humanitarian space as well as the political and economic situation in neighbouring countries. The Refugees Amendment Act, enacted in 2008, enforced control measures and the deterrence of 'new' asylums and limited refugees' rights in the same way as those of South Africans. It also instated the use of biometrics as a way to crack down on fraudsters (Wilkinson 2016).

At that time, the South African regime was heralded as one of the most progressive in the world. While the Refugees Act was being promulgated, a parallel process was taking place with respect to the Immigration Act of 2002, which was to become the post-apartheid immigration policy pronouncement. The policy process of the Act started with the Green Paper in 1997, followed by the White Paper of 1999, then the draft Bill of 2000 and finally the Immigration Act of 2002. The Act categorised as well as designated various permits for skilled migrants. It also expanded enforcement and border controls and monitoring of the Act by ordinary citizens. Since 2002 the Immigration Act has undergone various amendments, the first of which was the Immigration Amendment Act of 2004, which allowed, among others, more powers to the Minister and the Department of Home Affairs (DHA). It dissolved the liaison committee and civil society participation, as well as the anti-corruption unit. It also reduced the number of quota permits available, decreased the appeal window and enforced the deterrence of newcomers. The second amendment of the Immigration Act came in 2007, taking forward the African agenda that favoured cross-border trade and the relaxation of repatriation deposits of students.

The Immigration Amendment Act of 2011 redefined temporary residence permits introduced critical work visas and added a condition of capital injection with respect to business visas. It also redefined migration policy, while the Refugees Act of 2011 reduced the asylum transit visa window from 14 to 5 days as well as the registration of asylum seekers. The amendments came with increased prison sentences for those contravening the Act. In 2016 the Border Management Authority Bill was enacted by parliament. The Bill made provisions for the establishment, organisation and regulation of a single state entity to oversee the movements of people and goods through South African borders. The Bill was repealed by the National Council of Provinces in 2019. The latest amendment, in 2017, of the Refugees Amendment Act limited the rights of asylum seekers to work. Also in 2017, proposed amendments to the White Paper on International Migration were aimed at delinking residency from citizenship and introducing a point system as a way of ensuring skilled migration. At the regional level, a policy on migration has also been formulated, although member countries have been slow to ratify it. In 2005 the Southern Africa Development Community (SADC) adopted a policy instrument in the form of the facilitation of movement of persons, which sanctioned visa-free travel between SADC member states for a maximum period of 90 days. The protocol was a sign of goodwill and harmonisation of permanent and temporary residence policy on work permits. From about 2005–2009, there were large-scale deportations of Zimbabweans from South Africa. In 2010 the South African government ran a dispensation project in respect of Zimbabweans, which led to the documentation of about 250,000 migrants from that country.

10.3 Environmental Characteristics of Climate Change in Southern Africa

The increase in greenhouse emissions over the past decades has brought about extreme weather conditions. The effects of global environmental change have been witnessed in the form of ever-increasing incidences of natural disasters, for example, high sea levels, drought, extreme rainfall, earthquakes, floods, tsunamis and tropical cyclones. These environmental challenges have been experienced by many countries, especially several African countries, including the studied countries in this chapter. Scholars have deduced that these environmental challenges can and have influenced internal (rural–rural, rural–urban) and international migration or can displace people into camps, which can result in permanent migration (Greenpeace 2017; Abebe 2014; Chagué-Goff et al. 2012; Warner et al. 2010).

Over the past decade, southern African countries experienced tropical cyclones Favio, Eline, Kenneth, Dineo and Idai; South Africa experienced severe flooding in 2019. The characteristics of extreme weather conditions and the impacts felt by each country are different. Like many other countries in southern Africa that have been exposed and affected by climate change, Zimbabwe and Mozambique were heavily hit by cyclones Idai and Kenneth in 2019, which displaced many people from their households to camps and forced some people to migrate internationally. Some of the displaced persons remain in camps and are faced with an uncertain future (Al Jazeera 2020). Cyclones Idai and Kenneth were not the only environmental challenges that Zimbabwe has been confronted with over the past years. Droughts have hit many parts of Zimbabwe and have had an enormous impact on the country's agricultural sector.

South Africa has also experienced environmental challenges in the past and recently. High temperatures leading to droughts and low rainfall are some of many environmental challenges that South Africa is facing. KwaZulu-Natal (KZN), a coastal province, is prone to flooding because of high rainfall and storms. In 2019 urban and rural areas in KZN like the eThekwini Metropolitan Municipality, the Umshwati Local Municipality, the Ugu District Municipality and the King Cetshwayo District Municipality experienced high rainfall, which resulted in flooding and the displacement of many households. In the 2018–2019 annual rainfall season, The City of Cape Town witnessed low rainfall levels, which impacted the livelihoods of individuals, farmers and many others. Wine farmers were hit the hardest, and less wine was produced that season.

Zambia has experienced similar challenges in respect of intra-seasonal rainfall and drought. In a study on the extent of climate change and variability over the years in Kabwe and Livingstone, in Zambia, and the impact it has on rural-rural migration, Makondo and Thomas (2020) established that the climate of both areas has changed, with Livingstone experiencing more droughts and unreliable rainfall and Kabwe experiencing changes in rainfall and the duration of the rainy season (Makondo and Thomas 2020). One could assume that rural–rural migration would be the best adaptation strategy to cope with such changes. However, the study concluded that rural–rural migration is not safe from climate change risk (Makondo and Thomas 2020). Southern African countries are not the only countries in Africa that are experiencing an increase of migration due to challenges related to environmental change. North African countries are experiencing a high volume of seasonal migrants, along with increasing sea levels, drought and desertification, which will result in an increase in seasonal migrants not only from the countries of origin but also to the destination countries (Podesta 2019).

The International Organization for Migration (IOM) estimates that, by 2050, 200 million people may be displaced due to environmental challenges globally (Warner et al. 2010). According to the *Global report on internal displacement* (IDMC 2019a), at the end of 2018, 17.2 million people were displaced within their own countries due to environmental challenges and disasters. This figure further increased by the end of June 2019, with 7 million new people having been internally displaced due to environmental challenges and disasters within their own countries (IDMC 2019b). From the 17.2 million people that were displaced, 1.1 million were displaced due to geophysical and 16.1 million people were further displaced due to weather-related conditions, including floods, storms, droughts, wildfires, landslides and extreme temperatures.

The impacts of environmental challenges are already felt by vulnerable countries, and it is difficult and challenging for many of them to recover from these effects. The Global Climate Risk Index was analysed with regard to the studied countries, which include South Africa, Malawi, Mozambique, Swaziland and Zimbabwe (Table 10.1). The Global Climate Risk Index analyses the impacts of weather-related loss in different countries and regions (World Risk Report 2019). According to the Long-term Climate Risk Index (CRI) between 1999 and 2018, Mozambique was among the top 20 vulnerable countries globally (e.g. storms, floods and heat) (Eckstein et al. 2019). Zimbabwe was in the 36th position, which indicates it is less vulnerable to extreme weather conditions, despite both countries having been hit by cyclones recently. South Africa and Malawi, ranked 79th and 80th, respectively, show more resilience compared with Mozambique and Zimbabwe. Data on Swaziland was not available at the time of writing.

How well are countries prepared to adapt to extreme weather conditions varies from one country to another. According to the World Risk Index 2019, Mozambique has the least capacity to adapt to extreme weather conditions, followed by Zimbabwe. Swaziland is ranked third and Malawi fourth (Table 10.2). From the studied countries, South Africa is ranked the best country to adapt to extreme weather conditions.

The Notre Dame Global Adaptation Index (ND-GAIN 2015) came to similar conclusions regarding the studied countries' ability to adapt to extreme weather conditions. In 2015 Mozambique, Malawi and Swaziland were among the most vulnerable and less prepared of the case study countries. Interestingly, Zimbabwe was ranked better than Mozambique, Malawi and Swaziland. Again, South Africa was considered less vulnerable to the impacts of extreme weather and as being more able to adapt to extreme weather conditions. Brzoska and Fröhlich (2016) identify four types of climate change patterns or four types of migrants:

(i) Ecological-economic migrants

In this case, migration is adopted as a livelihood strategy to generate an income for the household or to diversify household income. It is not an individual decision to migrate, but a household decision. The effects of environmental risks/challenges have the potential to have an impact on these groups.

(ii) Climate disaster refugees

Refugees are people who have had to move from their country because of harm to seek protection in other countries (e.g. political instability) (White 2019). In the case of climate change, refugees are normally forcedly removed from their habitat due to environmental risks caused by climate change and are placed in a safer environ-

Table 10.1 Long-term Climate Risk Index (CRI) for 1999–2018

Country	Rank in 2018
South Africa	79
Malawi	80
Mozambique	14
Swaziland	–
Zimbabwe	36

Source: Eckstein et al. (2019)

Table 10.2 World Risk Index 2019 ranking for the studied countries

Country	Rank
South Africa	92
Malawi	55
Mozambique	46
Swaziland	90
Zimbabwe	52

Source: Bündis Entwicklung and UNU-EHS (2019)

ment. These people may choose to return to their normal habitat at a later stage.

(iii) Permanent climate refugees

Environmentally induced disasters can permanently destroy individuals' habitats. Farmers mostly fall into this category. Their livestock mostly depends on natural resources for survival, and if the resources get affected, they usually seek alternative natural resources for their livestock.

(iv) Climate-affected migrants

Land and water are one of the most important natural resources to the livelihoods and as a result migrants' populations such as herders of livestock's or farmers are therefore affected by changing environmental conditions, especially the herders in some parts of Africa and Asia. They end up changing their migration pattern or even forced to give up their lifestyles. Table 10.3 shows a summary of the types of migrants.

This study explores the applicability of some of the typologies mentioned above to understand the situation of victims of Idai and Kenneth in Chimanimani and Port St Johns. The victims of these cyclones are likely to be climate disaster refugees, or they can be climate-affected migrants. The applicability of any of the two typologies is dependent on timeframes related to the length of their stay in temporary shelters. The discussion of the findings will focus on this aspect.

Table 10.3 Types of migrants

Type	Distance travelled	Permanence	Agency	Economic activity at destination
Ecological-economic migrants	Direction and distance depend on 'risk capital' and economic opportunities	Temporary (seasonal, life cycle)	Individuals, but often group-decided; predominately young males	Seeking to be self-supporting; primary aim is to secure remittances
Climate disaster refugees	Short term (refugee camps, relatives)	Not permanent, shorter periods	Groups	Dependent on external support
Permanent climate refugees	Direction and distance depend on 'risk capital', external support and economic opportunities	Not permanent, shorter periods, permanent	Groups	Seeking to be self-supporting; tapping external support
Climate-affected migrants	Rerouting of migratory patterns	Depends on conditions	Groups	Seeking to be self-supporting

Source: Brzoska and Fröhlich (2016)

10.4 Materials and Methods

The case study research design was adopted with the aim of determining the link between migration and extreme weather events. Chimanimani, Zimbabwe, and Port St Johns, South Africa, were purposively selected due to cyclones and floods that occurred in the case study areas. Data collection was based on a household survey, fieldwork observations and secondary analysis of online data portals. A total of 220 household questionnaires were administered in Chimanimani (Zimbabwe) regarding cyclone Idai and 190 household questionnaires in Port St Johns (South Africa) regarding the 2019 floods. The questionnaires were administered on an offline question-pro data collection tool during the day and later uploaded on the question-pro online platform. Alternatively, the questionnaire survey links were sent to the affected individuals in Port St Johns to complete online. The study deployed the question-pro tool to collect and analyse data on both cyclone Idai and floods. The data was analysed to assess the impact of 2019 floods and cyclones Idai and Kenneth on migration patterns in Southern Africa.

The study was carried out in the affected southern African countries. The study focused on areas/communities that were greatly affected by cyclone Idai and the floods in Port St Johns in 2019 (Fig. 10.1). The case study areas were Chimanimani, Zimbabwe (cyclone Idai), and Port St Johns, South Africa (floods).

The respondents had to answer the following questions to assess the likelihood that they would migrate as a result of cyclone Idai and the 2019 floods: (a) If another flood or cyclone were to happen, where are you likely to move to? (b) Among your family or friends, who has migrated or moved due to floods or cyclones and where did they migrate or move to? and (c) If another flood or cyclone were to happen, what would be another driving factor for you to move or migrate?

10.5 Results and Discussion

Three questions were included in the household questionnaire survey to understand if cyclone Idai that occurred in Zimbabwe and the 2019 floods in South Africa influenced internal and international migratory patterns. The responses to the three questions are presented in Figs. 10.2, 10.3, 10.4, 10.5, and 10.6.

The results presented in Figs. 10.2 and 10.3 show that the effect of natural disasters is more significant for in-migration than out-migration. Most respondents from both Chimanimani (91%) (Fig. 10.2) and Port St Johns (83%) (Fig. 10.3) prefer to move internally (within their own country) despite recurrent flooding events. This finding is supported by Drabo and Mbaye (2015), who suggest that residents of low- and middle-

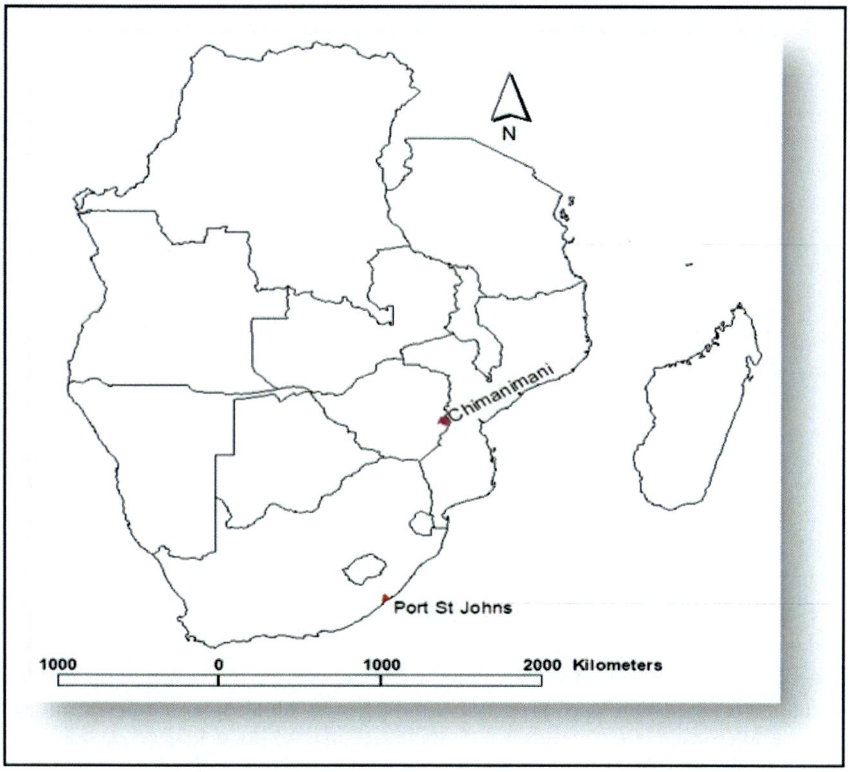

Fig. 10.1 Map of the study area. Source: Authors

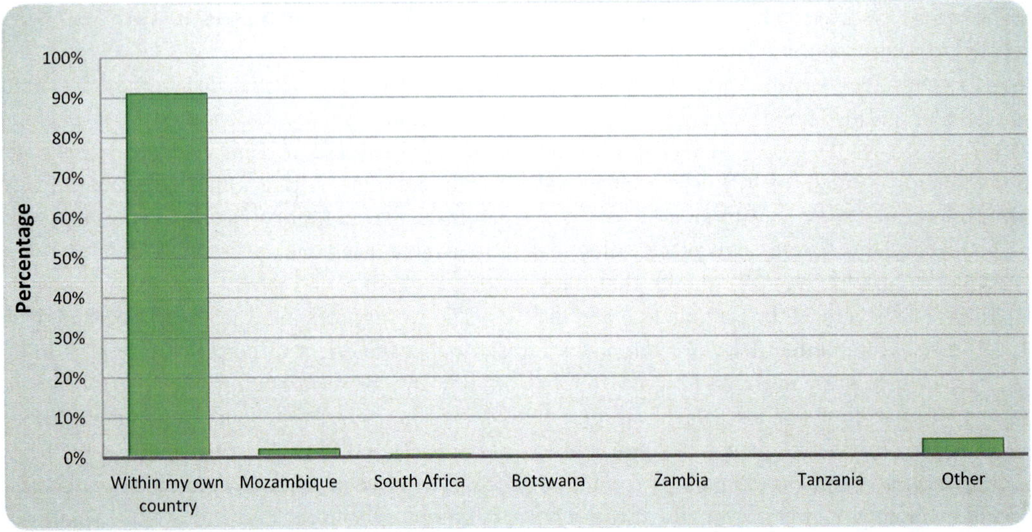

Fig. 10.2 Countries to which respondents from Chimanimani, Zimbabwe, are likely to migrate. Source: Zimbabwe fieldwork 2019 ($n = 220$)

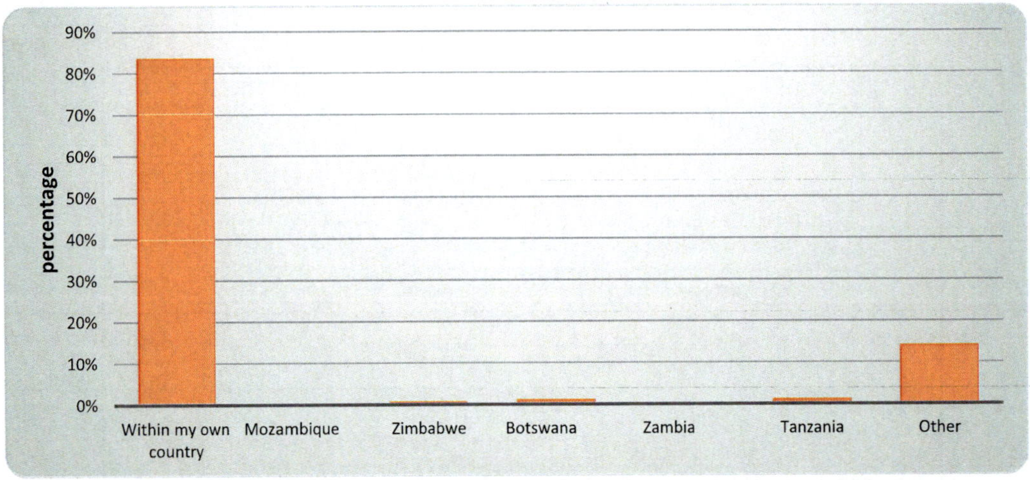

Fig. 10.3 Countries to which respondents from Port St Johns, South Africa, are likely to migrate. Source: Eastern Cape, South Africa, fieldwork 2019 ($n = 197$)

income countries do not automatically move to other countries in the wake of disasters, primarily because they have no alternatives at their disposal to do so, other than their rich counterparts, who are able to financially deal with the impacts of disasters without migrating.

Typically, wealthy people who reside in disaster-prone areas have home insurance that covers damage caused by natural disasters. Nonetheless, it is not only the financial capacity to migrate that influences cross-border disaster displacement. For instance, Cattaneo and Peri (2016) point out that despite the economic status of an individual or a country, migration can be influenced by other factors such as vulnerability. Beine and Parsons (2015) state that there is a certain variability within developing countries, which indicates that the degree of vulnerability

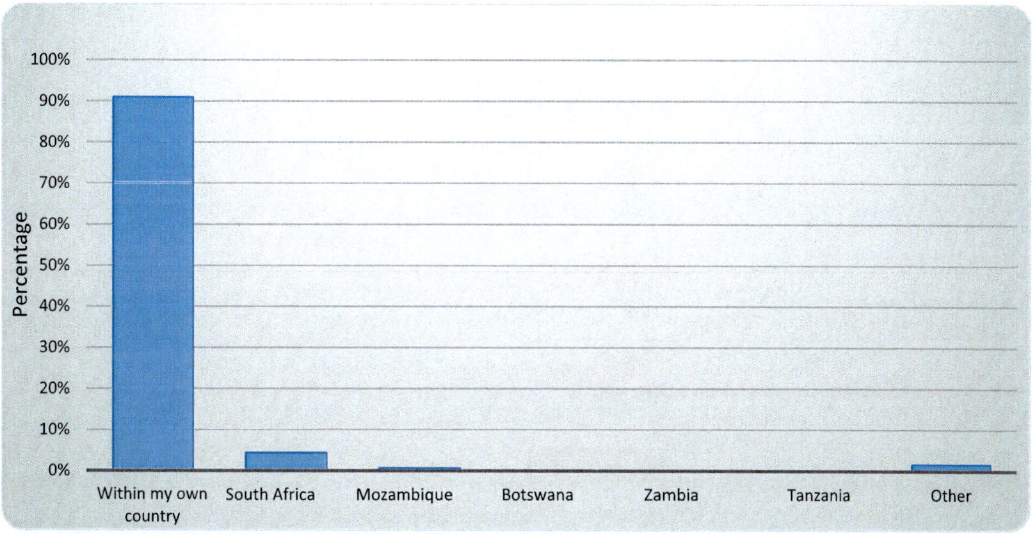

Fig. 10.4 Countries to which Chimanimani respondents' relatives or friends have migrated due to cyclone Idai. Source: Zimbabwe fieldwork 2019 ($N = 220$)

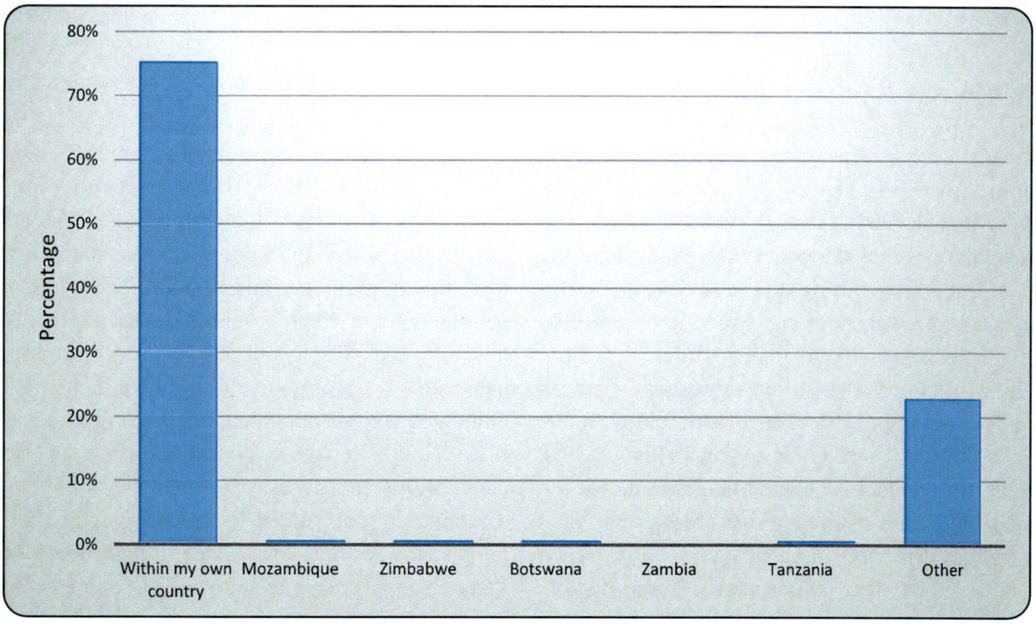

Fig. 10.5 Countries to which Port St Johns respondents' relatives or friends have migrated due to floods. Source: Eastern Cape, South Africa, fieldwork 2019 ($N = 197$)

and the extent of a disaster also affect the relationship between disasters and migration. The severity of a disaster is the main factor that influences the movement of people outside their country of residence. Basically, the more severe the disaster, the greater the likelihood that an individual will migrate. This is especially true in instances where a disaster hits an entire country as opposed to a specific area.

In addition, the decision not to migrate during disasters may be associated with challenges that migrants face in host countries. These challenges

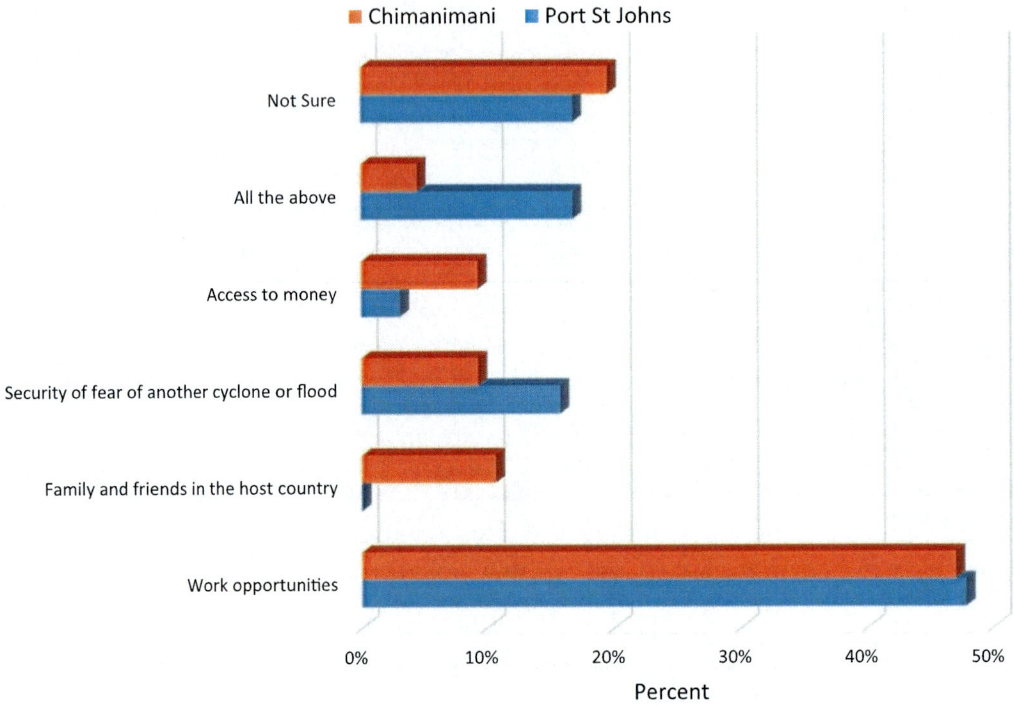

Fig. 10.6 Other reasons for respondents to migrate to other countries. Source: Fieldwork 2019

include access to resources, services and opportunities, which is key to their resilience. Wilson and Paradise (2016) remark that locals received evacuation assistance earlier and more efficiently than foreigners during the 2011 Japan earthquake. Moreover, migrants typically do not possess legal rights to be in a particular country, which often restricts their movement even in times of natural disasters such as flooding events. According to Clark-Ginsberg and Hunt (2016), illegal migrants in Dhaka, Bangladesh, choose not to evacuate disaster-prone areas even when warnings for upcoming disasters are issued because they fear deportation. These factors together contribute to the risk exposure of migrants during and after natural disasters, further undermining their well-being. Figures 10.2 and 10.3 also show that, though the figures are relatively low, respondents from Chimanimani and Port St Johns prefer to migrate to 'other countries' (4% and 14%), followed by Mozambique (2% and 0%), Zimbabwe (1% and 1%), Botswana (0% and 1%), Tanzania (0% and 1%) and Zambia (0% and 0%) respectively. Kim-Blanco's (2016) view that migration decisions are often mediated by economic, social and political drivers supports this result. For example, people are not likely to migrate to politically unstable and poor countries because migration has been a way for people to improve their lives. Other existing factors that trigger migration in many Southern Africa countries are access to clean water and access to grazing land. When migrants are confronted by environmental challenges like cyclones and floods, it is difficult for them to adapt, especially if they have limited or no access to grazing land.

Figure 10.4 shows that although respondents from Chimanimani have had a few relatives and friends who have moved to other countries due to cyclone Idai, most of them (91%) remained within the country while only 4% migrated to South Africa, 2% migrated to other countries, 1% migrated to Mozambique and 0% migrated to

Botswana, Tanzania and Zambia, respectively. This result can be attributed to the fact that most respondents from Chimanimani who were affected by cyclone Idai were accommodated in camp tents in the area. Although Zimbabweans have been known to migrate to various neighbouring countries, including South Africa, Mozambique and Botswana (Idemudia et al. 2013), this scenario of Chimanimani shows that the economic crisis of previous years was a major driver of cross-border migration in comparison with natural disasters. This does not negate the fact that rates of international migration in Africa have slightly increased in recent years due to climatic shocks (Ruohong et al. 2016). Although this is the case, the proportion of Africans living in other countries such as European countries and America has increased since 2000 (BBC News 2019). This is relatively a factor of the historical role of labour recruitment in these countries in relation to colonial ties. Another explanation relates to the requirements and the costs of immigration visas to enter other African countries, which are relatively higher compared to those of other countries outside Africa (Madowo 2018). Additional explanations could be linked to natural disasters. African countries, more than any other countries in the world, are considered to be the most vulnerable to weather events such as floods and droughts, which influences immigration preferences.

Port St Johns is no different in respect of the findings. As shown in Fig. 10.5, an estimated 75% of the respondents from Port St Johns indicated that despite recurring flooding, their relatives and friends who have been affected by floods have not moved to other countries. A total of 23% of the respondents indicated that some of their relatives or friends have migrated to 'other countries' after the floods while 1% have migrated to Mozambique, 1% to Zimbabwe, 1% to Botswana, 1% to Tanzania and 0% to Zambia, respectively. This result resonates with Cattaneo and Peri's (2016) view that despite pressing circumstances, African rural societies are not keen to migrate since people are still rooted by traditional, social and cultural ties, and the migratory trends within these societies have not significantly changed compared to other continents. However, it does not imply that African societies do not migrate within their own countries. Despite economic reasons to migrate to urban areas, there are also other reasons for rural to urban migration.

The Nansen Initiative (2015) identifies other root causes of rural migration, and these include rural poverty, food security, inequality, limited access to social protection and depletion of natural resources. One participant from Port St Johns highlighted the high poverty in the area, stating that her neighbour migrated to the city for better access to municipality resources. Lack of service delivery is a problem in many municipalities in South Africa, including those in rural areas. Figure 10.6 (n = 417 responses) shows various reasons that are pull factors for respondents from Chimanimani and Port St Johns to migrate to other countries. The main reasons given for migration to other countries by respondents from both countries are work opportunities.

Forty-eight percent of the respondents from Port St Johns gave work opportunities as their primary motivation to go to other countries, followed by those who are not sure (17%), those who indicated various reasons (17%), those who indicated security or fear of another cyclone or flood (16%) and those who indicated access to money (3%). Zero percent of the respondents indicated family and friends in the host country as a motivation to go to other countries. In respect of Chimanimani, respondents selected work opportunities at 47% as the main pull factor to other countries, followed by not sure at 19%, family and friends in the host country at 11%, security or fear of another cyclone or flood and access to money, both at 9%, and 'all the above' at 4%. These findings suggest that people are more likely to move to other countries in search of job opportunities than to flee the effects disasters (Bohra-Mishra et al. 2014). These findings further emphasise that job opportunities will always be a pull factor for migration, especially where there is high poverty and a depletion of natural resources caused by the effects of climate change.

10.6 Conclusion

People move to areas of safety, especially areas where emergency aid is provided, at the beginning and soon after a cyclone or floods. However, sometimes after the emergency aid has dried up, communities are left to rebuild and return to their 'normal' lives. The literature shows that a common adaptation strategy for rural households affected by disasters or other economic hardships is temporary migration to urban areas for a short-term stay. While communities may not migrate permanently as a result of floods or cyclones, temporary migration will continue to be an adaptation strategy for them. Other households may engage in informal cross-border trade as a form of livelihood. In the aftermath of cyclones such as Idai and Kenneth and the floods in Port St Johns, it would be hardly surprising to learn of increases in new arrivals in nearby urban areas or the nearest border towns.

The findings of the study were based on a survey comprising household questionnaires. The results showed that even though migration decisions are influenced by economic, social and political drivers, people generally prefer to move internally despite the reoccurrence of disasters. In concurrence with the literature, the study also revealed that most people in the study areas do not have a disposable income and therefore cannot move to other countries. The results further indicated that regardless of the economic status of an individual or a country, people will not necessarily migrate to other countries because there are several factors that influence migration. The study showed that localised migration is likely to be more common, in the sense that people are prepared to move to areas that are safer from extreme weather events within their own countries.

In conclusion, this chapter demonstrated that migration can be induced by cyclones and floods that have taken place in southern Africa. It further offered insight into the causal effects of displacement and migration, as well as the link between migration and climate change and the impact of climate change on the affected areas. The study also looked at patterns of migration due to cyclones (Idai and Kenneth) and floods.

Based on the analysis provided above, it is evident that cyclones and floods have put southern Africa and other parts of the world in a state of crisis in that individuals and communities have been forcefully displaced, either temporarily or permanently. Response measures to address the problem of internal or external migration need to be tailored to fit the socio-economic reality of southern Africa, given the rural setting of most settlements, which have limited access to necessities and information.

References

Abebe, M. A. (2014). Climate change, gender and migration in Eastern Africa. *Washington Journal of Environmental Law and Policy, 4*, 104–137.

Al Jazeera/Tendai Marima. (2020). *Cyclone Idai: One year on, Zimbabwe survivors loosing hope*. Retrieved November 13, 2020, from https://www.aljazeera.com/news/2020/3/15/cyclone-idai-one-year-on-zimbabwe-survivors-losing-hope.

BBC News. (2019). *African migration: Five things we've learnt*. Retrieved September 10, 2020, from https://www.bbc.com/news/world-africa-47705944.

Beine, M., & Parsons, C. (2015). Climatic factors as determinants of international migration. *The Scandinavian Journal of Economics, 117*, 723–767.

Biermann, F., & Boas, I. (2010). Preparing for a warmer world. Towards a global governance system to protect climate refugees. *Global Environmental Politics, 10*(1), 60–88.

Black, R., Stephen, R., Bennett, G., Thomas, S. M., & Beddington, J. R. (2011). Mobility as adaptation. *Nature, 478*, 447–449.

Bohra-Mishra, P., Oppenheimer, M., & Hsiang, S. M. (2014). Nonlinear permanent migration response to climatic variations but minimal response to disasters. *Proceedings of the National Academy of Sciences, 111*(27), 9780–9785.

Brzoska, M., & Fröhlich, C. (2016). Climate change, migration and violent conflict: vulnerabilities, pathways and adaptation strategies. *Migration and Development, 5*(2), 190–210.

Cattaneo, C., & Peri, G. (2016). The migration response to increasing temperatures. *Journal of Development Economics, 122*, 127–146.

Chagué-Goff, C., Andrew, A., Szczuciński, W., Goff, J., & Nishimura, Y. (2012). Geochemical signatures up to the maximum inundation of the 2011 Tohoku-oki tsunami - Implications for the 869AD Jogan and other palaeotsunamis. Sedimentary Geology, 282, 65–77. https://doi.org/10.1016/j.sedgeo.2012.05.021

Clark-Ginsberg, A., & Hunt, D. (2016). *We are human too! Concern Worldwide's efforts to reduce risks for the homeless migrants of Dhaka, Bangladesh*. Retrieved from www.concern.net/insights/pavement-dwellersstreets-today-fighting-better-tomorrow

De Haan, A. (2002). Migration and livelihoods in historical perspective: a case study of Bihar, India. *Journal of Development Studies, 38*(5), 115–142.

Drabo, A., & Mbaye, L. M. (2015). Natural disasters, migration and education: an empirical analysis in developing countries. *Environment and Development Economics, 20*, 767–796.

Eckstein, D., Künzel, V., Schäfer, L., & Winges, M. (2019). *Briefing paper: Global Climate Risk Index 2020*. Bonn: Germanwatch.

Greenpeace. (2017). *Climate change, migration and displacement. The underestimated disaster*. Retrieved November, 11, 2020 https://www.greenpeace.de/sites/www.greenpeace.de/files/20170524-greenpeace-studie-climate-change-migration-displacement-engl.pdf.

https://reliefweb.int/sites/reliefweb.int/files/resources/WorldRiskReport-2019_Online_english.pdf

https://gain.nd.edu/our-work/country-index/

Hassani-Mahmooei, B., & Parris, B. W. (2012). Climate change and internal migration patterns in Bangladesh: an agent-based model. *Environment and Development Economics, 17*(6), 763–780.

Idemudia, E. S., Williams, J. K., & Wyatt, G. E. (2013). Migration challenges among Zimbabwean refugees before, during and post arrival in South Africa. *Journal of Injury & Violence Research, 5*(1), 17–27. https://doi.org/10.5249/jivr.v5i1.185.

Ionesco, D., Mokhnacheva, D., & Gemenne, F. (2017). *The atlas of environmental migration*. New York: Routledge.

Internal Displacement Monitoring Centre (IDMC). (2019a). *GRID 2019. Global report on internal displacement*. Retrieved November 11, 2020, from https://www.internal-displacement.org/sites/default/files/publications/documents/2019-IDMC-GRID.pdf.

Internal Displacement Monitoring Centre (IDMC). (2019b). *Mid-year figures. internal displacement from January to June 2019*. Retrieved February 3, 2020, from https://www.internal- displacement.org/sites/default/files/publications/documents/2019-IDMC-GRID.pdf.

Kim-Blanco, P. 2016. *Internal migration and natural disasters in Mexico. A spatial modeling approach*. MA thesis. Unpublished.

Madowo, L. (2018). *Why is it so hard for Africans to visit other African countries?* BBC Africa. Retrieved from https://www.bbc.com/news/world-africa-45677447

Makondo, C. C., & Thomas, D. S. G. (2020). Seasonal and intra-seasonal rainfall and drought characteristics as indicators of climate change and variability in Southern Africa: a focus on Kabwe and Living stone in Zambia. *Theoretical and Applied Climatology, 140*, 271–284.

Marino, E., & Lazrus, H. (2015). Migration or forced displacement?: the complex choices of climate change and disaster migrants in Shishmaref, Alaska and Nanumea, Tuvalu. *Human Organization, 74*(4), 341–350.

Piguet, E., Pécoud, A., & De Guchteneire, P. (2011). Migration and climate change: an overview. *Refugee Survey Quarterly, 30*(3), 1–23.

Podesta, J. (2019). *The climate crisis, migration, and refugees*. Retrieved February 19, 2020, from https://www.brookings.edu/research/the-climate-crisis-migration-and-refugees/.

Quencez, M. (2012). Floods in Bangladesh and migration to India. *The State of Environmental Migration, 2011*, 57–71, N° 06/12.

Raleigh, C., Jordan, L., & I. (2017). Assessing the impact of climate change on migration and conflict. The World Bank, Washington, DC.

Ruohong, C., Shuaizhang, F., Oppenheimer, M., & Pytlikova, M. (2016). Climate variability and international migration: the importance of the agricultural linkage. *Journal of Environmental Economics and Management, 79*, 135–151.

Rwamatwara, E. (2005). Forced migration in Africa: a challenge to development*Stchiproben*. *Stchiproben Wiener Zeitchrift Für Kritische Afrikastudien, 8*, 173–191. Retrieved from http://stichproben.univie.ac.at/fileadmin/user_upload/p_stichproben/Artikel/Nummer08/10_Rwamatwara.pdf.

The Nansen Initiative. (2015). *Disasters, climate change and human mobility in Southern Africa: consultation on the draft protection agenda*. Retrieved October 12, 2020, from https://disasterdisplacement.org/wp-content/uploads/2015/07/230715_final_background_paper_south_africa_regional_consultation_screen.pdf.

USAID. (2019). *Fact Sheet Idai—3rd April 2019. 3rd April 2019*, 1–6.

UN Department of Economic and Social Affairs (DESA). (2019). *The number of international migrants reaches 272 million, continuing an upward trend in all world regions, says UN*. Retrieved June 2, 2020, from https://www.un.org/development/desa/en/news/population/international-migrant-stock-2019.html.

UNHCR. (2017). *UNHCR and climate change, disasters, and displacement*. Geneva.

Van Onselen, C. (1976). *Chibaro: African mine labour in Southern Rhodesia, 1900–1933*. London: Pluto Press.

Warner, K., Hamza, M., Oliver-Smith, A., Renaud, F., & Julca, A. (2010). Climate change, environmental degradation and migration. *Natural Hazards, 55*, 689–715.

Wentzel, M. (2003). *Historical and contemporary dimensions of migration between South Africa and its neighbouring countries*. (Unpublished paper delivered at HSRC migration workshop, Pretoria, 17–20 March).

White, G. (2019). "Climate refugees"—a useful concept? *Global Environmental Politics, 19*(4), 133–138.

Wilkinson, E. (2016). *Climate change, migration and the 2030 Agenda for Sustainable Development*. Briefing Paper. Retrieved November 11, 2020, from https://www.odi.org/publications/10655-climate-change-migration-and-2030-agenda-sustainable-development.

Wilson, B. S., & T. R. Paradise. (2016). *Refugee inclusion in earthquake casualty estimation: a case study in south-east Turkey*. Retrieved November 11, 2020, from https://rm.coe.int/migrants-in-drr-web-final/1680716585.

World Risk Report. (2019). *Focus supply*. Retrieved December 17, 2020, from https://reliefweb.int/sites/reliefweb.int/files/resources/WorldRiskReport-2019_Online_english.pdf.

11 Impact of Floods on Access to Drinking Water: A Focus on 2019 Floods in Magalasi Village in Chikwawa District, Malawi

Miriam Kalanda Joshua, Ruth Kalinga Chirwa, Cosmo Ngongondo, Maurice Monjerezi, Evance Mwathunga, and Raymond Kasei

Abstract

Access to drinking water remains a challenge in many rural areas in Sub-Saharan Africa. This situation is further compounded by frequent disasters such as floods, which affect water supply infrastructure. However, few scholarly works have examined access to drinking water for rural households vulnerable to floods in Southern Africa. This study analysed the extent to which the 2019 flooding induced by Cyclone Idai affected households' access to water in Magalasi Village in Chikwawa District in Malawi. Data was collected and analysed following the mixed methods research design in which both quantitative and qualitative approaches were used. The results show that during the floods, households relocated to a nearby school which has solar powered tap water. However, the increased numbers at the camp resulted in a decline in the daily minimum water per capita of 20 L/c/d in about 60% of the households as each tap is meant to carter for 250 people only. Furthermore, the use of the taps was restricted to 8 h (6 am–2 pm during weekdays and 8 am–5 pm during weekends), which coincided with times for farming activities. Consequently, households had to fetch water from a village about 3 km away. However, the alternative water sources were also inaccessible during severe flooding conditions, with intermittent and contaminated water. Considering the health challenges associated with consumption of insufficient and contaminated water, this study proposes construction of more resilient deep boreholes for continuous water supply.

M. K. Joshua (✉) · R. K. Chirwa · C. Ngongondo
E. Mwathunga
Department of Geography and Earth Sciences, University of Malawi, Chancellor College, Zomba, Malawi
e-mail: cngongondo@cc.ac.mw; emwathunga@cc.ac.mw

M. Monjerezi
Department of Chemistry, University of Malawi, Chancellor College, Zomba, Malawi
e-mail: mmonjerezi@cc.ac.mw

R. Kasei
Climate Change and Food Security Department, University for Development Studies, Tamale, Ghana

Keywords

Human rights · Water · SDG 6.1 · Cyclone Idai · Flooding · Vulnerable · Households

11.1 Introduction

Water is a human right such that everyone is entitled "to sufficient, safe, acceptable, physically accessible and affordable water for personal and domestic uses" (UN Economic and Social

G. Nhamo, L. Chapungu (eds.), *The Increasing Risk of Floods and Tornadoes in Southern Africa*, Sustainable Development Goals Series, https://doi.org/10.1007/978-3-030-74192-1_11

Council 2002, p. 2). However, over 600 million people in developing countries do not have access to water (WHO and UNICEF 2014; Pullan et al. 2014). About half of the population without access live in the Sub-Saharan Africa (SSA) (Loftus 2015; Stoler 2012; WHO and UNICEF 2017b) and more specifically in rural areas (WHO and UNICEF 2013; Pullan et al. 2014). The population without access to drinking water is an annual averaged estimate, which does not take into account seasonal patterns and access patterns during extreme precipitation events. Many studies agree that climate risks such as floods have adverse effects on water infrastructure (Anderson et al. 2010; Rosinger 2018) with consequences on seasonal availability of water for domestic use (Tucker et al. 2014). Hence, if drinking water access estimates are inclusive for example of access during flooding conditions, then the proportions of households without access would be large. This means that the current proportions of water accessibility are underestimated. The situation is a challenge for SDG 6 in the sense that "ensuring universal access to safe and affordable drinking water by 2030, requires that we invest in adequate infrastructure, provide sanitation facilities and encourage hygiene at every level" (United Nations 2015, p. 14). At the same time, the continent is faced with frequent climate shocks which result in disasters such as floods.

Floods, considered the leading cause of natural disaster deaths globally (Doocy et al. 2013), are normally a result of the overflowing of the normal confines of a stream or other body of water or the accumulation of water over areas that are not normally submerged. Various characteristics of the climate system such as rainfall intensity, duration, amount, timing and drainage basin conditions can affect the spatial extent and intensity of floods (Bates et al. 2008; Kundzewicz et al. 2014). Tropical cyclones are among well-known triggers of flooding events globally (Whitfield 2012), due to the intense rainfall that normally accompanies them. Under a future warmer climate, the global frequency of tropical cyclones intensity is expected to decrease, although the more intense tropical cyclones are projected to increase (Sugi et al. 2017). Low-lying rural areas in tropical regions especially those in developing countries have been reported to be the most flood-prone areas, largely due to poverty, limited infrastructures, and access to resources and health-care services (Shimi et al. 2010; Kumar et al. 2016) and in general, to low levels of flood protection (Tanoue et al. 2016). In Bangladesh, Shimi et al. (2010) argued that over 70 million people are annually affected by tropical cyclone-induced floods, thereby affecting their access to basics such as safe water. Similar flood impacts on accessibility to basics such as water supply and transportation infrastructure have been reported by among others: Kirsch et al. (2012) in India, Chakraborty et al. (2016) in Bangladesh, Twisa and Buchroithner (2019) and in Tanzania, Mabuku et al. (2016) in Zambia and Eastern Zambezi region of Namibia and Bryndal et al. (2017) in the Kasiniczanka River basin in Poland.

Over southern Africa, tropical cyclones originating from around the Madagascar and Mozambique Channel in South West Indian Ocean often results in heavy rainfall and consequently flooding once they make landfall over the sub-continent. Rakotobe et al. (2016) estimated that the southern Africa region experiences an average of 3–4 cyclones in each cyclone year between November and April. Furthermore, Christensen et al. (2007) and Lasolle (2012) show an increasing frequency of cyclones in the South West Indian Ocean area associated with extreme precipitation events, specifically, floods in Southern Africa. The increased frequency of cyclones over Southern Africa has immediate negative impacts on: human health leading to loss of life; livelihoods and socio-economy; and destruction of vital infrastructure such as water supply and sanitation (Fitchett and Grab 2014; Rakotobe et al. 2016, Lequechane et al. 2020). For instance, the occurrence of the category 4 Cyclone Idai in March 2019 is reported to have affected 1.85 million people in Mozambique alone, which was accompanied with the destruction of about 122,700 houses, vital infrastructure for water supply, electricity, transportation, communication and power grid (Lequechane et al.

2020 and references therein). Cambaza et al. (2019) reported that when cyclone Cyclone Kenneth made landfall over Northern Mozambique in April 2019, over 374,000 were affected and needed about USD104 million worth of assistance. This was coupled with the destruction of over 40,000 houses and water, sanitation and vital infrastructure. Chanza et al. (2020) provide an account of the impacts of Cyclone Idai in the low-lying Risitu Valley in Zimbabwe based on the local peoples' experiences. The study by Chanza et al. (2020) reveals the extent of the losses of valuables, livelihoods, and vital infrastructure incurred by the people due to the Cyclone Idai, which were also compounded by some deficiencies in the disaster response by the local authorities. However, a key lesson from the Lisitu Valley study is the role that was played by local "heroes" and "Samaritans," in the disaster response process which helped in minimising threats to life and livelihoods.

Lumbroso et al. (2008) argued that although the magnitudes of the flooding events are a contributing factor, the livelihoods of the people in Southern Africa region are highly affected due to their high vulnerability compounded by socio-economic, institutional and political factors. The is further collaborated by Smiley and Hambati (2019) in their study on floods impact on drinking water access in Dar es Salaam, Tanzania, who found that it is the poorest who experience the worst from flooding. This is contrary to the situation in many developed countries which have resilient measures in place to mitigate from the effects of flooding on people's livelihoods (Tanoue et al. 2016). However, the development and implementation of effective and resilient mitigation measures should be holistic and informed by an understanding of the experiences of communities that have been affected by the flooding events (Badi 2017). Access to water is one critical challenge that most communities affected by floods face. However, there are very few scholarly works that have examined access to drinking water for rural households vulnerable to floods in Malawi and Southern Africa, a situation that is generally prevalent in many developing countries (Chambers 2012). This gap in scholarly works on access to drinking water can also be partly attributed to seasonal bias as most surveys are undertaken in the dry season (Wright et al. 2012; Kostyla et al. 2015). Yet, such estimates are central in providing the right picture of access and guiding policy and interventions addressing persistent inequalities in the SSA (Pullan et al. 2014). Sustainable Development Goal #6.1 aims to achieve universal access to drinking water (WHO and UNICEF 2017a; UNICEF and WHO 2019). Thus, drinking water should be accessible to all people at all times when needed and free from contamination (UN Economic and Social Council 2002). In addition, SDG 6 further calls for water to be within a collection distance of 30 min if the source is not located within the dwelling unit. On the other hand, such efforts in ensuring access to potable water within a 30-min collection time also contributes towards meeting SDG 13 targets, in the wake of increased flooding due to climate change, as well as SDG 3 with the risks to human life that are associated with floods. Hence, understanding how human right to water is met during extreme precipitation or flooding conditions is very important to contribute to the formulation of resilient coping strategies. In the wake of Cyclone Idai, some studies have been undertaken to assess the impacts the cyclone had on human livelihoods in some of the affected areas in Mozambique (e.g. Lequechane et al. 2020; Cambaza et al. 2019) and Zimbabwe (e.g. Chanza et al. 2020). However, documented literature on similar studies is very scanty for Malawi. Hence, this study contributes to this gap using the case of Magalasi Village in Chikwawa District in southern Malawi.

The main aim of this study was, therefore, to analyse the extent to which households affected by the 2019 flooding in the wake of Tropical Cyclone Idai had their human right to water respected in terms of sufficiency, safe and continuous supply of drinking water within 1000 m return distance. Thus, the study assessed flooding implications on drinking water access during and in the aftermath of Cyclone Idai in Chikwawa rural using Magalasi Village as a case study. Specifically, the study (i) analysed water sufficiency; (ii) explored safety of water consumed;

(ii) examined continuity of water supply in water sources; (iv) analysed distance travelled to water sources; and (v) assessed affordability of water during and in the aftermath of the Cyclone Idai. The assessment was based on the provisions of water as a human right in the Malawi National Water Policy (Government of Malawi 2007) and other legislations.

11.2 Study Area, Data and Methods

11.2.1 Description of the Study Area

The study was conducted in Magalasi Village, located in Chikwawa District in Southern Malawi (Fig. 11.1). Magalasi Village is located in the lower Shire River Valley, a floodplain which is one of the most flood-prone areas in Malawi (Winsemius et al. 2015). Chikwawa District was one of most heavily affected districts by both the 2015 and 2019 flooding events (Government of Malawi 2015, 2019b). The area lies in a typically semi-arid region, with average annual rainfall ranging from as low as 170–967.6 mm and maximum temperatures averaging 37.6 °C in December and 27.6 °C in July (Joshua et al. 2016). Although the national post disaster report includes effects of flooding on water and sanitation in the district, it only focuses on drinking water quality (Government of Malawi 2019b). Consequently, this leaves out gaps on most of the standards included in human right to water: sufficiency, physical accessibility, continuity of access and affordability. Such standards are therefore covered in this study.

11.2.2 Data Collection and Analysis

Data collection was undertaken in the immediate aftermath of Cyclone Idai in March 2019. At the time of this study, Magalasi Village had a total of 144 households with an average household size of 6.5, according to information sourced from the Key Informants (KIs). The study employed a case study design using mixed (qualitative and quantitative) research methods in triangulation to collect data and information for further in-depth analyses. The case study design was applied in this study as the approach is considered more ideal and flexible in evaluative mixed research studies (Yin 2014; Cresswel 2013) undertaken immediately following an event of interest (Cyclone Idai triggered flooding in this case) which affected one or more individuals in a specific community (Magalasi Village). The triangulated qualitative data collection methods included key informant interviews, Focus Group Discussions (FGDs), document analysis and observations. Through triangulation, the study harnessed diverse viewpoints (Olsen 2004) to cast more light on the peoples' predicament during the Cyclone Idai-induced floods.

The study firstly purposively sampled key informant interviews ($n = 6$) to get an overview of the village background and the state of vulnerability to flood using a checklist. These included village headmen, one elderly resident, a village development committee member holding a health-related position, one district/regional water officer and two officers from non-governmental organizations (NGOs) carrying out programmes related to disasters in the area. In addition, a household survey was undertaken to interview randomly selected household heads ($n = 34$) on their experiences related to water access during Cyclone Idai. The sampled households represented about one-quarter of the 144 households in Magalasi Village. Furthermore, FGDs ($n = 2$) were conducted with six participants in each group who included men and women, the elderly and the youth who were disaggregated by wealth group (either wealthier or poorer). Further information was collected through observations in the village and document analysis of official documents related to drinking water service delivery.

The data from the qualitative methods were thematically analysed. The quantitative data from the household survey were analysed in excel to produce summarised estimates on service delivery standards based on the National Water Policy and related legislation provisions.

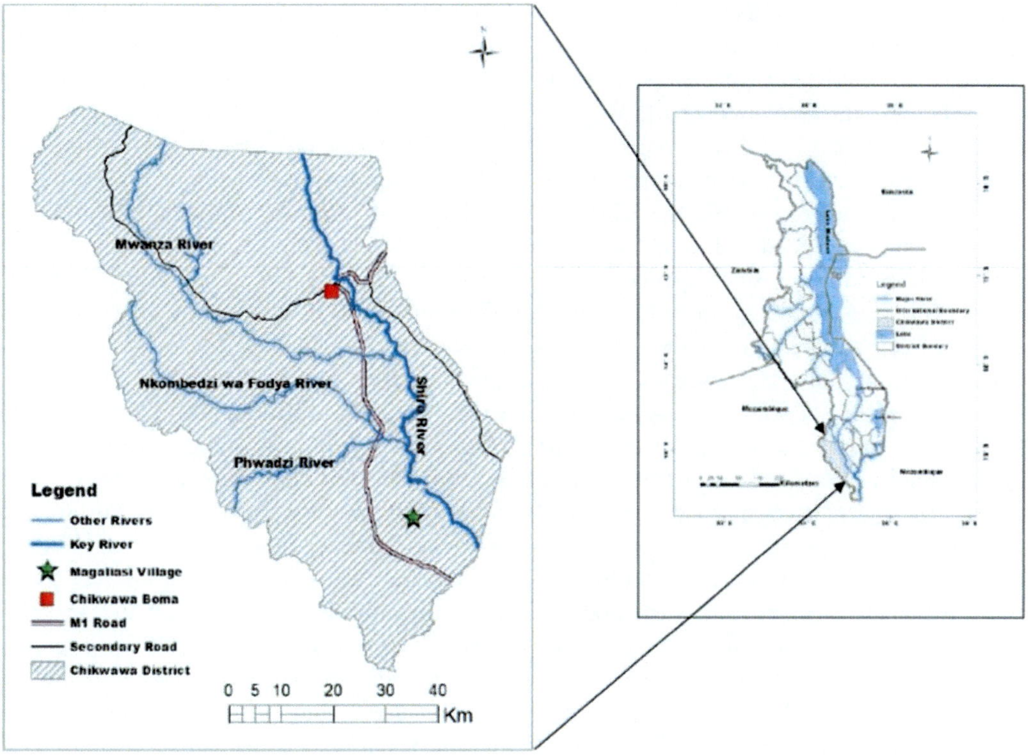

Fig. 11.1 Map of study area. Source: Authors, Data from Malawi Shapefiles in ArcGIS (2020)

11.3 Results and Discussions

11.3.1 Water Sufficiency and Continuity of Water Supply in Water Sources

This section analyses attainment of water sufficiency in Magalasi Village during and in the aftermath of Cyclone Idai using two criteria: household quantities used per capita per day and hours of water supply a day. The quantity per capita per day in each household was estimated by dividing the total drinking water collected by the household size.

The National water policy provides a very basic level of water delivery that drinking water should be available "within convenient distance" (Government of Malawi 2007, p. 5). This distance is defined as 1 km or 30 min round trip (Government of Malawi 2010b, 2010c). On the other hand, sufficient quantities is defined as a minimum of 20 L per capita per day (Government of Malawi 2010a). Using this criterion, about 44% of the households in Magalasi Village had sufficient quantities (≥20 L/c/d) and over half (56%) fetched insufficient water (<20 L/c/d) during the flooding conditions. In terms of continuity, water service delivery is expected to be continuous (24 h and 7 days a week). However, using these criteria, the focus group discussions (FGDs) revealed that drinking water supply was not continuous in the village and was limited to 8 h per day. This likely affected sufficiency of water collected.

11.3.2 Safety of Water

The study results show that the main source of water for all households during normal times is solar-powered piped water at Mchenga Primary School, which is within the village. However, the people do not access the water at any time as they desire because access rules restrict access to 8 h

(6 am–2 pm during weekdays and 8 am–5 pm during weekends). During the closure times, households fetch drinking water from a nearby village about 3 km away (radius). During the flooding conditions, all households from Magalasi Village and the surrounding villages moved to and camped at Mchenga Primary School. The taps remained their main source of water, and the access rules were still note relaxed. However, a borehole in the neighbouring village which has been an alternative source was also inaccessible as the route to the water source was impassable. Hence, during the closure times, households were forced to use water from unsafe sources. Studies show that intermittent piped water supply system is often prone to microbial contamination (Kumpel and Nelson 2013, 2016). This means that that almost all the water consumed in Magalasi Village during the flooding conditions was unsafe as the piped water may not be always safe from contaminants.

The National Water Policy emphasises provision "of water of acceptable quality" (Government of Malawi 2007 p3) using improved water sources as a proxy to drinking water quality. The improved water sources include piped water, boreholes, protected shallow wells and springs (Government of Malawi 2010c). However, many studies have criticised this measure because many improved water sources have been associated with contamination. Additionally, the World Health Organisation now recommends use of a water safety plan, a holistic framework for maintaining good quality water as it guides identification and management of hazards (WHO 2017). However, at the time of the study, the Government of Malawi was still using the proxy to drinking water service delivery (Government of Malawi 2019a). In addition, according to the district/regional water officer (key informant), the country was yet to have a national water safety plan. It did not also have a water safety plan for the rural areas. In this regard, the study used the main sources of water to assess safety of water used during the flooding conditions.

With regard to policy, the study established that the Malawi National Water Policy (Government of Malawi 2007) has specific provisions for managing disasters. Two of the National Water Policy objectives related disaster management aims to: "ensure timely provision of potable water and sanitation for vulnerable communities especially children, women and persons with disabilities during water-related disasters;" and "to provide basic requirements of potable water supply to all affected areas" (Government of Malawi 2007, p. 15). The management strategies include "providing emergency water supply systems including boreholes, hand-dug wells and portable treatment units" (Government of Malawi 2007). However, the study established that these emergency water supply strategies were not delivered by service providers during the flooding conditions. Furthermore, it was learnt from the FGDs that water-borne diseases (diarrhoea and cholera) were widely spread during the flooding season. Consumption of the unsafe water therefore explains this finding, confirming wide recognition of health-related challenges resulting from drinking unsafe water (Stoler 2012; Tarrass and Benjelloun 2012; Pullan et al. 2014).

11.3.3 Distance Travelled to Water Source

Physical accessibility is measured in two ways: Collection time within 1 km or 30 min round trip (Government of Malawi 2010b, 2010c) and number of water users per water source. According to the second criteria, one borehole should serve a maximum number of 250 people within 500 m radius (Government of Malawi 2010c) and a communal tap should serve a maximum of 250 people within the radius of 500 m (Government of Malawi 2010b, 2010c). The results show that during flooding conditions, households relocated and camped at a nearby school (Mchenga) which hosted 1100 people (Government of Malawi 2019c). According to the FGDs and participant observation, the school has solar powered pumped tap water fitted with three taps (Figs. 11.2 and 11.3). This means that each tap water in

Fig. 11.2 Solar panel and a tank providing water at Mchenga School in Magalasi Village. Source: Authors Fieldwork (2019)

Magalasi Village is expected to serve 360 people. There are 144 households with 936 people in Magalasi Village, suggesting that even during normal times, the three taps are still inadequate for the community.

The situation is therefore worse during the flooding conditions. Using the population threshold for a single tap, numbers at the camp exceeded the maximum number (250 people) required for a single tap. With a population of 1100 people, each tap served about 367 people, which is more than the recommended requirement. Although the taps were within 500 m radius from the camp, long waiting times/queues meant long collection time per trip. Furthermore, as stated in the previous section, the study established that the use of the taps was restricted to 8 h (6 am–2 pm during weekdays and 8 am–5 pm during weekends). The weekdays' access times coincide with times for farming activities. This situation consequently forced households to fetch water from a borehole in Phazi Village, about 6kms return trip (Fig. 11.4). This distance is beyond the recommended 1 km return trip. Additionally, the alternative option also became inaccessible during severe flooding conditions. Consequently, the daily minimum per capita of 20 L/c/d declined in about 60% of the households. Thus, physical accessibility standard was compromised during the flooding conditions.

11.3.4 Affordability of Drinking Water

This section analyses affordability of drinking water in Magalasi Village during the flooding conditions. The National Water Policy aspires affordable water to all users at all times (Government of Malawi 2007). All the household respondents and FGD indicated that the water was affordable to all because they did not pay user fees for water facilities. However, this finding relates to direct costs. The households incurred indirect costs to access water due to the long queues involved, thereby limiting time for other productive activities. Other indirect costs included money spent on water-borne diseases treatment and travel difficulties to water sources. These findings concur with study results in Guatemala on health-related implications/costs during extreme climate events, though this earlier study focused on drought conditions (Caruso et al. 2015; Sorenson et al. 2011). Hence, if all the indirect costs were considered, the water was unaffordable in most of the households.

Fig. 11.3 The three taps at Mchenga primary school closed, waiting for the time to be opened. Source: Authors Fieldwork (2019)

Fig. 11.4 Women from Magalasi Village queuing for water in Phazi Village. Source: Authors Fieldwork (2019)

11.4 Conclusions and Recommendations

Tropical cyclones affect community livelihoods and vital infrastructure such as water because of the floods that they are normally associated with. Resilient measures for coping with water related challenges that rural communities face in the aftermath of floods and other disasters can be developed and affected implemented by learning from the experiences of affected communities. This study assessed the effects of the 2019 Cyclone Idai-induced flooding on drinking water for households affected by flooding using mixed research methods. A case study of the flood-prone Magalasi Village in Chikwawa District in Southern Malawi was used. During the flooding conditions during and after Tropical Cyclone Idai, people from Magalasi Village who sought refuge at Mchenga School Camp had to deal with insufficient quantities of contaminated water, intermittently and from afar. The study findings show that the Camp's three (3) solar-powered taps were overstretched beyond the recommended number of 250 people per tap as they had to carter for the water needs of over 1100 people (360 per tap). In addition, the water supply access restrictions to the taps to 8 h per day affected the people especially during the weekdays when they also needed to attend to other livelihood chores such as farming. Consequently, the people were compelled to walk long distances (more than 3 km away) in search of alternative water sources. However, these sources were also not always accessible and in some cases also not safe, thereby exposing the people to risks of waterborne diseases due to the contaminated water. Although some of these challenges are prevalent during normal periods due to the semi-arid conditions, the situation became even worse during Cyclone Idai-induced flooding conditions.

Considering the health challenges associated with consumption of insufficient and contaminated water, this study proposes construction of secured deep boreholes which are less affected by the climate extremes and are likely to provide continuous water supply. Further, in cases where existing water infrastructure fail to carter for the water needs of all those affected by floods, relief interventions should include drilling of boreholes within the camp. In addition, there is need to provide basic at source water treatment kits to flood-affected communities depending on alternative water sources to reduce the risk of water-borne diseases. Furthermore, this study concludes that attainment of SDG 6.1 remains threatened until drinking water access challenges during flooding conditions receive adequate attention from policy makers and service providers.

References

Anderson, S., Gundel, S., & Vanni, M. (2010). *The impacts of climate change on food security in Africa: A synthesis of policy issues for Europe*. IIED.

Badi, G. (2017). *Enhancing UK flood mitigation measures through local communities flood knowledge*. Leeds: Leeds Beckett University.

Bates, B. C., Kundzewicz, Z. W., Wu, S. & Palutikof, J. P. (Eds). (2008). Climate change and water. *Technical paper of the Intergovernmental Panel on Climate Change*, pp 210. Geneva: IPCC Secretariat.

Bryndal, T., Franczak, P., Kroczak, R., Cabaj, W., & Kołodziej, A. (2017). The impact of extreme rainfall and flash floods on the flood risk management process and geomorphological changes in small Carpathian catchments: A case study of the Kasiniczanka river (outer Carpathians, Poland). *Natural Hazards, 88*, 95–120. https://doi.org/10.1007/s11069-017-2858-7.

Cambaza, E., Mongo, E., Anapakala, E., Nhambire, R., Singo, J., & Machava, E. (2019). Outbreak of cholera due to cyclone Kenneth in northern Mozambique. *International Journal of Environmental Research and Public Health, 16*, 2925. https://doi.org/10.3390/ijerph16162925.

Caruso, B. A., Sevilimedu, V., Fung, I. C.-H., Patkar, A., & Baker, K. K. (2015). Gender disparities in water, sanitation, and global health. *Lancet, 386*, 650–651.

Chakraborty, T. P., Enamul Kabir, A. H. M., & Ghosh, G. C. (2016). Impact and adaptation to cyclone AILA: Focus on water supply, sanitation and health of rural coastal community in the South West Coastal Region of Bangladesh. *Journal of Health and Environmental Research, 2*(3), 13–19. https://doi.org/10.11648/j.jher.20160203.11.

Chambers, R. (2012). Seasonal poverty: Integrated, overlooked and therefore opportunity. In S. Devereux, R. Sabates-Wheeler, & R. Longhurst (Eds.), *Seasonality, rural livelihoods and development*. Earthscan: London and New York.

Chanza, N., Siyongwana, P. Q., Williams-Bruinders, L. Gundu-Jakalasi, V., Mudavanhu, C., Sithole V. B., & Manyani, A. (2020). Closing the gaps in disaster management and response: drawing on local experiences with Cyclone Idai in Chimanimani, Zimbabwe. *International Journal of Disaster Science, 11*, 655–666. https://doi.org/10.1007/s13753-020-00290-x

Christensen, J. H., Hewitson, B., Bussuioc, A., Chen, A., Gao, X., Held, I., Jones, R., Koli, R. K., Kwon, W. T., Laprise, R., Rueda, V. M., Mearns, L., Menéndez, C. G., Räisänen, J., Rinke, A., Sarr, A., & Whetton, P. (2007). Regional climate projections. In S. Solomon, D. Qin, M. Manning, Z. Chen, M. Marquis, & K. B. Averyt (Eds.), *Climate change 2007: The physical science basis. Contribution of working group 1 to the fourth assessment report of the Intergovernmental Panel on Climate Change*. Cambridge: Cambridge University Press.

Cresswel, J. W. (2013). *Research design: Qualitative, quantitative and mixed research methods* (4th ed.). Sage Publications: London.

Doocy, S., Daniels, A., Murray, S., & Kirsch, T. S. (2013). The human impact of floods: A historical review of events 1980-2009 and systematic literature review. Edition 1. *PLOS Currents Disasters*. https://doi.org/10.1371/currents.dis.f4deb457904936b07c09daa98ee8171a.

Fitchett, J. M., & Grab, S. W. (2014). 66-year tropical cyclone record for South-East Africa: Temporal trends in a global context. *International Journal of Climatology, 34*, 3604–3615.

Government of Malawi. (2007). *National Water Policy* (2nd ed.). Lilongwe: Ministry of irrigation and water development.

Government of Malawi. (2010a). *Implementation guidelines for rural water supply and sanitation*. Lilongwe: Ministry of Irrigation and Water Development.

Government of Malawi. (2010b). *Implementation manual for piped and point water supply systems*. Lilongwe: Ministry of Irrigation and Water Development.

Government of Malawi. (2010c). *Water supply and sanitation district operational manual*. Lilongwe: Ministry of Irrigation and Water Development.

Government of Malawi. (2015). *Malawi 2015 floods post disaster needs assessment report*. Department of Disaster Management Affairs (DoDMA): Lilongwe.

Government of Malawi. (2019a). *2018 Malawi housing and population census: Main report*. Zomba: National Statistics Office.

Government of Malawi. (2019b). *Malawi 2019 floods post disaster needs assessment (PDNA)*. Lilongwe: Office of the President and Cabinet.

Government of Malawi. (2019c). *Ngabu ADD report*. Chikwawa: Ngabu ADD.

Joshua, M. K., Ngongondo, C., Chipungu, F., Monjerezi, M., Liwenga, E., Majule, A. E., Stathers, T., & Lamboll, R. (2016). Climate change in semi-arid Malawi: Perceptions, adaptation strategies and water governance. *Jàmbá: Journal of Disaster Risk Studies, 8*(3), a255. https://doi.org/10.4102/jamba.v8i3.255.

Kirsch, T. D., Wadhwani, C., Sauer, L., Doocy, S., & Catlett, C. (2012). Impact of the 2010 Pakistan floods on rural and urban populations at six months. *PLOS Currents Disasters*. https://doi.org/10.1371/4fdfb212d2432.

Kostyla, C., Baina, R., Cronk, R., & Bartram, J. (2015). Seasonal variation of fecal contamination in drinking water sources in developing countries: A systematic review. *Science of the Total Environment, 514*, 333–343.

Kumar, V., Cheng, S. Y. C., & Singh, A. K. (2016). Impact of flood on rural population and strategies for mitigation: A case study of Darbhanga district, Bihar state, India. *Contemporary Rural Social Work Journal, 8*(1).

Kumpel, E., & Nelson, K. L. (2013). Comparing microbial water quality in an intermittent and continuous piped water supply. *Water Research, 47*(14), 5176–5188.

Kumpel, E., & Nelson, K. L. (2016). Intermittent water supply: Prevalence, practice, and microbial water quality. *Environmental Science & Technology, 50*, 542–553.

Kundzewicz, Z. B., Kanae, S., Seneviratne, S. I., Handmer, J., Nicholls, N., Peduzzi, P., Mechler, R., Bouwer, L. M., Arnell, N., Mach, K., Muir-Wood, R., Brakenridge, G. R., Kron, W., Benito, G., Honda, Y., Takahashi, K., & Sherstyukov, B. (2014). Flood risk and climate change: Global and regional perspectives. *Hydrological Sciences Journal, 59*(1), 1–28. https://doi.org/10.1080/02626667.2013.857411.

Lasolle, D. (2012). *SADC policy paper on climate change: Assessing the policy options for SADC member states*. SADC Policy Analysis and Dialogue Programme.

Lequechane, J. M., Mahumane, A., Chale, F., Nhabomba, C., Salomão, C., Lameira, C., Chicumbe, S., & Baltazar, C. M. (2020). Mozambique's response to cyclone Idai: How collaboration and surveillance with water, sanitation and hygiene (WASH) interventions were used to control a cholera epidemic. *Infectious Diseases of Poverty, 9*(68). https://doi.org/10.1186/s40249-020-00692-5.

Loftus, A. (2015). Water (in)security: Securing the right to water. *The Geographical Journal, 181*, 350–356.

Lumbroso, D., Ramsbottom, D., & Spaliveiro, S. (2008). Sustainable flood risk management strategies to reduce rural communities' vulnerability to flooding in Mozambique. *Journal of Flood Risk Management, 1*, 34–42.

Mabuku, M. P., Senzanje, A., Mudhara, M., Jewitt, G., & Mulwafu, W. (2016). Rural households' flood preparedness and social determinants in Mwandi district of Zambia and eastern Zambezi region of Namibia. *International Journal of Disaster Risk Reduction*. https://doi.org/10.1016/j.ijdrr.2018.03.014.

Olsen, W. (2004). Triangulation in social research: Qualitative and quantitative methods can really be mixed. In Holborn, N. (Ed), *Development in Sociology*. Ormskirk: Causeway Press.

Pullan, R. L., Freeman, M. C., Gething, P. W., & Brooker, S. J. (2014). Geographical inequalities in use of improved drinking water supply and sanitation across sub-Saharan Africa: Mapping and spatial analysis of cross-sectional survey data. *PLoS Medicine, 11*, 1–17.

Rakotobe, Z. L., Harvey, C., Rao, N. S., Dave, R., Rakotondravelo, J. C., Randrianarisoa, J., Ramanahadray, S., Andriambolantsoa, R., Razafimahatratra, H., Rabarijohn, R. H., Rajaofara, H., Rameson, H., & MacKinnon, J. L. (2016). Strategies ofsmallholder farmers for coping with the impacts of cyclones: A case study from Madagascar. *International Journal of Disaster Risk Reduction, 17*, 114–122.

Rosinger, A. Y. (2018). Household water insecurity after a historic flood: Diarrhea and dehydration in the Bolivian Amazon. *Social Science & Medicine, 197*, 192–202.

Shimi, A. C., Parvin, G. A., Biswas, S., & Shaw, R. (2010). Impact and adaptation to flood: A focus on water supply, sanitation and health problems of rural community in Bangladesh. *Disaster Prevention and Management, 19*(3), 298–313.

Smiley, S. L., & Hambati, H. (2019). Impacts of flooding on drinking water access in Dar Es Salaam, Tanzania: Implications for the sustainable development goals. *Journal of Water, Sanitation and Hygiene for Development, 9*(2), 392–396. https://doi.org/10.2166/washdev.2019.168.

Sorenson, S. B., Morssink, C., & Campos, P. A. (2011). Safe access to safe water in low income countries: Water fetching in current times. *Social Science & Medicine, 72*, 1522–1526.

Stoler, J. (2012). Improved but unsustainable: Accounting for sachet water in post-2015 goals for global safe water. *Tropical Medicine & International Health, 17*, 1506–1508.

Sugi, M., Murakami, H., & Yoshida, K. (2017). Projection of future changes in the frequency of intense tropical cyclones. *Climate Dynamics, 49*, 619–632. https://doi.org/10.1007/s00382-016-3361-7.

Tanoue, M., Hirabayashi, Y., & Ikeuchi, H. (2016). Global-scale river flood vulnerability in the last 50 years. *Scientific Reports, 6*, 36021. https://doi.org/10.1038/srep36021.

Tarrass, F., & Benjelloun, M. (2012). The effects of water shortages on health and human development. *Perspectives in Public Health, 132*, 240–244.

Tucker, J., Macdonald, A., Coulter, L., & Calow, R. C. (2014). Household water use, poverty and seasonality: Wealth effects, labour constraints, and minimal consumption in Ethiopia. *Water Resources and Rural Development, 3*, 27–47.

Twisa, S., & Buchroithner, M. F. (2019). Seasonal and annual rainfall variability and their impact on rural water supply services in the Wami River Basin, Tanzania. *Water, 11*, 2055. https://doi.org/10.3390/w11102055.

UN Economic and Social Council. (2002). *General Comment No. 15. The right to water (arts. 11 and 12 of the International Covenant on Economic, Social and Cultural Rights)*. United Nations Economic and Social Council, Committee on Economic, Social and Cultural Rights.

United Nations. (2015). *Transforming our world: The 2030 agenda for sustainable development*. New York: United Nations.

UNICEF, & WHO. (2019). *Progress on household drinking water, sanitation and hygiene 2000-2017: Special focus on inequalities*. New York: United Nations Children's Fund (UNICEF) and World Health Organization (WHO).

Whitfield, P. H. (2012). Floods in future climates: A review. *Journal of Flood Risk Management, 5*(4), 336–365.

WHO. (2017). *Safely managed drinking water - thematic report on drinking water 2017*. Geneva: World Health Organisation.

WHO, & UNICEF. (2013). *Progress on sanitation and drinking-water, 2013 update*. Geneva: World Health Organization.

WHO, & UNICEF. (2014). *Progress on sanitation and drinking-water, 2014 update*. Geneva: World Health Organization.

WHO & UNICEF (2017a). *Meeting report: WHO/UNICEF joint monitoring Programme for water supply and sanitation task force on monitoring inequalities for the 2030*. Sustainable Development Agenda. WHO and UNICEF.

WHO, & UNICEF. (2017b). *Progress on drinking water, sanitation and hygiene: 2017 update and SDG baselines*. Geneva: World Health Organization (WHO) and the United Nations Children's Fund (UNICEF).

Winsemius, H. C., Jongman, B., Veldkamp, T. I. E., Hallegatte, S., Bangalore, M. W., & Ward, J. P. (2015). *Disaster risk, climate change, and poverty: Assessing the global exposure of poor people to floods and droughts*. Washington, DC: World Bank Group.

Wright, J. A., Yang, H., & Walker, K. (2012). Do international surveys and censuses exhibit 'dry season' bias? *Population, Space Place, 18*, 116–126.

Yin, R. K. (2014). *Case study research design and methods* (5th ed., 282 pages). Thousand Oaks, CA: Sage.

12 Vulnerability of Settlements to Floods in South Africa: A Focus on Port St Johns

Saul Ngarava, Leocadia Zhou, Abbyssinia Mushunje, and Petronella Chaminuka

Abstract

The objective of the study was to highlight the flood vulnerabilities exhibited in different settlements in South Africa, using the case of the Port St Johns. The study utilised a cross-sectional survey design and a purposive sample of 195 respondents. A multiplicative flood vulnerability index (FVI) was used to analyse the data. The results showed that the causes of flood vulnerabilities included environmental degradation, poor drainage, poor settlement planning, location of settlements on wetlands, slopes and waterways. This was exhibited at varying degrees in the high-density, medium-density, low-density, rural, rural (urban)-like and informal settlements. Informal settlements identified were located on waterways, while the rural settlements identified were on degraded land, as well as on wetlands and slopes. An FVI of 0.39 was obtained indicating that the study area was vulnerable to flooding. Furthermore, physical FVI had the highest loading of 0.47, followed by environmental FVI (0.40), social FVI (0.35) and economic FVI (0.33), respectively. Households in informal settlements showed high physical and social vulnerabilities, and were characterised by blocked drainages, inadequate storm water infrastructure and a high degree of social differentiation, which increased their vulnerability. Rural settlements exhibited high economic and environmental vulnerabilities, and were characterised by a high unemployment rate; low incomes; lack of alternative livelihoods; location on slopes and flood-prone areas, increasing their vulnerability to flooding. The chapter concludes that Port St Johns is vulnerable to flooding, with high physical and environmental vulnerabilities.

Keywords

Settlement · Flood vulnerability index · South Africa

12.1 Introduction

The United Nations Sustainable Development Goal (SDG) 11B endeavours to improve settlements that adopt and implement integrated poli-

S. Ngarava (✉) · L. Zhou
Risk & Vulnerability Science Centre,
University of Fort Hare, Alice, South Africa
e-mail: SNgarava@ufh.ac.za; LZhou@ufh.ac.za

A. Mushunje
Department of Agricultural Economics & Extension,
University of Fort Hare, Alice, South Africa
e-mail: AMushunje@ufh.ac.za

P. Chaminuka
Economic Analysis Unit, Agricultural Research
Council, Pretoria, South Africa
e-mail: ChaminukaP@arc.agric.za

G. Nhamo, L. Chapungu (eds.), *The Increasing Risk of Floods and Tornadoes in Southern Africa*,
Sustainable Development Goals Series, https://doi.org/10.1007/978-3-030-74192-1_12

cies relating to disaster resilience by 2020, as well as to develop and implement holistic disaster risk management in line with the Disaster Risk Reduction 2015–2030 Sendai Framework (UN 2020). In achieving this goal, there is need to integrate ecological, social and economic dimensions, due to the inseparability of natural and human systems (Jiménez-Aceituno et al. 2019). Human settlements present unique systems based on their economic, social and ecological dimensions. Any disruptions to such systems due to naturally occurring disasters such as flooding present vulnerability and compromises achievement of SDG 11.

From another perspective, SDG 13.1, which seeks to strengthen resilience and adaptive capacity to natural disasters, has a bearing on human settlements (UN 2020). Flooding is one of the most devastating and inevitable natural disasters whose intensity, frequency and geographical spread have increased over time (Berezi et al. 2019). Floods are a result of a large amount of water overflow beyond its normal confines (UN-SPIDER 2017). They are the most prevalent natural risk due to climate change, and are anticipated to occur more frequently (Nasiri and Shahmohammadi-kalalagh 2013). The exposure, susceptibility, resilience and adaptive capacity to natural disasters such as flooding (SDG 13) of human settlements (SDG 11) are put to the fore, given the disparities within and between different settlements, making assessments of these capacities essential in understanding variations in vulnerabilities (de Andrade and Szlafsztein 2018).

According to Bruwer et al. (2017), South Africa is primarily exposed to weather-related natural hazards, with flooding and fires topping the list. The country experienced more than 1800 flood events between the period 1647 and 2013 (Bruwer et al. 2017). Le Roux et al. (2019) highlighted that 5% of settlements in the country are at extreme risk of increased flooding, whilst 16% are at high risk, and 33% and 46% at medium and low risk, respectively. South Africa has an annual risk of flooding at 83.3%, and the population vulnerability is high due to economic factors and geographical location (Zuma et al. 2012). The most vulnerable provinces in South Africa are Limpopo, North-West and Eastern Cape (Le Roux and Van Huyssteen 2010). However, this does not spare other South African provinces from vulnerability and exposure to floods (Munyai 2017).

The April 2019 flooding in Port St Johns, Eastern Cape Province, was overshadowed by the flooding that also took place in Durban, KwaZulu-Natal Province, South Africa, during the same time. This period witnessed 85 deaths in Durban, several injuries, between 1000 and 1600 people displaced and more than R1 billion in infrastructural damage, where 190 mm of rainfall in 24 h was recorded in Port St Johns, a record amount in 33 years (Niehaus 2019). Low-lying rural and informal settlements such as Green Farm were left inaccessible due to the flash-flood within the area which caused the Mzimvubu river to burst its banks, combined with the topography, and high oceanic tides pushing the river water. Hundreds of residents, mainly children and women, were evacuated and accommodated at the town hall and churches (Daniel 2019; DCG 2019; News24 2019; SABC 2019).

Despite increased knowledge about vulnerability, flood risk is still widespread, raising questions about the efficacy and effectiveness of vulnerability assessments on flood mitigation and adaptation (Khan 2012; Nasiri and Shahmohammadi-kalalagh 2013). There are limited studies on vulnerability of communities in South Africa to climate change-induced flooding, with few focusing at local level in rural areas (Munyai 2017; Musyoki et al. 2016; Zuma et al. 2012). Platforms such as the Greenbook (Le Roux et al. 2019) have tried to consolidate this fragmented data and information through user interface platforms. However, consolidated documentation is scanty. Flood disasters are growing in frequency, and their destructive nature worldwide is due to a variety of environmental and human factors (Musyoki et al. 2016). The need to understand vulnerability of settlements to the floods cannot be overemphasised. Thus, this chapter examines settlement vulnerabilities to climate change-induced flooding, focusing on

Port St Johns, which experienced flooding in April, 2019.

12.2 Literature Review

12.2.1 Vulnerability and its Measurements

Vulnerability refers to "the characteristics and circumstances of a community, system or asset that makes it susceptible to damaging effects of a hazard" (UNISDR 2009). Vulnerability is embedded in the concept of risk, where Risk × Hazard = Vulnerability (Fernandez et al. 2016). Assessing vulnerability is a prerequisite for understanding society's exposure to natural hazards and is a requirement for designing effective risk and vulnerability reduction designs (Hamidi et al. 2020). Flood vulnerability is the major element in flood risk assessment and damage evaluation, with various methods in assessing flood vulnerability, providing guides in developing indicators in quantifying various types of vulnerabilities (Hamidi et al. 2020). These have transcended those that conceive vulnerability as a social or biophysical phenomenon, to those that emphasise its integrative nature (Mwale et al. 2015).

Cutter et al. (2008) developed the hazards of place model where they indicate that vulnerability is the end product of both social and biophysical dimensions, and is spatially different. Bohle (2001) utilised the double structure of vulnerability model where vulnerability had two aspects, i.e. external (hazard) and internal (capacity). Turner II et al. (2003) divided vulnerability into exposure, response and adaptation in a socio-ecological context. According to the pressure and release model, the progression of vulnerability can be explained through risk; which is a product of a natural hazard and vulnerability (Blaikie et al. 2004). Birkmann (2006) and Bogardi and Birkmann (2005) conceptualised the Borgadi, Birkman and Cardona (BCC) framework indicating that vulnerability should be linked in a socio-economic and environmental sphere, and is assessed through exposure, susceptibility and coping strategies. Birkmann (2006) also identified that vulnerability can be measured in five spheres, namely environmental, institutional, economic, social and physical. The MOVE (Methods for the Improvement of Vulnerability Assessment in Europe) framework was proposed by Birkmann et al. (2013), conceptualising the multi-dimensionality of vulnerability and integrating exposure, susceptibility and resilience as key, under various thematic dimensions such as institutional, environmental, socio-economic and physical.

Flood vulnerability depends on various factors which include political, social, economic and environmental (Nasiri and Shahmohammadi-kalalagh 2013), making it difficult to measure as these are not homogenous, spatially and temporally (Pandey et al. 2010). However, vulnerability assessments have fallen into four categories, namely vulnerability index system, vulnerability curve method, disaster loss and modelling methods (Dapeng Huang et al. 2012; Nasiri et al. 2016; Nasiri and Shahmohammadi-kalalagh 2013).

Even though various definitions of vulnerability can be established, Fernandez et al. (2016) and Tate (2011) summarised them into three classes, namely exposure, sensitivity (susceptibility or resistance) and adaptive capacity (recovery potential or resilience). Furthermore, four dimensions are required in vulnerability assessment, namely (a) the environmental dimension—potential impacts on the natural environment and ability of ecosystems to cope and recover from hazard impacts; (b) the social dimension—the presence of human beings, individuals or communities, and their capacity to cope, resist and recover from hazard impacts; (c) the economic dimension accounting for the potential impacts of hazards on economic assets; and (d) the physical dimension—potential of physical impact on the built environment (Fernandez et al. 2016; Mwale et al. 2015).

12.2.2 Settlement Exposure, Susceptibility and Resilience to Flooding

Exposure is the possibility of people and/or physical items to be affected (Hamidi et al. 2020). Nasiri and Shahmohammadi-kalalagh (2013) identified exposure as the scope that human settlements and people's lives are positioned in within flood-risk areas. It is the "presence of people; livelihoods; environmental services and resources; infrastructure; or economic, social and cultural assets in places that could be adversely affected" (Mwale et al. 2015). Exposure indicators mainly relate to countable or density of exposed assets (de Andrade and Szlafsztein 2018; Röthlisberger et al. 2017). Other indicators include climate, flood duration, flood frequency, topography, experience, population, land use and urbanisation, amongst others (Hashim et al. 2018; Jha and Gundimeda 2019; Karmaoui and Balica 2019).

Sensitivity is the "predisposition of elements at risk to suffer harm" (Mwale et al. 2015). It is when a system has exposed elements due to weak socio-economic conditions, increasing the likelihood of being harmed by external shock (Balica et al. 2009; Hamidi et al. 2020). Sensitivity indicators refer to quantitative socio-economic factors such as population, educational level, gender, poverty, attitudes, building type, household income and access to services, amongst others (de Andrade and Szlafsztein 2018; Hummel et al. 2016).

Resilience refers to the adaptive capacity of a community or social-ecological system in response to an identified hazard (Birkmann et al. 2013; Hamidi et al. 2020; Nasiri and Shahmohammadi-kalalagh 2013). It is "the ability of a system and its component parts to anticipate, absorb, accommodate, or recover from the effects of a hazardous event" (Mwale et al. 2015). Indicators for adaptive capacity include educational levels, employment status, dependency ratio, income, income sources, assets, water storage facilities, tenure, insurance, risk perception, the capacity to shelter, emergency services, access to electricity, collective action, social networking, communication, warning systems, experience, flood recovery time, drainage system, land use, awareness and preparedness, and can be qualitative in measurement (Hamidi et al. 2020; Jha and Gundimeda 2019; Karmaoui and Balica 2019).

12.2.3 Conceptual Framework

The study adapted the expanded vulnerability framework as utilised by Hashim et al. (2018) as well as Turner II et al. (2003), where the local setting such as a village constitutes a unit of analysis (Fig. 12.1). The framework is adaptable to qualitative and quantitative designs. Vulnerability exists in a multi-faceted human-environment system at different temporal, functional and spatial scales. Interaction between exposures, sensitivities and resilience factors potentially affects the existing system, providing insights on the vulnerability.

In this instance, as depicted in Fig. 12.2, interactions between exposures, sensitivities and resilience in an existing system provide insights into the vulnerability of an area (Hashim et al. 2018; Turner II et al. 2003).

12.3 Materials and Methods

This section focuses on the methods that were used in collecting the data that was used in the study. It comprises the description of the study area, the research design as well as the analytical methods utilised.

12.3.1 Study Area

The study was carried out in Port St Johns (PSJ), located in the Eastern Cape (EC) Province, South Africa (Fig. 12.3).

Port St Johns (PSJ) is one of five local municipalities in the OR Tambo District Municipality (DM), with a 55 km shoreline in the Eastern Cape Province, South Africa (OR Tambo DM 2016). Port St Johns is a category B municipality, mean-

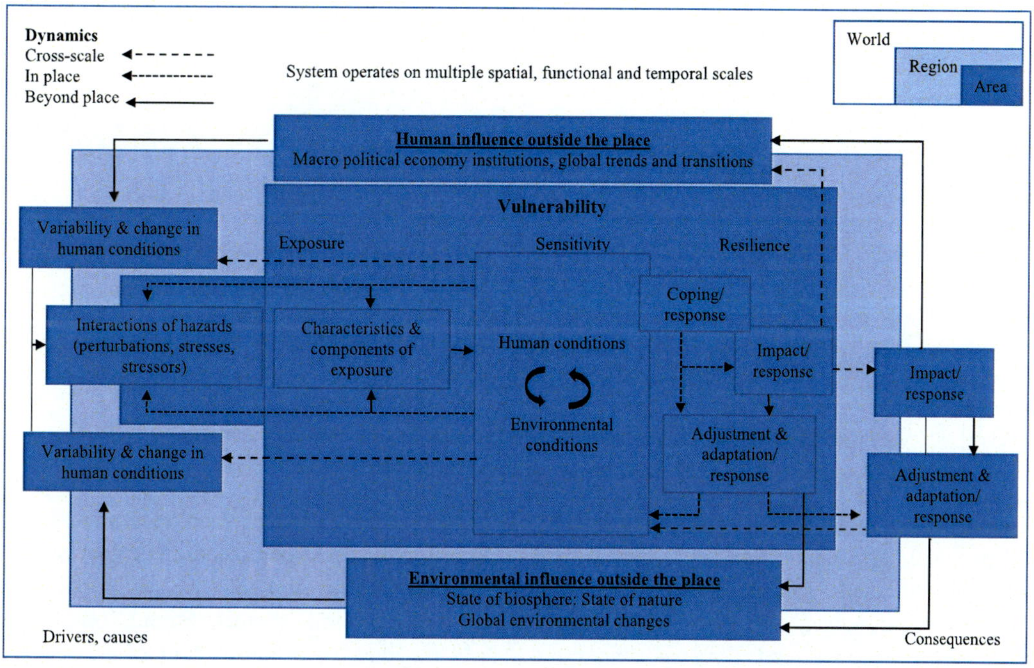

Fig. 12.1 Interaction among available factors within the expanded vulnerability framework. Source: Adapted from Hashim et al. (2018) and Turner II et al. (2003)

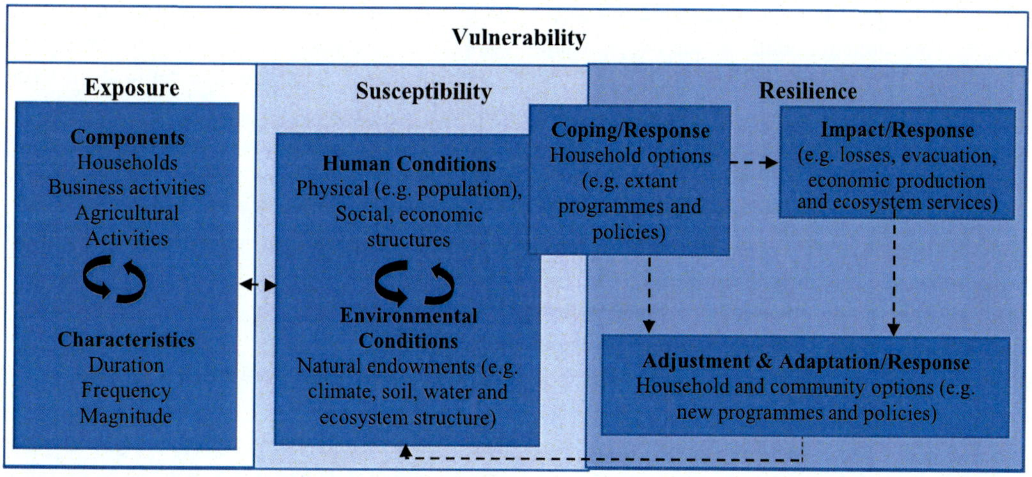

Fig. 12.2 Details of vulnerability factors in the vulnerability framework. Source: Hashim et al. (2018) and Turner II et al. (2003)

ing it shares municipality executive and legislative authority (Worklaw 2019). PSJ comprises urban settlements and approximately 130 rural areas/villages. Ingquza Hill Local Municipality borders it in the north and Nyanden Local Municipality to the south. PSJ has 20 wards, with a total area of 1239 km^2 (OR Tambo DM 2016). PSJ Local Municipality has a population of 168,000 (projected to be 177,000 in 2021), at 11.44% of the DM's total. The local municipality has an unemployment rate of 50.3%, with 33,951 households, and a 53.8% female population (OR Tambo DM 2016; Port St. Johns LM 2018). In terms of gross value added (GVA) in the PSJ

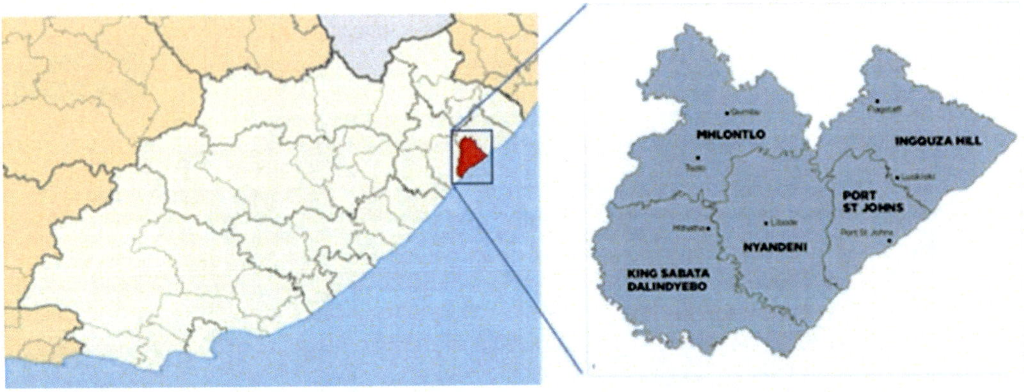

Fig. 12.3 Study site. Source: Municipalities of South Africa (2020) and Wikipedia (2016)

economy, government and community, social and personal services account for 57.7% whilst wholesale and retail contribute 16% with finance, insurance, real estate and banking at 13%. Agriculture, fisheries and forestry contribute 1.35% to the local municipality's GVA, and 5.44% of its total employment (Le Roux et al. 2019).

Port St Johns, settlement vulnerabilities to flooding, environmental and service access were the highest elements in the local municipality's vulnerability to flooding in the urban areas. In the rural areas, the highest elements were socio-economic factors, services and growth pressure (Le Roux et al. 2019). The southern half of the local municipality has a low flooding hazard, whilst increased flooding hazard exists in the northern parts of the local municipality. Furthermore, there is a higher risk along the coast line, as well as along the Mzimvubu river channel (Le Roux et al. 2019).

12.3.2 Study Design

This study adopted a structural paradigm as highlighted by Stăngă and Grozavu (2012), focusing on the vulnerability of political and socio-economic structures to the human exploitation of nature or lack of resources. The basic aspect of proper risk assessment is to identify the vulnerability factors, usually divided into two large categories: internal factors, which are the basis of coping capacity and external factors (exposure to stress/perturbation or to societal structures) (Turner II et al. 2003) or biophysical and socio-economic factors (Füssel 2007). The study aimed at index construction of the various settlements in the Port St Johns area based on the vulnerability framework (Fig. 12.2). Each vulnerability dimension has indicators associated with measurable variables. These variables and indicators were selected based on relevance, representativeness, accuracy and replicability (de Andrade and Szlafsztein 2018). In acquiring the indicators and subsequent index, quantitative research methods were utilised. A cross-sectional survey was conducted through questionnaire administration from the first to the 18th of December, 2019.

12.3.3 Sampling Procedure

The study utilised the non-probability purposive sampling method. The initial Ward selection was based on purposive sampling, where 4 out of the 20 wards within PSJ LM were selected based on the flooding that occurred within that area in April 2019. Wards 1, 4, 6 and 10 were selected, and with the assistance of the Ward councillors, the affected households were identified. A total sample size of 195 respondents was reached. A semi-structured questionnaire was utilised in data collection, with enumerators translating into the local Xhosa dialect during administration. Ethical clearance[1] for the study was obtained observing ethical considerations such as discontinuity, refusal to participate, and anonymity.

12.3.4 Analytical Framework

According to Hashim et al. (2018) and Fernandez et al. (2016), a flood vulnerability assessment is utilised in determining the environmental, economic, social and physical vulnerabilities to flood effects at a particular time. A flood vulnerability index (FVI) can be computed with various methods such as derivative and correlation methods, fuzzy comprehensive assessments, fuzzy similarity method, simple fuzzy classification, factor analysis and statistical process control (Hashim et al. 2018). This study used a multiplicative method in constructing the FVI. The flood vulnerability indicators were combined to compute the FVI model using the following equation (Hamidi et al. 2020; Hashim et al. 2018; Karmaoui and Balica 2019):

$$FVI_i = \left(\frac{E*S}{R}\right) \text{ ; where } i = PV, SV, EcV \text{ and } EnV \quad (12.1)$$

where E is Exposure, S is Susceptibility, R is Resilience, PV is Physical Vulnerability, SV is Social Vulnerability, EcV is Economic Vulnerability and EnV is Environmental Vulnerability. The mean total FVI, which lies between 0 and 1, is computed as follows:

$$\mu(FVI) = \frac{\left(\frac{E*S}{R}\right)_{PV} + \left(\frac{E*S}{R}\right)_{SV} + \left(\frac{E*S}{R}\right)_{EcV} + \left(\frac{E*S}{R}\right)_{EnV}}{4} \quad (12.2)$$

The data was initially normalised to convert into non-dimensional units using the following normalisation formula which interpolates the minimum and maximum of obtained data (Hashim et al. 2018; Nazeer and Bork 2019):

$$Z_{ij} = \frac{X_{ij} - \min\left(X_{ij}\right)}{\max\left(X_{ij}\right) - \min\left(X_{ij}\right)} \quad (12.3)$$

$$Z_{ij} = \frac{\max\left(X_{ij}\right) - X_{ij}}{\max\left(X_{ij}\right) - \min\left(X_{ij}\right)} \quad (12.4)$$

where X_{ij} represents the value of j indicators (j = 1, 2, …, n) of the ith household (i = 1, 2, …n), Max (X_{ij}) is the maximum value, Min (X_{ij}) is the minimum value and Z_{ij} is the matrix corresponding to the normalised score in which its scaled value lies between 0 and 1. Eq. (12.3) was utilised when the indicator had a direct relationship with vulnerability, whilst Eq. (12.2) was used when the indicator had an indirect relationship with vulnerability (Nazeer and Bork 2019). Microsoft Excel was used to analyse the data. Table 12.1 shows the indicators that were used in the study.

In interpreting the values of the FVI, the designations in Table 12.2 were utilised (Hashim et al. 2018):

Index construction has however been criticised for being subjective, both in variable selection and weighting, the inaccessibility or unavailability of certain variables, aggregation problems at different scales, as well as validation difficulty. Different combinations of the variables may also produce several vulnerability assessments (Fernandez et al. 2016).

12.4 Results and Discussion

12.4.1 Demographics

Figure 12.4 shows that in all the settlements that were targeted, most of the respondents were owners of their residences, followed by tenants. According to Fernandez et al. (2016), landlords are more inclined to pursuing construction changes in their houses and have insurance and access to information about financial support during recovery relative to tenants. Mogammad (2006) added to this by indicating that tenants cannot upgrade their dwellings without the authority of the owner. Owners can flood-proof

[1]Ethical Clearance Certificate: UNISA 2019_CRERC_FA.

Table 12.1 Indicators utilised in creating the flood vulnerability index

Factors	Indicators		
	Exposure	Susceptibility	Resilience
Physical	Nature of settlement; lack of drainage facilities; poorly designed houses	Connection to the national grid; Mobile phone network availability; witnessed such damage since 2019; damage to homes; damage to domestic water supply; damage to sanitation and hygiene	Tenure
Economic	Access to land	Floods causing blackouts	Employment status; household income; income; lack of alternative livelihoods
Environmental	Residing in flood-prone areas	Damage to agricultural land; damage to grazing land; damage to vegetation	The extent of damage to forests/plantation
Social	Settled in vulnerable areas	Household structure; household size; number of smart phones; availability of radio; availability of TV; language preference; time for displaced people to be accommodated; communication package clear; enough time to prepare; sex, age, ethnicity; household characteristics; social class; lack of information	Deaths during flooding

Table 12.2 Interpretation of the vulnerabiliy indixes

Index value	Description
0.75–1.00	**Very high vulnerability** An area has a very high vulnerability to floods. Either the physical, social, economic, environmental or all aspects are very highly vulnerable to floods
0.50–0.75	**High vulnerability to floods** An area has a high vulnerability level to floods. Either the physical, social, economic, environmental or all aspects tend to be highly vulnerable to floods.
0.25–0.50	**Vulnerable to floods** An area has a moderate vulnerability level to floods. Either the physical, social, economic, environmental or all aspects are moderately vulnerable to floods.
0.01–0.25	**Low vulnerability to floods** An area has a low vulnerability level to floods. Either the physical, social, economic, environmental or all aspects are lowly vulnerable to floods.
<0.01	**Very low vulnerability to floods** An area has a very low vulnerability level to floods. Either the physical, social, economic, environmental or all aspects are very low vulnerable to floods.

Source: Hashim et al. (2018)

their dwellings as a mitigation strategy or immediately repair flood-damaged houses. However, home owners may be more vulnerable if the flood damage is extensive, with high switching costs, relative to tenants who have low switching costs to other dwellings (Imran et al. 2019; Mogammad 2006).

Figure 12.5 shows causes of settlement vulnerability, as perceived by the respondents, across different settlement types. In terms of vulnerability due to environmental degradation, half of the respondents were from rural settlements, followed by 31% and 12.5% from informal and rural (urban) like settlements, respectively. Of the respondents that identified the location of settlements on waterways as a cause of vulnerability, 48% were from informal settlements, while 40% were from rural settlements. Flooding is exacerbated by land degradation which is related to sensitivity to flood vulnerability (Davis-Reddy and Vincent 2017). Environmental degradation is more common in rural settlements, relative to urban settlements, because of infrastructure such as paving, hence it causes more vulnerability to flooding. Furthermore, human settlement expansion in rural settlements has exacerbated degradation, leading to vulnerability to flooding (Davis-Reddy and Vincent 2017; Olorunfemi 2011). Location in waterways is exacerbated by flood plains which tend to attract human settle-

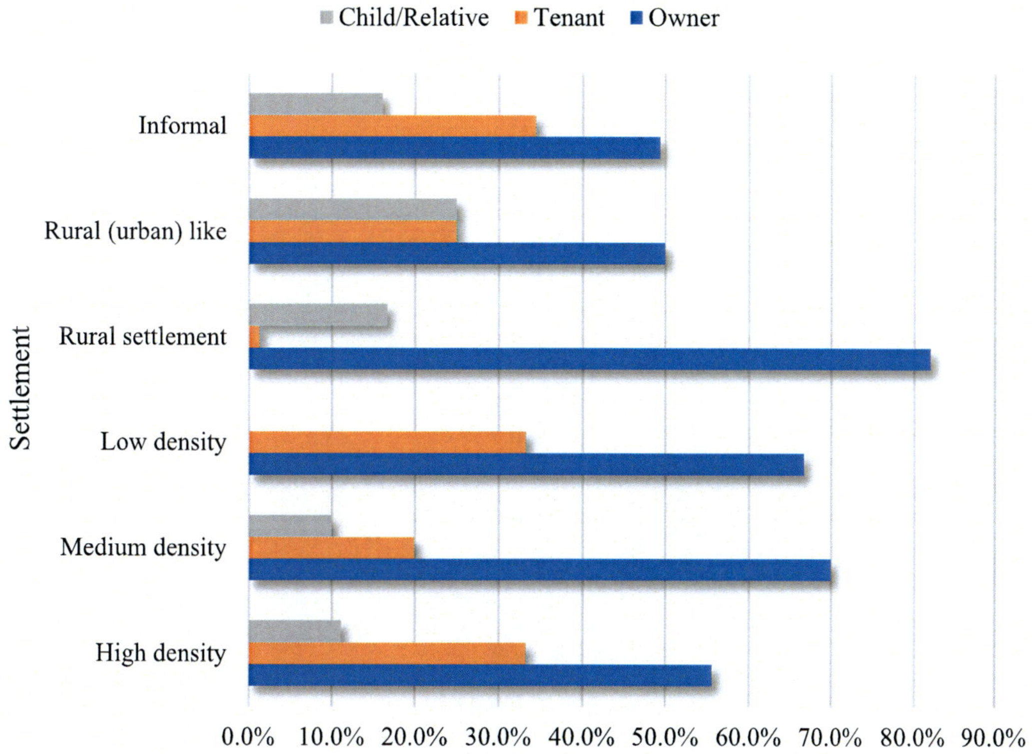

Fig. 12.4 Household ownership in settlements in the study area. Source: Survey data 2019

ments and development as they are flat, fertile and located near water.

More respondents identified location on slopes and wetlands as causes of flood vulnerability in rural settlements than informal settlements. Poor settlement planning as well as poor drainage were equally identified in both rural and informal settlements but not in other settlements (Fig. 12.5).

12.4.2 Flood Vulnerability Index

The study area is vulnerable to flooding as indicated by the FVI of 0.39 (Table 12.3). In all dimensions of the FVI, the study area was vulnerable to flooding. Physical vulnerability had the highest index followed by environmental, social and economic, respectively. Le Roux et al. (2019) highlighted that Port St Johns had a very high socio-economic vulnerability. The economic and physical vulnerabilities were identified to be medium, whilst the environmental vulnerability was low. The differences might be due to the aggregation of their findings, where the results were a composite of the whole local municipality, relative to specific areas as targeted by the current study. Furthermore, the variables utilised in compiling the indices were different to those utilised by Le Roux et al. (2019).

Figure 12.6 shows that of the households that were very highly vulnerable to flooding, 48% were in informal settlements whilst 35% and 13% were in rural settlement and medium density areas, respectively. Sixty percent of the households that were highly vulnerable to flooding were also in informal settlements, with 19% found in rural settlements, whilst both high-density and medium-density areas had 9%. Fifty-two percent of households that were vulnerable to flooding were from rural settlements, whilst 61% of households that had little vulnerability were also from rural settlements. Kotzee and Reyers (2016), however, indicated that resilience

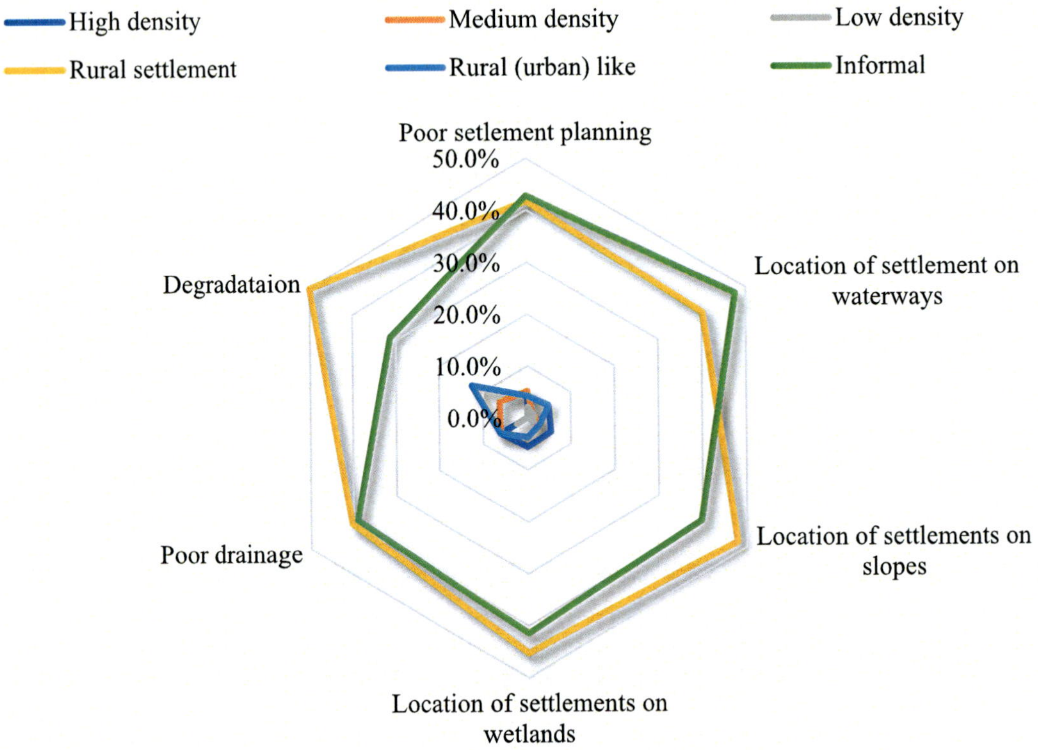

Fig. 12.5 Causes of settlement vulnerability to flooding. **Source:** Survey data

Table 12.3 Flood vulnerability index of Port St Johns

Port St Johns' flood vulnerability index		
***FVI* component**	***FVI* Value**	***FVI* Designation**
$FVI_{Physical}$	0.47	Vulnerable to flooding
FVI_{Social}	0.35	Vulnerable to flooding
$FVI_{Economic}$	0.33	Vulnerable to flooding
$FVI_{Environmental}$	0.40	Vulnerable to flooding
FVI_{Total}	**0.39**	**Vulnerable to flooding**

Source: Survey data 2019

(which tends to reduce vulnerability) is higher in urban areas relative to rural areas and as such will likely contribute to the reduction in vulnerability.

Nasiri and Shahmohammadi-kalalagh (2013) highlighted that urban areas are densely populated making them vulnerable to flood impacts. Furthermore, it is the poor rural-urban migrants who occupy informal settlements in flood-prone areas and flood plains, with high population density, poor shelter and low adaptive capacity (Few 2003). Informal settlements are mainly characterised by geographical (Musyoki et al. 2016), housing, economic and tenure vulnerabilities (Usamah et al. 2014). In this case, this tends to outweigh the resilience experienced through trust, social cohesion, community involvement and communication (Usamah et al. 2014). The findings were however contrary to Stăngă and Grozavu (2012) who indicated that large settlements such as informal settlements develop services and activities that improve adaptation to cope with crisis. In this case, small rural villages will have reduced ability to cope with crisis and hence are more vulnerable.

Informal settlements had the highest physical vulnerability to flooding, followed by rural settlements, with the exception of very low vulnerability (Fig. 12.7). Most low-density areas had very low physical vulnerability to flooding. Le Roux et al. (2019) identified that rural settlements had higher physical vulnerability than urban settlements. Factors such as blocked drainage, poorly designed houses, lack of connectivity, washing

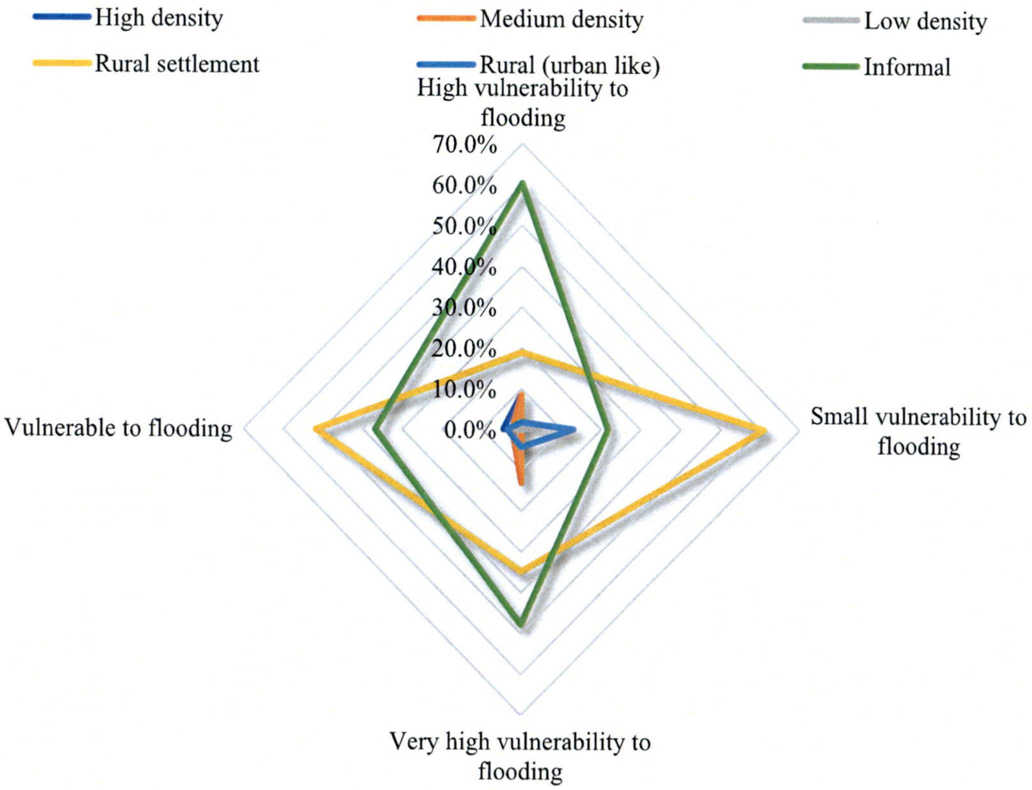

Fig. 12.6 Vulnerability of the different settlements in Port St Johns. **Source:** Survey data

away of unpaved roads, as well as inadequate sewage and storm-water infrastructure have had an impact on vulnerability to flooding in informal settlements (Mogammad 2006; Musungu et al. 2016; Olorunfemi 2011).

Rural settlements exhibited economic vulnerability relative to informal settlements (Fig. 12.8). Furthermore, rural (urban) like, low, medium and high density showed low vulnerability. Le Roux et al. (2019) highlighted similar results where rural settlements exhibited higher economic vulnerabilities than urban ones. This is due to the high unemployment rate within the area, at 50.3%, and low household incomes as well as lack of alternative livelihoods, which are all exacerbated in rural settlements (Port St. Johns LM 2018).

Rural settlements also exhibited environmental vulnerability to flooding relative to informal settlements as shown in Fig. 12.9. That was contrary to the study by Le Roux et al. (2019) who found that urban settlements were more environmentally vulnerable than rural settlements. This was supported by Olorunfemi (2011) who highlighted that informal settlement location on environmentally sensitive areas such as flood plains, slopes and wetlands has exacerbated environmental degradation and vulnerability. However, due to the physiography of rural settlement location on slopes, it tends to increase vulnerability to flooding mainly through degradation (Musyoki et al. 2016). Furthermore, rural settlements are mainly agricultural-based, and flood plains are attractive due to the fertile soils and proximity to water, making such settlement vulnerable when there is exacerbated degradation through damage to agricultural land, grazing land as well as forests and plantations (Musyoki et al. 2016). Figure 12.10 illustrates how informal settlements exhibited social vulnerability relative to rural settlements.

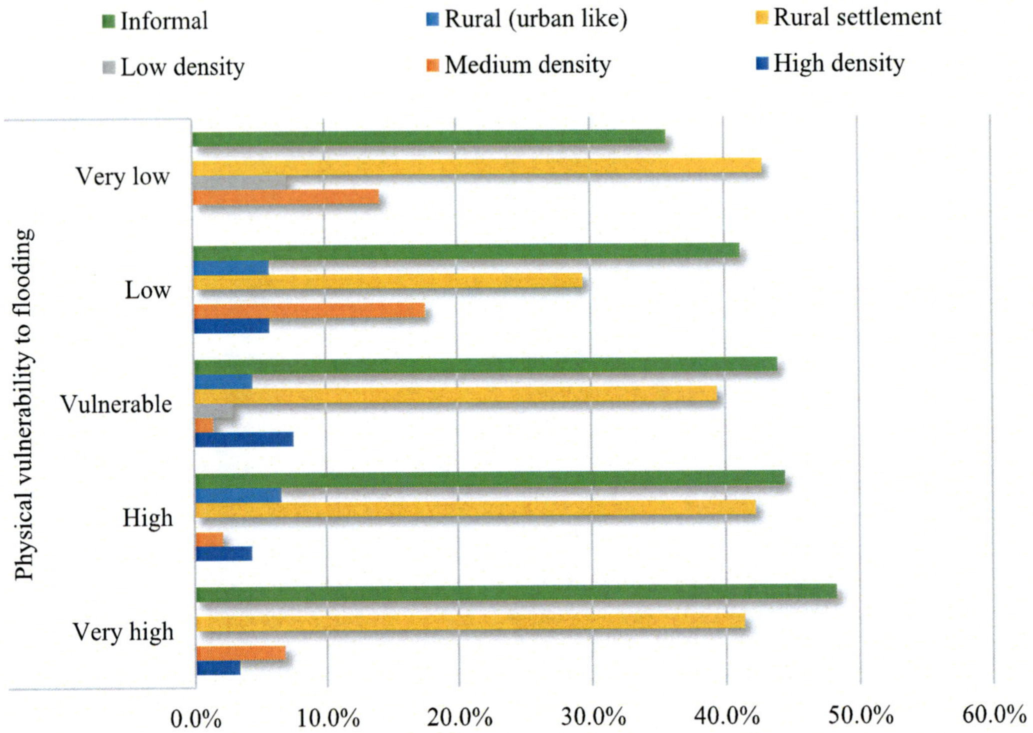

Fig. 12.7 Physical vulnerability of settlements to flooding. Source: Survey data 2019

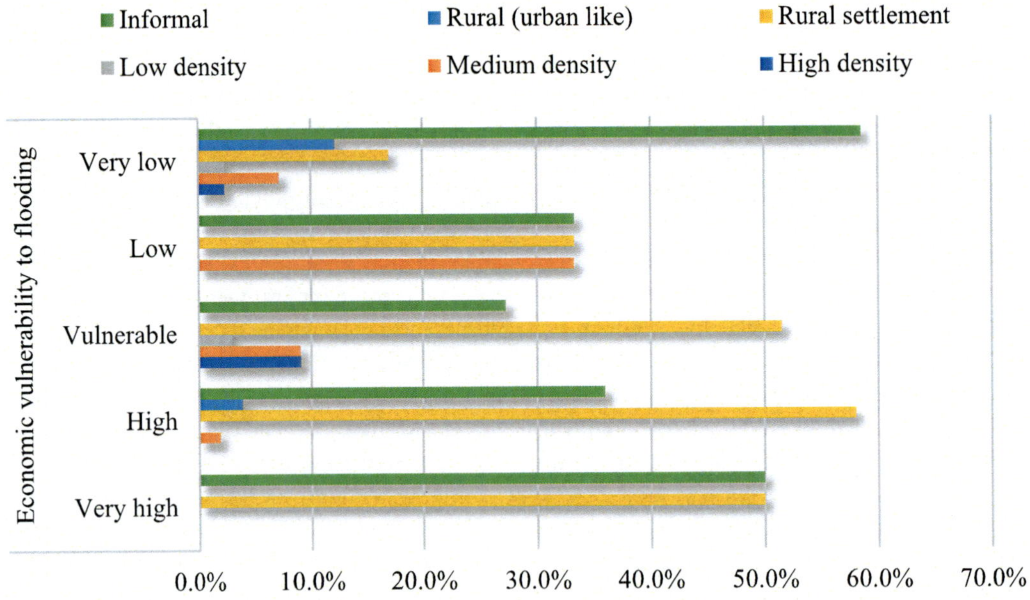

Fig. 12.8 Economic vulnerability of settlements to flooding. Source: Survey data 2019

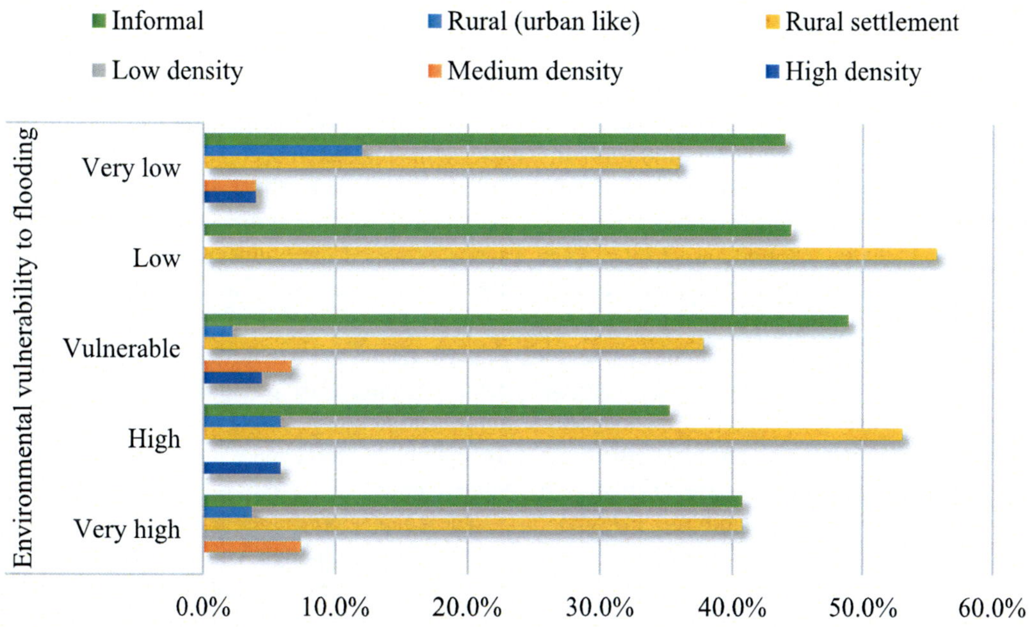

Fig. 12.9 Environmental vulnerability of settlements to flooding. **Source:** Survey data

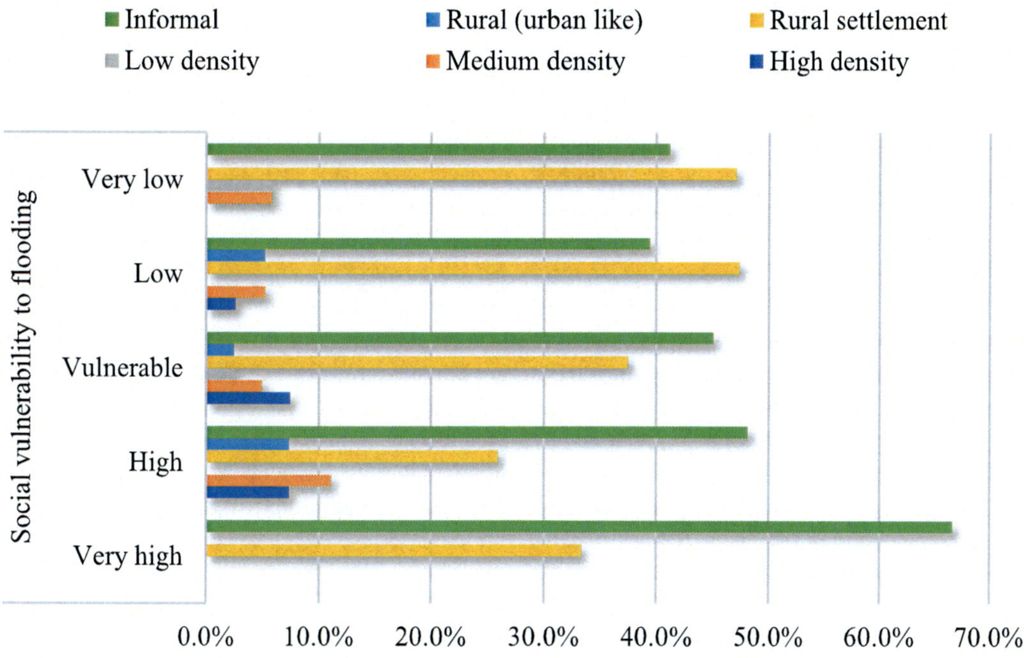

Fig. 12.10 Social vulnerability of settlements to flooding. Source: Survey data 2019

There was also a gradual increase in vulnerability for medium- as well as high-density settlements. These findings were contrary to Le Roux et al. (2019) who identified that rural settlements had higher social vulnerability than urban settlements. This was mainly based on access to services such as education, housing and health. However, informal settlements are characterised by high poverty levels, overcrowding, social problems, impermanent and often neglected structure of dwellings, low level of service provision and high degree of social differentiation. There is also an increased number of injuries, ill-health and deaths due to flooding in informal settlements as well as damage and loss of assets (Mogammad 2006; Olorunfemi 2011). Less social vulnerability in rural settlements can be attributed to the strong network of social ties (social capital), with greater community interaction and stronger bonds which can increase recovery and response both before and after flooding. This is unlike in informal settlements with isolated households and limited networks to draw upon during flooding, exhibiting low social resilience (Kotzee and Reyers 2016).

12.5 Conclusion

Flooding has an impact on attainment of SDG 11B (improve settlements) and SDG 13.1 (strengthening resilience and adaptive capacity to natural disasters). There are disparities in human settlements based on the exposure, susceptibility, resilience and adaptive capacity to flooding. Assessments of these capacities are essential in understanding variation in flood vulnerabilities amongst various human settlements. A significant proportion of settlements in South Africa are at risk of flooding, exemplified by the flooding which occurred in Port St Johns in 2019. However, there has been little documentation on the vulnerabilities of different settlements to flooding in South Africa.

This study has shown that in terms of flood vulnerabilities, most of the causes that were highlighted included degradation, which was mostly identified in rural settlements. Location on wetlands as well as on slopes was also mostly identified in the rural settlements. Location of settlements on waterways was mostly identified in the informal settlements, whilst poor drainage was identified in both the rural and informal settlements. High-density, medium-density, low-density as well as rural (urban)-like settlements were low in the flood vulnerabilities. The overall FVI showed that Port St Johns is vulnerable to flooding, and in all the FVI dimensions (physical, social, economic and environmental) PSJ exhibits its vulnerability to flooding. Physical vulnerability had the highest impact on flood vulnerability, followed by environmental. Informal settlements exhibited high vulnerabilities, whilst rural settlements exhibited low vulnerabilities to flooding. The other settlements (high-density, medium-density, low-density and rural (urban)-like) exhibited low vulnerabilities. Informal settlements had the highest physical and social vulnerabilities to flooding, whilst rural settlements had the highest economic and environmental vulnerabilities.

The chapter concludes that Port St Johns is vulnerable to flooding, with high physical and environmental vulnerabilities. Furthermore, informal settlements were found to be the most vulnerable to flooding, followed by rural settlements. Informal settlements had high physical and social vulnerabilities, whilst rural settlements had high economic and environmental vulnerabilities. Therefore, informal settlements are characterised by blocked drainages; impermanent, neglected and poorly designed houses; lack of connectivity; inadequate storm water infrastructure; high poverty; overcrowding; social problems; low levels of service provision; limited networks; and a high degree of social differentiation, which tend to increase their physical and social vulnerability to flooding. Rural settlements are characterised by a high unemployment rate; low incomes; lack of alternative livelihoods; location on slopes and flood-prone areas such as flood plains, which tend to increase their economic and environmental vulnerability to flooding.

References

Balica, S., Douben, N., & Wright, N. G. (2009). Flood vulnerability indices at varying spatial scales. *Water Science Technology, 60*, 2571–2580. https://doi.org/10.2166/wst.2009.183.

Berezi, O. K., Obafemi, A. A., & Nwankwoala, H. O. (2019). Flood vulnerability assessment of communities in the flood prone areas of Bayelsa State,

Nigeria. *International Journal of Geology and Earth Sciences, 5*(3), 19–37. Retrieved from https://www.researchgate.net/profile/Hycienth_Nwankwoala2/publication/335971218_Flood_Vulnerability_Assessment_of_Communities_in_the_Flood_Prone_Areas_of_Bayelsa_State_Nigeria/links/5d875130458515cbd1b1916e/Flood-Vulnerability-Assessment-of-Communities-in.

Birkmann, J. (2006). Measuring vulnerability to promote disaster-resilient societies: Conceptual frameworks and definitions. In J. Birkmann (Ed.), *Measuring vulnerability to natural hazards: Towards disaster resilient societies* (pp. 7–54). New York: United Nations Press.

Birkmann, J., Cardona, O. D., Harbat, A. H., Pelling, M., Schneiderbauer, S., Kienberger, S., Keiler, M., Alexander, D., Zeil, P., & Welle, T. (2013). Framing vulnerability, risk and societal responses: The MOVE framework. *Natural Hazards, 67*, 193–211. https://doi.org/10.1007/s11069-013-0558-5.

Blaikie, P., Wisner, B., Cannon, T., & Davis, I. (2004). *At risk: Natural hazards, people's vulnerability and disasters*. London: Routledge Taylor & Francis Group.

Bogardi, J., & Birkmann, J. (2005). Vulnerability assessment: The first step towards sustainable risk reduction. In D. Malzahn & T. Plapp (Eds.), *Disasters and society: From hazard assessment to risk reduction* (pp. 75–82). Berlin: Logos Verlag Berlin.

Bohle, H. G. (2001). Vulnerability and criticality: Perspectives from social geography. In E. Dyck (Ed.), *IHDP update 2/2001, international human dimensions programme on global environmental change* (pp. 3–5). Bonn, Germany: UNU-IHDP.

Bruwer, A., van Staden, M., Le Roux, A., & van Niekerk, W. (2017). Understanding the social and economic implication of global change. In J. Mambo & K. Faccer (Eds.), *South African risk and vulnerability atlas* (2nd ed.). Pretoria, South Africa: African Sun Media.

Cutter, S. L., Barnes, L., Berry, M., Burton, C., Evans, E., Tate, E., & Webb, J. (2008). A place-based model for understanding community resilience to natural disasters. *Global Environmnetal Change, 18*, 598–606. https://doi.org/10.1016/j.gloenvcha.2008.07.013.

Daniel, L. (2019, April 23). *Port St Johns floods: Residents evacuated amid heavy rainfall. The south African*. Retrieved from https://www.thesouthafrican.com/news/port-st-johns-floods-evacuation-news-updates-april-2019/

Dapeng Huang, R. Z., Huo, Z., Mao, F. E., & Zheng, W. (2012). An assessment of multidimensional flood vulnerability at the provincial scale in China based on the DEA method. *Natural Hazards, 64*, 1575–1586.

Davis-Reddy, C. L., & Vincent, K. (2017). *Climate Risk and Vulnerability: A handbook for Southern Africa* (2nd ed.). Pretoria, South Africa: CSIR. https://doi.org/10.1201/b17047-5.

DCG. (2019). *Minister Zweli Mkhize visits area affected by floods in Eastern Cape. Pretoria, South Africa*. Retrieved from https://www.gov.za/speeches/minister-zweli-mkhize-see-first-hand-damages-caused-recent-rains-eastern-cape-27-apr-2019

de Andrade, M. M. N., & Szlafsztein, C. F. (2018). Vulnerability assessment including tangible and intangible components in the index composition: An Amazon case study of flooding and flash flooding. *Science of the Total Environment, 630*, 903–912. https://doi.org/10.1016/j.scitotenv.2018.02.271.

Fernandez, P., Mourato, S., Moreira, M., & Pereira, L. (2016). A new approach for computing a flood vulnerability index using cluster analysis. *Physics and Chemistry of the Earth, 94*, 47–55. https://doi.org/10.1016/j.pce.2016.04.003.

Few, R. (2003). Flooding, vulnerability and coping strategies: Local responses to a global threat. *Progress in Development Studies, 3*(1), 43–58. https://doi.org/10.1191/1464993403ps049ra.

Füssel, H. M. (2007). Vulnerability: A generally applicable conceptual framework for climate change research. *Global Environmental Change, 17*, 155–167.

Hamidi, A. R., Zeng, Z., & Khan, M. A. (2020). Household vulnerability to floods and cyclones in Khyber Pakhtunkhwa, Pakistan. *International Journal of Disaster Risk Reduction, 46*(January), 101496. https://doi.org/10.1016/j.ijdrr.2020.101496.

Hashim, M. S., Hassan, S., & Bakar, A. A. (2018). Developing a household flood vulnerability index: A case study of Kelantan. *Scholars Journal of Economics, Business and Management, 5*(7), 575–589. https://doi.org/10.21276/sjebm.2018.5.7.4.

Hummel, B., Cutter, S., & Emrich, C. (2016). Social vulnerability to natural hazards in Brazil. *Brazil International Journal of Disaster Risk Science, 7*, 111–122.

Imran, M., Sumra, K., Mahmood, S. A., & Sajjad, S. F. (2019). Mapping flood vulnerability from socioeconomic classes and GI data: Linking socially resilient policies to geographically sustainable neighborhoods using PLS-SEM. *International Journal of Disaster Risk Reduction, 41*(July), 101288. https://doi.org/10.1016/j.ijdrr.2019.101288.

Jha, R. K., & Gundimeda, H. (2019). An integrated assessment of vulnerability to floods using composite index—A district level analysis for Bihar, India. *International Journal of Disaster Risk Reduction, 35*(October 2018), 101,074. https://doi.org/10.1016/j.ijdrr.2019.101074.

Jiménez-Aceituno, A., Peterson, G. D., Norström, A. V., Wong, G. Y., & Downing, A. S. (2019). Local lens for SDG implementation: Lessons from bottom-up approaches in Africa. *Sustainability Science*, (0123456789). https://doi.org/10.1007/s11625-019-00746-0.

Karmaoui, A., & Balica, S. (2019). A new flood vulnerability index adapted for the pre-Saharan region. *International Journal of River Basin Management*, 1–15. https://doi.org/10.1080/15715124.2019.1583668.

Khan, S. (2012). Vulnerability assessments and their planning implications: A case study of the Hutt Valley, New Zealand. *Natural Hazards, 64*, 1587–1607.

Kotzee, I., & Reyers, B. (2016). Piloting a social-ecological index for measuring flood resilience: A composite index approach. *Ecological Indicators, 60*, 45–53. https://doi.org/10.1016/j.ecolind.2015.06.018.

Le Roux, A., & Van Huyssteen, E. (2010). Socio-economic landscape: The south African socioeconomic and settlement landscape. In *South Africa risk and vulnerability atlas* (pp. 15–21). Pretoria: CPD Print.

Le Roux, A., van Huyssteen, E., Arnold, K., & Ludick, C. (2019). *Green Book. The vulnerabilities of South Africa's settlements*. Pretoria, South Africa. Retrieved from https://pta-gis-2-web1.csir.co.za/portal/apps/GBCascade/index.html?appid=280ff54e54c145a5a765f736ac5e68f8.

Mogammad, A. B. (2006). *Analysing urban flood risk in low-cost settlements of George, Western Cape, South Africa: Investigating physical and social dimensions*. University of Cape Town.

Municipalities of South Africa. (2020). *Port ST Johns local municipality*. Retrieved October 27, 2020, from https://municipalities.co.za/map/1038/port-st-johns-local-municipality

Munyai, R. B. (2017). *An assessment of community flood vulnerability and adaptation: A case study of Greater Tzaneen local municipality*. South Africa: University of Venda.

Musungu, K., Drivdal, L., & Smit, J. (2016). Collecting flooding and vulnerability information in informal settlements: The governance of knowledge production. *South African Geographical Journal, 98*(1), 84–103. https://doi.org/10.1080/03736245.2015.1117013.

Musyoki, A., Thifhulufhelwi, R., & Murungweni, F. M. (2016). The impact of and responses to flooding in Thulamela Municipality, Limpopo Province , South Africa. *Jàmbá: Journal of Disaster Risk Studies, 8*(2), 1–10. https://doi.org/10.4102/jamba.v8i2.166.

Mwale, F. D., Adeloye, A. J., & Beevers, L. (2015). Quantifying vulnerability of rural communities to flooding in SSA: A contemporary disaster management perspective applied to the lower Shire Valley, Malawi. *International Journal of Disaster Risk Reduction, 12*, 172–187. https://doi.org/10.1016/j.ijdrr.2015.01.003.

Nasiri, H., Mohd Yusof, M. J., & Mohammad Ali, T. A. (2016). An overview to flood vulnerability assessment methods. *Sustainable Water Resources Management, 2*(3), 331–336. https://doi.org/10.1007/s40899-016-0051-x.

Nasiri, H., & Shahmohammadi-kalalagh, S. (2013). Flood vulnerability index as a knowledge base for flood risk assessment in urban area. *Journal of Novel Applied Sciences, 2*(8), 269–272.

Nazeer, M., & Bork, H. R. (2019). Flood vulnerability assessment through different methodological approaches in the context of north-west Khyber Pakhtunkhwa, Pakistan. *Sustainability (Switzerland), 11*(23). https://doi.org/10.3390/su11236695.

News24. (2019, April 22). *Hundreds evacuated over flooding as hevy rainfall batters Port St Johns. News24*. Retrieved from https://www.news24.com/SouthAfrica/News/hundreds-evacuated-over-flooding-as-heavy-rain-batters-port-st-johns-20190422

Niehaus, J. (2019). *Message from the NSTF Executive Director: No longer "safe as houses."* Pretoria, South Africa. Retrieved from http://www.nstf.org.za/wp-content/uploads/2019/05/EDs-May-2019.pdf

Olorunfemi, F. B. (2011). *Managing flood disasters under a changing climate: Lessons from Nigeria and South Africa (2011 No. 1)*. Ibadan, Nigeria.

OR Tambo DM. (2016). *Coastal management Programme 2016–2021*. Mthatha, Eastern Cape Province.

Pandey, A. C., Singh, S. K., & Nathawat, M. S. (2010). Waterlogging and flood hazards vulnerability and risk assessment in Indo Gangetic plain. *Natural Hazards, 55*, 273–289.

Port St. Johns LM. (2018). *Port St Johns LM Final IDP*. Port St Johns. Retrieved from https://ia801004.us.archive.org/31/items/opment_plans_2019-20_02_local_municipalities_ec154_port_st_johns_ec154_port_st_johns_idp_2018-19_pdf/Documents/01. Integrated development plans/2019-20/02. Local municipalities/EC154 Port St Johns/EC154 Port St Joh.

Röthlisberger, V., Zischg, A. P., & Keiler, M. (2017). Identifying spatial clusters of flood exposure to support decision making in risk management. *Science of the Total Environment, 598*, 593–603.

SABC. (2019, April 23). *Heavy floods in Port St Johns. SABC Online*. Retrieved from https://www.sabcnews.com/sabcnews/heavy-floods-in-port-saint-johns/

Stăngă, I. C., & Grozavu, A. (2012). Quantifying human vulnerability in rural areas: Case study of Tutova Hills (Eastern Romania). *Natural Hazards and Earth System Science, 12*(6), 1987–2001. https://doi.org/10.5194/nhess-12-1987-2012.

Tate, E. C. (2011). *Indices of social vulnerability to hazards: Model uncertainty and sensitivity*. University of South Carolina.

Turner, B. L., II, Kasperson, R. E., Matson, P. A., McCarthy, J. J., Corell, R. W., Christensen, L., Eckley, N., Kasperson, J. X., Luers, A., Martello, M. L., Polsky, C., Pulsipher, A., & Schiller, A. (2003). A framework for vulnerability analysis in sustainability science. *Proceedings of the National Academy of Sciences, 100*, 8074–8079.

UN. (2020). *Sustainable Development Goal 11: Make cities and human settlements inclusive, safe, resilient and sustainable*. Retrieved March 16, 2020, from https://sustainabledevelopment.un.org/sdg11

UNISDR. (2009). *Terminology on disaster risk reduction*. Geneva, Switzerland.

UN-SPIDER. (2017). *Space-based information for disaster management and emergency response*. Natural Hazards, Defitions. Retrieved March 16, 2020, from http://www.un-spider.org/risks-and-disasters/natural-hazards.

Usamah, M., Handmer, J., Mitchell, D., & Ahmed, I. (2014). Can the vulnerable be resilient? Co-existence of vulnerability and disaster resilience : Informal settlements in the Philippines. *International Journal of Disaster Risk Reduction, 10*, 178–189. https://doi.org/10.1016/j.ijdrr.2014.08.007.

Wikipedia. (2016). *Map of the Eastern Cape with Port St Johns highlighted*. Retrieved October 27, 2020, from https://en.m.wikipedia.org/wiki/File:Map_of_the_Eastern_Cape_with_Port_St_Johns_highlighted_(2016).svg

Worklaw. (2019). *Chapter 7 of the constitution of the Republic of South Africa local government*. Retrieved March 17, 2020, from https://www.worklaw.co.za/SearchDirectory/Legislation/conchapter7.asp

Zuma, B. M., Luyt, C. D., Chirenda, T., & Tandlich, R. (2012). Flood disaster management in South Africa: Legislative framework and current challenges. In *International Conference on Applied Life Sciences (ICALS)* (pp. 127–132). Turkey, September 10–12: ISALS.

Check for updates

Impacts of Floods on Livestock Production in Port St Johns, South Africa

13

Saul Ngarava, Leocadia Zhou, Abbyssinia Mushunje, and Petronella Chaminuka

Abstract

The study highlights the effects of floods on livestock production in South Africa, using cross-sectional survey data from 195 purposively selected respondents in Port St Johns Local Municipality. The results indicate that amongst all livestock, small stocks were the most affected by flooding, especially amongst households settled in flood-prone areas. This resulted in reduced agricultural production for sheep, poultry, pig, goat and cattle producers. The flooding affected the availability of meat, with beef, mutton and chicken experiencing reduced availability, whilst the availability of goat and pork meat increased. Flooding also caused mutton to be relatively expensive, whereas the other livestock meat products were relatively cheaper. The availability and the affordability of dairy products decreased. The flooding caused partial damage to agricultural and grazing land, taking between 1 and 2 weeks to restore vegetation after the flooding. It was perceived that environmental degradation was the leading cause of flooding that affected cattle production. Climate change and excessive rainfall were the leading causes of flooding affecting sheep and chickens, respectively. The packaging of the flood warning message was explicit and timeously communicated for sheep producers. The study concludes that flooding affected sheep supply and demand more disproportionally than other livestock species. It recommends the promotion of flood-tolerant vegetation to ease shocks in livestock production. There is also a need to strengthen infrastructure against flooding.

Keywords

Flood · Livestock · Sustainable development goals · South Africa

S. Ngarava (✉) · L. Zhou
Risk & Vulnerability Science Centre,
University of Fort Hare, Alice, South Africa
e-mail: SNgarava@ufh.ac.za; LZhou@ufh.ac.za

A. Mushunje
Department of Agricultural Economics & Extension,
University of Fort Hare, Alice, South Africa
e-mail: AMushunje@ufh.ac.za

P. Chaminuka
Economic Analysis Unit, Agricultural Research
Council, Pretoria, South Africa
e-mail: ChaminukaP@arc.agric.za

13.1 Introduction and Background

Floods and storms account for the greatest share of natural disasters' effects on economic damage and death toll (Carrera et al. 2015). They have directly taken nearly 100,000 lives, affect-

G. Nhamo, L. Chapungu (eds.), *The Increasing Risk of Floods and Tornadoes in Southern Africa*, Sustainable Development Goals Series, https://doi.org/10.1007/978-3-030-74192-1_13

ing 1.4 billion people worldwide over the last decade ending 2016 (Mahmood et al. 2016). According to Deen (2015), floods tend to undermine agricultural systems by affecting infrastructure, contaminating water bodies, increasing human and livestock disease susceptibility as well as causing human and livestock loss. This tends to disrupt yields, food security, income generation and poverty alleviation, undermining the achievement of Sustainable Development Goals (SDGs) 1 and 2 which relate to poverty alleviation and zero hunger (UN 2015, 2019). This is exacerbated by the effect of floods in eroding poor people's assets and further undermining their livelihoods in terms of social networks, infrastructure, housing and labour productivity (IPCC 2014; Shisanya and Mafongoya 2016).

An estimated 1.3 billion people globally rely on livestock for their livelihoods, where 600 million directly own livestock, contributing 5% of the total food energy and 25% of dietary protein (Ngarava 2019). Livestock is a potential vehicle in achieving SDGs 1 and 2 (UN 2015, 2019). In South Africa, 43% of the 2.3 million agro-based households practise livestock production. The sub-sector contributes 46.3% to the country's agricultural GDP, with nearly three million subsistence farmers and 240,000 smallholder farmers, producing an estimated 900,000 tonnes of meat annually (CIA World Fact Book 2004; IDC 2016). The sector employs nearly 500,000 people, with 2.1 million directly or indirectly dependent on it for their livelihoods, with wages amounting to R6.1 billion from 38,500 commercial farms in 2016 (IDC 2016; Ngarava 2019). Consumption of livestock products within South Africa has been increasing, especially amongst the black middle class due to changes in lifestyle and consumer preferences, globalisation, urbanisation, income growth and high population growth (NAMC 2013; Ndoro et al. 2014). South Africa consumes 25–50 g/capita/day of red meat, with the country being a net exporter of beef and sheep meat (ARC 2016; Ngarava 2019). However, there is limited investigation on the effects of hazards such as flooding on the livestock sector in South Africa.

There has been limited documentation (Khandlhela and May 2006; Musyoki et al. 2016) on how rural communities in South Africa are vulnerable to climate change-induced flooding. Even online platforms such as the Greenbook (Le Roux et al. 2019) have also been limited on how flooding affects livestock production in South Africa. Flood disasters are growing in frequency worldwide due to a variety of environmental and human factors, and are becoming more frequent and destructive as well bringing huge economic losses to property and services (Mahmood et al. 2016; Musyoki et al. 2016). Come the year 2050, the South African livestock sector could be adversely impacted by climatic changes and environmental factors (Ngarava 2019; Theron 2015). Natural disasters and floods in particular are becoming more frequent and destructive (Musyoki et al. 2016). Hence, the objective of this study was to outline the effect of natural hazards such as flooding on livestock production, providing insights as well as policy recommendations. The study utilised the case of Port St Johns, which was vulnerable to flooding in April 2019.

Rojas-downing et al. (2017) and Turpie and Visser (2015) predict that climate change-induced flooding will influence changes in agricultural production levels, shifts in production length and time as well as physical damage to agricultural land. This can have devastating effects on the supply of livestock and their products. Food security is compromised due to the loss of livestock through floods. Livestock production activities are also impacted causing loss of economic earnings. The infrastructure used for livestock production is compromised, increasing the cost of production and ultimately prices (Biswas et al. 2019; Jacob 2011; Turpie and Visser 2015). Price increases of livestock and their products, also compounded by the stress and shocks on incomes, affect the demand. Potential impacts would include changes in production and feed (crop and forage) quality, food quality, availability of water, diseases and biodiversity for both livestock and feed. This affects the production and supply of livestock and their products. Flooding affects the form and structure of roots, leaf growth rate change and decreased total yield of veld grasses.

Flooding also damages grazing fields, where excess water destroys grass, increasing the unavailability of quality feed (Biswas et al. 2019). In some instances, the vegetation is shifted from grassland to forests, having an impact on the availability of feed (Turpie and Visser 2015). The risk of livestock mortality increases due to reduced grain supply and exposure to water-borne diseases following a flood (Deen 2015). Williams et al. (2016) exemplified this by highlighting that the widespread seasonal rainfall that causes flooding is responsible for the Rift Valley Fever (RVF) in South Africa. This forces producers to shift production from livestock to crop (Turpie and Visser 2015), ultimately influencing the supply of livestock and their products. Salinisation caused by the introduction of sea water into fresh water aquifers would transform the chemical and biological components of the water through heavy metals which influence livestock production (Rojas-downing et al. 2017). Furthermore, salinisation could have an effect on animal metabolism, fertility, digestion, respiratory and nervous system as well as skeletal, excretory and cardiovascular impairment. Flooding further precipitates actions of vector-borne pests such as mosquitoes, ticks and flies, with a high probability of the emergence of new diseases (Rojas-downing et al. 2017). If the flood does not result in drowning and death of the animals, the factors mentioned above will lead to the slowing down of supply and demand of livestock and their products in the event of flooding.

Flood-affected people prefer selling their livestock to coping with flooding (Mahanta and Das 2017). Some producers will dispose of the livestock due to difficulty in migrating during the flooding relative to income. However, Deen (2015) highlighted that the falling livestock prices associated with flooding compel households not to sell their livestock, despite the reduced income and consumption, in order to retain their livestock even though there is reduced productivity. These different scenarios have varying effects on the supply of livestock and ultimately the industry. Flooding has also been observed to destroy infrastructure which is essential in the demand and supply of livestock and their products. It tends to destroy physical infrastructure such as bridges, utilities and road networks (Turpie and Visser 2015). It disrupts service delivery, leading to loss of livestock and results in communities being cut-off from market facilities (Lahiff et al. 2008). In the 2010–2011 flooding in Limpopo Province, South Africa, Musyoki et al. (2016) identified that damage to the roads hampered access to major towns, with more than 100 villagers affected in accessing markets and shopping facilities, having a bearing on the demand and supply of livestock products. In Vietnam, McElwee et al. (2017) observed that during the 2003 flooding, income losses were almost 100% mainly due to the floods sweeping away the investments that had been made in animal production. In Pakistan, Deen (2015) highlighted that minimal loss of livestock was envisaged because of prior information on the flooding, with arrangements made to transport livestock. This is through observation that mortality risk increased through the underdevelopment of marketing systems.

In Africa, the net revenue per farm loss ensured by flooding is likely to be greater for large producers than for small producers. This will also be the case for the value of livestock owned as well as net revenue per livestock value (Turpie and Visser 2015). In Limpopo Province, South Africa, Musyoki et al. (2016) observed the destruction of agricultural land and livestock in the flooding that occurred in 2010–11. Khandlhela and May (2006) also observed that the 2000 floods in Limpopo Province, South Africa, resulted in 17% of the sampled households suffering a loss of crops and livestock. Furthermore, livestock, mainly goats and cattle were lost during the floods, with 5% of the sampled households resorting to selling their animals in order to respond to the floods. It is therefore observed that flooding has devastating effects on livestock production. This aspect has, however, not been adequately investigated in South Africa.

South Africa is exposed to weather-related natural hazards such as flooding and fires, with the country experiencing more than 1800 flood events between the period 1647 and 2013 (Bruwer et al. 2017). Le Roux et al. (2019) highlighted that 5% of settlements in the country are at extreme risk of increased flooding, whilst 16% are at high risk, and 33% and 46% at medium and low risk, respectively. The country has an 83.3% annual risk of flooding, and the population vulnerability is high due to economic factors and geographical location (Zuma et al. 2012). The most vulnerable provinces in South Africa are the Eastern Cape, KwaZulu-Natal, the North-West and Limpopo (Le Roux and Van Huyssteen 2010). In the southern parts of the Eastern Cape Province, Jacobs (2017) found that between 1902 and 2013, there were 45 floods events. The top five events with the highest 24-h rainfall amounts during this time were from East London (1970), Port Elizabeth (1968), East London (2002), Port Elizabeth (1981) and East London (2005) with 1101.4 mm, 455 mm, 317.2 mm, 316 mm and 273 mm, respectively (Jacobs 2017).

The April 2019 flooding in Port St Johns, Eastern Cape Province, as well as Durban, KwaZulu-Natal Province, South Africa, witnessed 85 deaths, an unknown number of injuries, between 1000 and 1600 people displaced and more than R1 billion in infrastructural damage, where 190 mm of rainfall in 24 hours was recorded in Port St Johns, a record amount in 33 years (Niehaus 2019). Low-lying rural and informal settlements such as Green Farm were left inaccessible due to the flash flood within the area which caused the Mzimvubu river to burst its banks, combined with the topography, and high oceanic tides pushing the river water. Hundreds of residents, mainly children and women, were evacuated and accommodated at the town hall and churches (Daniel 2019; DCG 2019; News24 2019; SABC 2019).

13.2 Methodology

13.2.1 Study Site

The study was carried out in Port St Johns (PSJ), located in the Eastern Cape (EC) Province, South Africa (Fig. 13.1).

The EC Province is the second largest province in South Africa, after the Northern Cape, with an area of 168,966 km^2 in the South East of the country. It consists of two metropolitan municipalities, six district municipalities subdivided into 37 local municipalities (Mandleni 2011; Masifunde and Zingisa 2011; Ngarava 2019). According to Ngarava (2019), 67% of the province's population reside in the former homeland areas, with limited infrastructural and economic development. Livestock, as well as their products, are the largest agricultural sub-sector in the province, with an estimated R2.4 billion annual gross income, which is 70% of the province's total income (IDP 2016a). Forty-five percent of the households in the Eastern Cape Province exclusively own livestock, with 37.2% of households practising mixed farming. From the 323,763 households involved with livestock in the province, 5% exclusively own cattle, whilst 67% own cattle with a combination of other livestock, that is, sheep, goats, pigs and poultry. In 2016, the Eastern Cape Province had 2.8 million cattle, 7.6 million sheep, 3.2 million goats, ½ a million pigs and 3.8 million chickens (Ngarava 2019; StatsSA 2016). The province has 22% of the country's total cattle, with 65% of these owned by communal farmers, accounting for between 80% and 90% of the livestock asset value (ComMark Trust 2009; Ngarava 2019). Livestock production within the province has been limited by lack of information and access to markets as well as control of production conditions. This has been exacerbated by poor veld management, limited access to credit and finance, insufficient and dilapidated infrastructure as well as low skills levels (ComMark Trust 2009; IDP 2016a, b, c; Ngarava 2019).

Port St Johns (PSJ) is a category B local municipality (sharing municipal executive and legislative authority with a category C municipality), with 130 rural areas/villages, which is located in the OR Tambo District Municipality (DM), Eastern Cape Province, South Africa (OR Tambo DM 2016b; Worklaw 2019). The local municipality has 20 wards on an area of 1239 km^2, with a population of 168,000 of which 53.8% are female (OR Tambo DM 2016b). Twenty-three percent of PSJ's population has no schooling, with the local municipality having an average

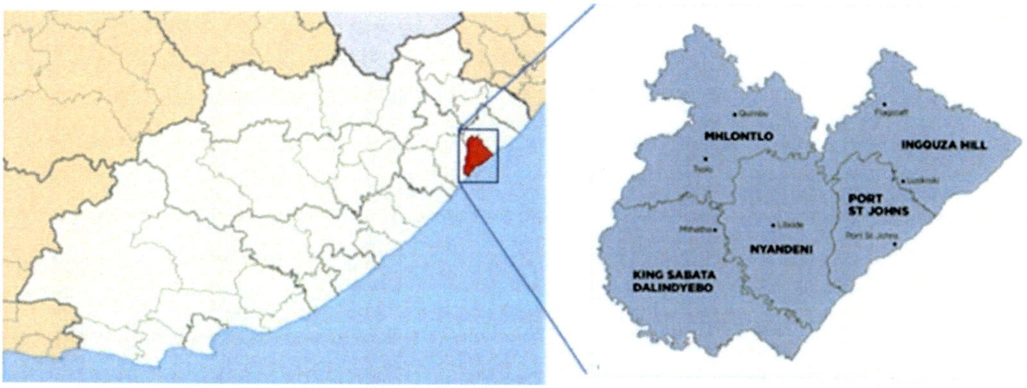

Fig. 13.1 Study site. Source: Municipalities of South Africa (2020) and Wikipedia (2016)

weighted household income of R2773 (OR Tambo DM 2016c). The local municipality has a R752 million gross value added (GVA), representing 4.2% of the OR Tambo District Municipality. Furthermore, of the 16,073 labour force, 7983 are employed, with 64,855 being economically inactive (OR Tambo DM 2016c). Agriculture, Fisheries and Forestry contribute 1.35% to the local municipality's GVA, and 5.44% of its total employment (Le Roux et al. 2019). Port St Johns has 32,000 cattle and 16,950 sheep. The local municipality has an annual demand of 3981 tonnes of meat, with R163 801 expenditure on red meat (OR Tambo DM 2016a, d).

13.2.2 Research Design

The cross-sectional survey design was adopted as the strategy for data collection and analysis. Four wards (1, 4, 6 and 10) were purposively selected based on the flooding that occurred within these areas in April, 2019. A total sample size of 195 respondents was utilised. A semi-structured questionnaire was utilised in data collection between the first and 18th of December 2019. Ethical clearance[1] was obtained observing ethical considerations such as discontinuity, refusal to participate and anonymity. Descriptive statistics and cross-tabulations were used to analyse the data.

[1] Ethical Clearance Certificate: UNISA 2019_CRERC_FA

13.3 Results and Discussion

Figure 13.2 shows that 90.4% of the respondents were from non-urban settlements, whilst 9.6% were from urban areas. This was also reflective in the proportion of the identified impact of flooding on the various livestock species, where more than 80% was observed for each of the identified livestock species. This is due to the location of livestock production which is mainly practised in rural areas. In this instance, the informal and rural areas constituted the non-urban areas. According to Deen (2015), OR Tambo DM (2016a) as well as Elum et al. (2017), the poorest households are the most vulnerable to flooding. McElwee et al. (2017, p. 480) highlighted that it is the "poorer households that spend more time in ex-ante actions to protect livestock, especially in low-cost actions such as tying down livestock pens during flooding". Non-urban areas are characterised by the most poor and vulnerable, who also tend to own livestock as livelihood strategy. Figure 13.2 also shows that there were 16.7% of respondents in urban areas who highlighted that flooding had an impact on sheep production, followed by 9.1% and 8.3% for cattle and goats, respectively.

Figure 13.3 shows that most of the respondents were in agreement that livestock were

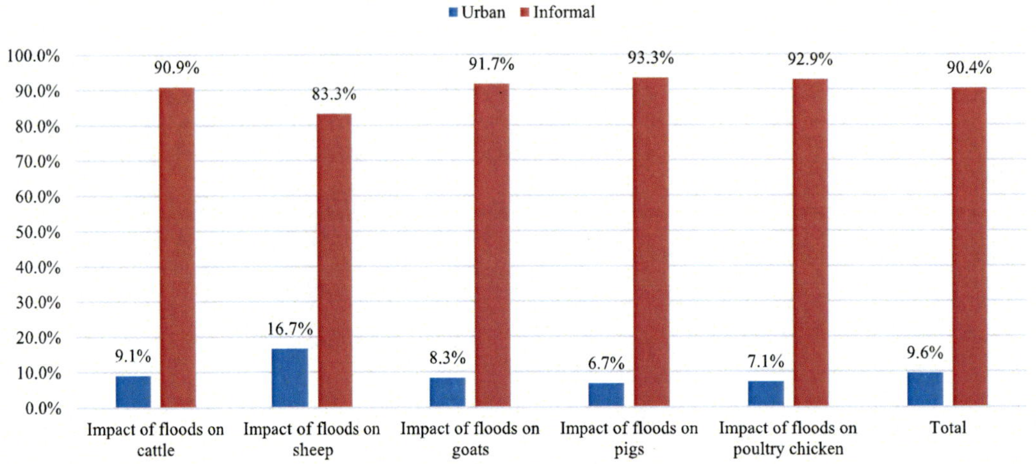

Fig. 13.2 Settlement location vs impact of flooding on various livestock species in Post St Johns. Source: Survey data (2019)

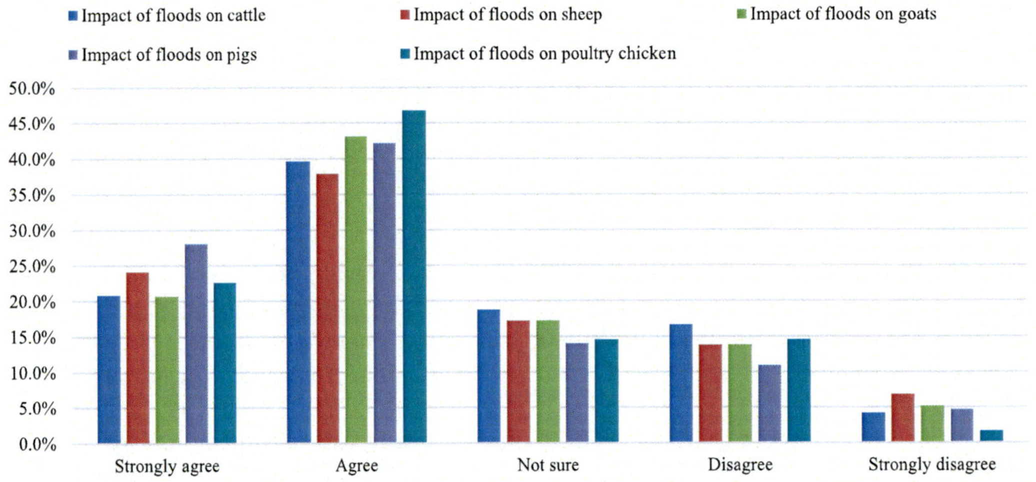

Fig. 13.3 Perceptions of settling in flood vulnerable areas vs impact of flooding on various livestock species in Post St Johns. Source: Survey data (2019)

affected by flooding if people settled in flood vulnerable areas in Post St Johns. The perceptions were that pigs were most affected, followed by poultry, goats, sheep and lastly cattle.

Figure 13.4 shows the perception of decreased agricultural production as a result of flooding and the various livestock species that were affected by the floods. It is shown that 75% of the respondents agreed that, as a result of flooding, there was a decrease in agricultural production of sheep, followed by poultry, pigs, goats and cattle, respectively. This also supports the notion that there was more impact on small stock relative to large stock.

13.3.1 Demand-Side Impact of Flooding on Livestock Production in Port St Johns

Figure 13.5 shows the availability of meat products during and after the flooding in April, 2019. More respondents indicated that beef was readily available during the flooding relative to after.

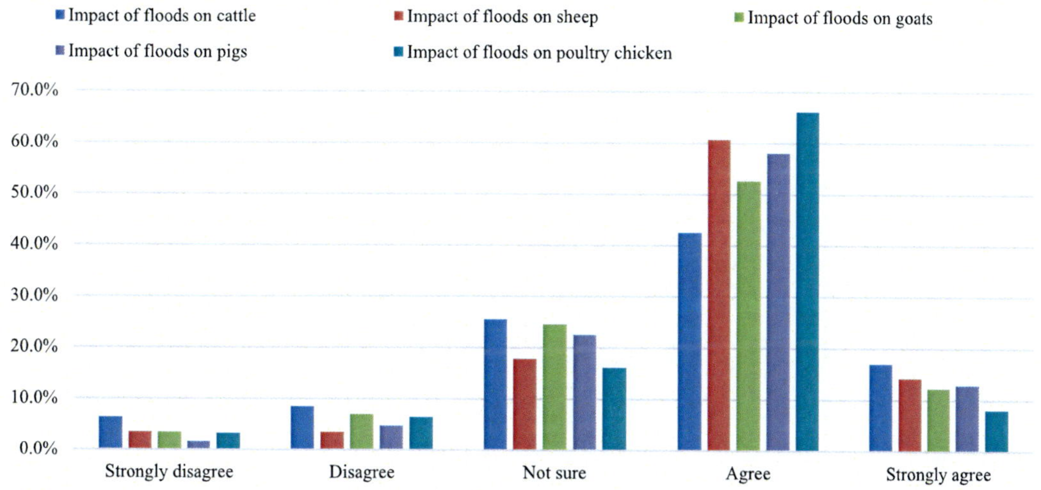

Fig. 13.4 Perceptions of decreased agricultural production resulting from flooding vs impact of flooding on various livestock species in Post St Johns. Source: Survey data (2019)

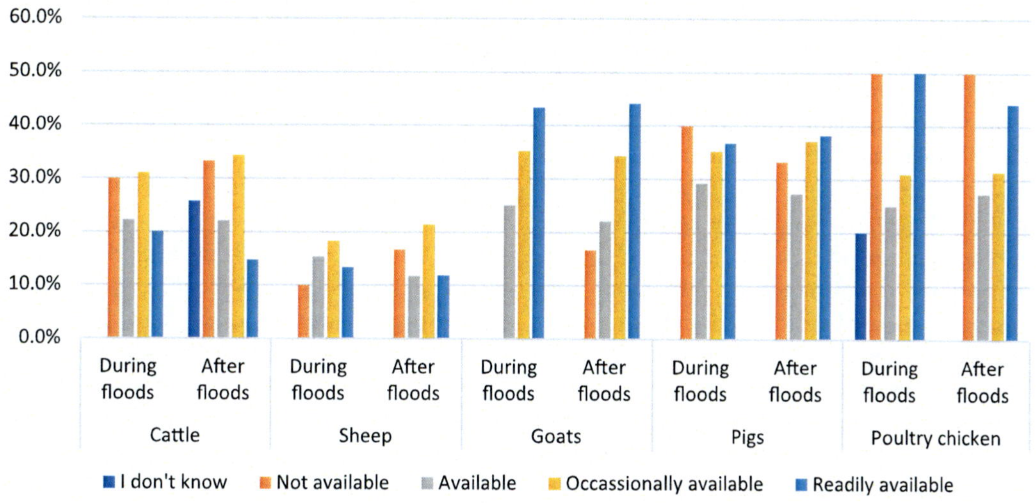

Fig. 13.5 Availability of meat products during and after the flooding in Post St Johns. Source: Survey data (2019)

This was also the trend for mutton and poultry. Thus, for beef, mutton and chicken, availability decreased, possibly as a result of the flooding. However, the availability of goat and pork meat increased relative to the flooding. Flooding will tend to affect animal health and lead to the emergence of diseases, impacting productivity and availability (AGRA 2014). However, AU-IBAR (2015) highlighted that due to flooding and because of globalisation, cheaper imported livestock products find their way into the affected areas. This can maintain the availability of livestock and their products, but could negatively impact on local production and be a threat to livestock development.

There was a reduction in the number of respondents that regarded beef, goat, pork and chicken as being expensive, during and after the flooding (Fig. 13.6). Mutton was the only meat type which had an increased number of respondents regarding it as expensive. Thus, mutton was relatively more expensive than the other livestock products. Alavian et al. (2009), as well as Pentz (2009) identified that results of climate change

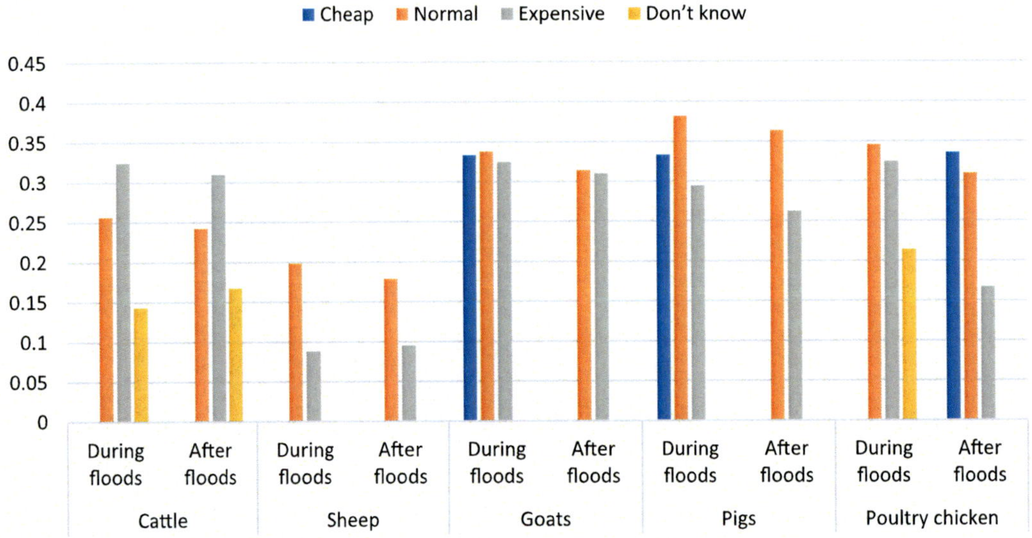

Fig. 13.6 Affordability of meat products during and after the flooding in Post St Johns. Source: Survey data (2019)

such as flooding can lead to a decrease in food production, creating market imbalances and pushing prices upwards. This can have the effect of shifting income, capital and human resources, and thus an economy-wide adjustment and impact (AGRA 2014; Alavian et al. 2009). In a study in South Africa, Nethengwe (2007) found that there was a sequence in the disposal of livestock in a crisis, where chicken and goats were disposed first, then cattle later if the crisis persisted. This can have localised effects on the relative supply of various livestock species.

The number of respondents who could readily acquire cattle and goat dairy products decreased during and after the flooding (Fig. 13.7). The number of respondents that found the cattle and goat dairy products expensive also increased during this period. Sheep dairy product availability did not change during and after flooding, whilst the number of respondents that found the sheep dairy products expensive increased. The results show that there was a relative decrease in the availability and affordability of dairy products after the flooding. This can be explained by the disruptions of transportation systems caused by flooding, which may hinder effective delivery, especially given that dairy products are transported daily (AGRA 2018).

13.3.2 Supply-Side Impact of Flooding on Livestock Production in Port St Johns

Land is a resource allowing the practice of livestock production (Ngarava 2019). Table 13.1 shows that most of the livestock producers had access to land. Furthermore, there was partial damage to agricultural land caused by flooding. It further took between 1 and 2 weeks to restore the agricultural land after the flooding had occurred. In a study in South Africa, Jacob (2011) identified that agriculture was the most affected by foods, following infrastructure and rehabilitation. Flooding will tend to cause physical damage to agricultural land affecting supply of livestock and their products (Rojas-downing 2017; Turpie and Visser 2015). Table 13.1 also shows that nearly half of all respondents indicated that there was partial damage to grazing land and vegetation, also taking between 1 and 2 weeks to recover. Flooding has an impact on the form and structure of roots and leaf growth rate, damaging grazing fields and decreasing yield of veld grasses (Biswas et al. 2019).

Table 13.2 shows that 28.6% of the respondents identified that environmental degradation caused flooding for cattle production, whilst 69% identified that climate change caused flooding for

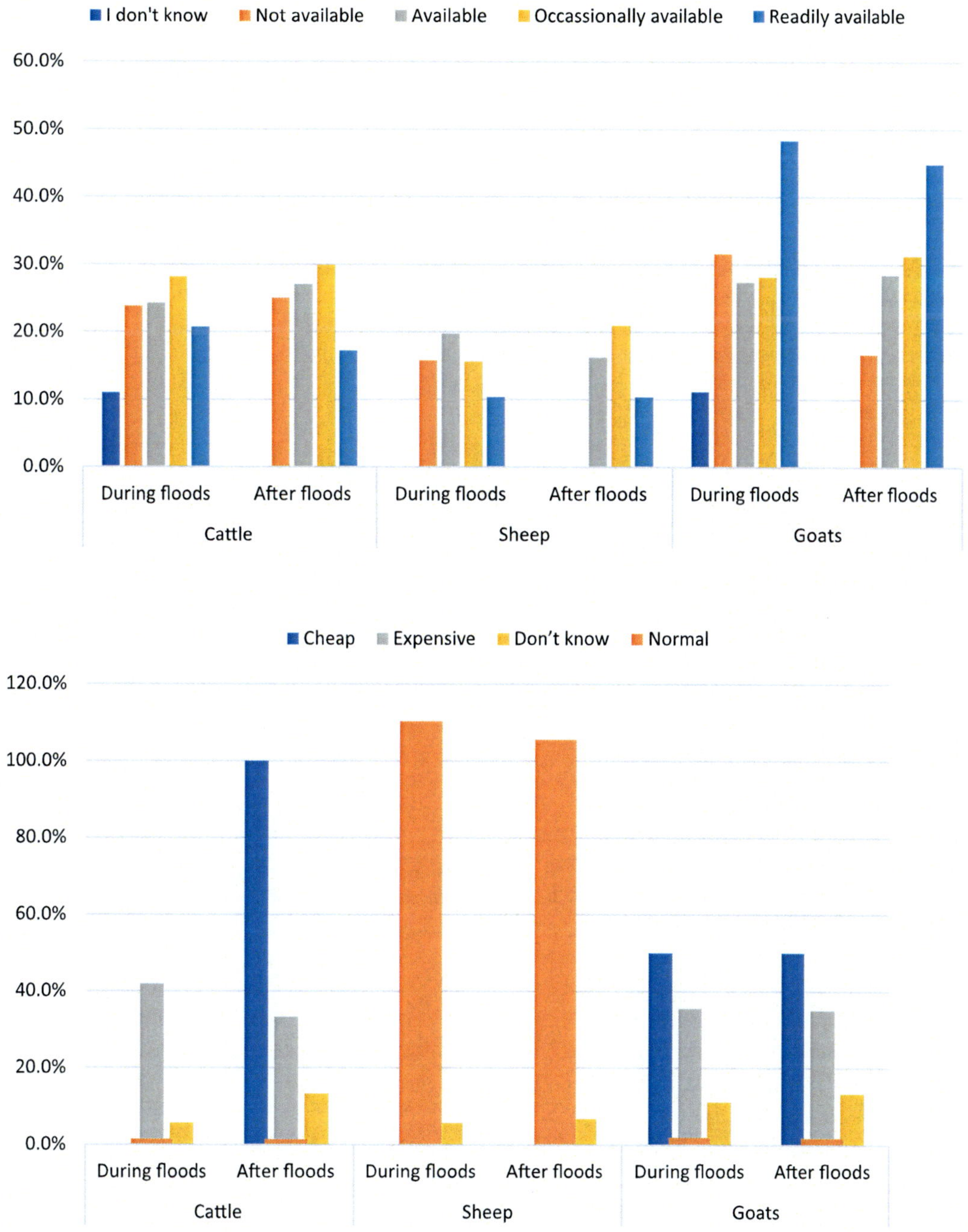

Fig. 13.7 Availability and affordability of dairy products during and after the flooding in Post St Johns. Source: Survey data (2019)

sheep production with 63.5% in agreement that it was excessive rainfall that caused flooding for poultry. Flooding is exacerbated by land degradation which is related to sensitivity to flood vulnerability (Davis-Reddy and Vincent 2017). The findings are contingent upon the livestock husbandry systems practised. Cattle production in the area is mostly through extensive grazing, and thus environmental degradation would have a more profound effect. Excessive rainfall would

Table 13.1 Access and damage to land vs impact of flooding on various livestock species in Post St Johns

	Access to land		Own arable land		Own grazing land		
Type of livestock affected by flooding	Yes	No	Yes	No	Yes	No	
Cattle	67,3%	32,7%	44,9%	55,1%	22,4%	77,6%	
Sheep	51,7%	48,3%	27,6%	72,4%	20,7%	79,3%	
Goat	58,6%	41,4%	29,3%	70,7%	13,8%	86,2%	
Pigs	54,7%	45,3%	25,0%	75,0%	9,4%	90,6%	
Chickens	71,4%	28,6%	41,3%	58,7%	14,3%	85,7%	
Extent of damage to agricultural land							
Type of livestock affected by flooding	**Complete damage**	**Partial damage**	**No damage**	**Do not know**	**Not applicable**		
Cattle	12,2%	40,8%	16,3%	10,2%	20,4%		
Sheep	6,9%	44,8%	13,8%	6,9%	27,6%		
Goat	6,9%	46,6%	10,3%	13,8%	22,4%		
Pigs	6,3%	46,0%	9,5%	12,7%	25,4%		
Chickens	9,7%	50,0%	12,9%	6,5%	21,0%		
Duration to restore agricultural land							
Type of livestock affected by flooding	**No longer usable**	**1–2 weeks**	**2–4 weeks**	**5–6 weeks**	**7–8 weeks**	**2+ months**	**Not applicable**
Cattle	4,2%	29,2%	2,1%	–	–	25,0%	39,6%
Sheep	–	34,5%	13,8%	–	–	6,9%	44,8%
Goat	3,4%	24,1%	10,3%	–	1,7%	13,8%	46,6%
Pigs	1,6%	28,6%	9,5%	1,6%	1,6%	9,5%	47,6%
Chickens	1,6%	24,6%	9,8%	6,6%	3,3%	16,4%	37,7%
Extent of damage to grazing land							
Type of livestock affected by flooding	**Complete damage**	**Partial damage**	**No damage**	**Do not know**	**Not applicable**		
Cattle	8,7%	32,6%	30,4%	8,7%	19,6%		
Sheep	3,8%	50,0%	15,4%	7,7%	23,1%		
Goat	1,9%	43,4%	17,0%	15,1%	22,6%		
Pigs	–	41,7%	16,7%	13,3%	28,3%		
Chickens	3,3%	41,7%	16,7%	13,3%	25,0%		
Duration to restore grazing land							
Type of livestock affected by flooding	**No longer usable**	**1–2 weeks**	**2–4 weeks**	**5–6 weeks**	**7–8 weeks**	**2+ months**	**Not applicable**
Cattle	4,3%	17,4%	6,5%	4,3%	–	15,2%	52,2%
Sheep	–	25,0%	17,9%	–	–	10,7%	46,4%
Goat	1,8%	18,2%	10,9%	1,8%	–	10,9%	56,4%
Pigs	–	18,0%	11,5%	0,0%	–	9,8%	60,7%
Chickens	1,6%	14,8%	13,1%	1,6%	–	9,8%	59,0%
Extent of damage to vegetation							

(continued)

Table 13.1 (continued)

	Access to land		Own arable land		Own grazing land		
Type of livestock affected by flooding	Yes	No	Yes	No	Yes	No	
Type of livestock affected by flooding	**Complete damage**	**Partial damage**	**No damage**	**Do not know**	**Not applicable**		
Cattle	4,3%	53,2%	14,9%	10,6%	17,0%		
Sheep	3,6%	53,6%	14,3%	14,3%	14,3%		
Goat	1,8%	60,7%	10,7%	10,7%	16,1%		
Pigs	1,6%	59,7%	14,5%	14,5%	9,7%		
Chickens	1,6%	54,1%	11,5%	16,4%	16,4%		
Duration to restore vegetation							
Type of livestock affected by flooding	**No longer usable**	**1–2 weeks**	**2–4 weeks**	**5–6 weeks**	**7–8 weeks**	**2+ months**	**Not applicable**
Cattle	4,3%	17,4%	8,7%	–	6,5%	13,0%	50,0%
Sheep	3,6%	28,6%	17,9%	–	7,1%	3,6%	39,3%
Goat	–	23,6%	16,4%	–	5,5%	14,5%	40,0%
Pigs	1,6%	21,0%	16,1%	–	6,5%	12,9%	41,9%
Chickens	1,6%	19,7%	16,4%	3,3%	–	13,1%	45,9%

Source: Survey data (2019)

Table 13.2 Perceptions on causes of flooding vs impact of flooding on various livestock species in Post St Johns

	Environmental degradation		Climate change		Excessive rainfall	
Type of livestock affected by flooding	Yes	No	Yes	No	Yes	No
Cattle	28,6%	71,4%	67,3%	32,7%	63,3%	36,7%
Sheep	10,3%	89,7%	69,0%	31,0%	51,7%	48,3%
Goat	19,0%	81,0%	67,2%	32,8%	58,6%	41,4%
Pigs	14,1%	85,9%	64,1%	35,9%	59,4%	40,6%
Poultry chicken	17,5%	82,5%	60,3%	39,7%	63,5%	36,5%

Source: Survey data (2019)

disallow chicken foraging and confinement to shelter, and thus would be more pronounced for poultry production.

Table 13.3 shows that location on slopes had a direct relationship with complete damage of agricultural land for 62.5% of the respondents. Location on wetlands was identified to cause complete damage of agricultural land by 43.8%, whilst 75% and 25% of the respondents identified a relationship between poor drainage and degradation as causes of flooding with complete damage to agricultural land. Location in waterways is exacerbated by flood plains which tend to attract human settlements and development as they are flat, fertile and located near water. Nethengwe (2007) posited that individuals and governments have pursued land-use decisions that develop flood zones. This promotes settlement patterns in flood-risk areas such as flood plains, higher topography and poor vegetation cover. This will result in livestock overgrazing grasslands and overuse of marginal land (Nethengwe 2007). According to OR Tambo DM (2016a) farming in a flood plain increases the susceptibility to flooding. Table 13.3 also shows that 20% of the respondents identified that location on slopes had a relationship with complete damage to grazing land, whilst 40% of the

Table 13.3 Perceptions on the extent of damage to land vs causes of damage to agricultural and grazing land as a result of flooding in Port St Johns

	Location on slopes		Location on wetlands		Poor drainage		Degradation	
Extent of damage to agricultural land	Yes	No	Yes	No	Yes	No	Yes	No
Complete damage	62,5%	37,5%	43,8%	56,3%	75,0%	25,0%	25,0%	75,0%
Partial damage	52,3%	47,7%	40,0%	60,0%	50,8%	49,2%	13,8%	86,2%
No damage	31,3%	68,8%	21,9%	78,1%	31,3%	68,8%	6,3%	93,8%
Do not know	50,0%	50,0%	26,9%	73,1%	46,2%	53,8%	7,7%	92,3%
Not applicable	29,8%	70,2%	28,1%	71,9%	54,4%	45,6%	8,8%	91,2%
Extent of damage to grazing land								
Complete damage	20,0%	80,0%	40,0%	60,0%	60,0%	40,0%	20,0%	80,0%
Partial damage	64,4%	35,6%	51,1%	48,9%	57,8%	42,2%	13,3%	86,7%
No damage	27,8%	72,2%	13,9%	86,1%	38,9%	61,1%	8,3%	91,7%
Do not know	48,5%	51,5%	27,3%	72,7%	48,5%	51,5%	9,1%	90,9%
Not applicable	33,8%	66,2%	29,4%	70,6%	51,5%	48,5%	11,8%	88,2%

Source: Survey data (2019)

respondents indicated that location on wetlands had a relationship with complete damage to grazing land. The respondents identified that poor drainage and degradation had a 60% and 20% relationship, respectively, with complete damage to grazing land caused by flooding.

13.3.3 Institutional Impact of Flooding on Livestock Production in Port St Johns

Table 13.4 indicates that 78.7% of the respondents indicated that poultry producers never received flooding communication. This was also the case for 77.6% and 71.4% of goat and cattle producers, respectively. Only 24.1% of sheep producers had received clear and straightforward communication concerning the flooding. Table 13.4 further shows that 88.7% of the pig producers never had enough time to avoid the flooding, whilst only 17.2% and 15.3% of the sheep and poultry producers had enough time to prepare to avoid the flooding. The message to avoid flooding was clearly communicated to 24.1% of the sheep producers, as well as 14.3% and 12.9% of the cattle and pig producers, respectively. The results reflect that the packaging of the flood warning was more explicit and timeously communicated for sheep producers. For all livestock producers, between 30.4% and 39.3% of the respondents identified that it took between 1 and 2 weeks to repair the road infrastructure that had been damaged by the floods.

Alavian et al. (2009) highlighted that poor maintenance and deterioration of existing infrastructure increase vulnerability to flooding. Flooding causes loss of economic earnings, damages and compromises infrastructure used in livestock production, increases the cost of production and that of livestock. AGRA (2014) indicated that although food may be available during or even after flooding, disruptions of transportation systems may hinder effective delivery. This also applies to input transportation for the various agribusinesses along with the livestock value chain. The 2011 floods in South Africa also witnessed damage in infrastructure, costing the farmers $284 million, and increasing in food prices as supplies were affected (Conway-Smith 2011).

13.4 Conclusion

The livestock sector has been instrumental in the economies of third world countries in achieving Sustainable Development Goals 1 and 2. Livestock demand, supplies and institutional arrangements have been affected by climate change-induced flooding. There has been an increase in the occurrence and destructive nature of floods in South Africa, as witnessed in Port St Johns in April 2019. However, there has been little documentation as to how communities in South Africa are vulnerable to climate change-induced flooding. This study established that the poorest households located in non-urban areas are heavily dependent on livestock. Thus, they are the most affected if livestock affected by flooding. It emerged that small stocks were the most affected by flooding in the study site, especially when people settled in flood-prone areas. This resulted in reduced agricultural production of sheep, poultry, pig, goat and cattle producers, respectively. The flooding affected availability of meat products, with beef, mutton and chicken experiencing reduced availability, whilst availability increased for goat and pork. Flooding also caused mutton to be relatively more expensive, whereas the other livestock meat products were relatively cheaper. For dairy products, the availability as well as the affordability decreased, indicating a negative effect caused by the floods, possibly through infrastructural destruction. Ownership of land limited livestock production. The flooding caused partial damage to agricultural and grazing land, taking between 1 and 2 weeks to restore vegetation after the flooding. Environmental degradation was the leading cause of flooding for cattle production, whilst climate change and excessive rainfall were the leading causes for sheep and chickens, respectively. The packaging of the flood warning message was more explicit and timeously communicated for sheep producers.

Table 13.4 Communication vs impact of flooding on various livestock species in Post St Johns

Packaging of communication of flooding in a clear and easy manner						
Type of livestock affected by flooding	Yes	No	Never got the warning			
Cattle	18,4%	10,2%	71,4%			
Sheep	24,1%	6,9%	69,0%			
Goat	13,8%	8,6%	77,6%			
Pigs	16,1%	12,9%	71,0%			
Chickens	13,1%	8,2%	78,7%			
	Enough time to prepare to avoid flooding					
Type of livestock affected by flooding	**Yes**	**No**				
Cattle	12,8%	87,2%				
Sheep	17,2%	82,8%				
Goat	13,0%	87,0%				
Pigs	11,3%	88,7%				
Chickens	15,3%	84,7%				
Message communicate clearly to avoid the flood disaster						
Type of livestock affected by flooding	**Yes**	**No**	**Not applicable**			
Cattle	14,3%	26,5%	59,2%			
Sheep	24,1%	24,1%	51,7%			
Goat	12,3%	24,6%	63,2%			
Pigs	12,9%	27,4%	59,7%			
Chickens	6,5%	22,6%	71,0%			
Duration to recover road infrastructure after the flooding						
Type of livestock affected by flooding	**No longer usable**	**1–2 weeks**	**2–4 weeks**	**5–6 weeks**	**2+ months**	**Not applicable**
Cattle	2,2%	32,6%	4,3%	2,2%	19,6%	39,1%
Sheep	3,6%	39,3%	10,7%	3,6%	7,1%	35,7%
Goat	0,0%	30,4%	8,9%	3,6%	23,2%	33,9%
Pigs	1,6%	35,5%	11,3%	4,8%	16,1%	30,6%
Chickens	1,7%	31,7%	16,7%	8,3%	23,3%	18,3%

Source: Survey data (2019)

Flooding within Port St Johns appears to affect sheep production more than any other livestock species. This is because of reduced sheep production, availability and affordability of mutton and the effect of climate-change-induced flooding on sheep production. This is despite sheep production having an easy, clear and timeous flood communication and warning. The study recommends that waterlogged-tolerant vegetation should be promoted to circumvent the reduction in vegetation and grazing land that occurs due to flooding. This could ease the shocks in livestock production. However, a balance between flood tolerance of the plants as well as its palatability to the various livestock species needs to be reached. There is also a need to strengthen the infrastructure against flooding such as the communication infrastructure and thus maintaining market access and supporting farmers to cope with changes to agricultural and grazing land conditions.

References

AGRA. (2014). *Africa agriculture status report: Climate change and smallholder agriculture in sub-Saharan Africa. Nairobi, Kenya.* Retrieved from https://agra.org/wp-content/uploads/2018/09/AASR-2018.pdf

AGRA. (2018). *Africa agriculture status report 2018. Nairobi, Kenya.* Retrieved from https://agra.org/wp-content/uploads/2018/09/AASR-2018.pdf

Alavian, V., Qaddumi, H. M., Dickson, E., Diez, S. M., Danilenko, A. V., Hirji, R. F., Puz, G., Pizarro, C., Jacobsen, M., & Blankespoor, B. (2009). *Water and climate change: Understanding the risks and making climate-smart investment decisions.* Washington, D.C.: The World Bank. https://doi.org/10.1007/s10103-012-1221-1.

ARC. (2016). *Annual beef bulletin.* Irene, South Africa.

AU-IBAR. (2015). *Livestock development strategy for Africa (LiDeSA) 2015–2035: The roadmap to a successful livestock sector-Transforming Afdrica's livestock sector for accelerated equitable growth.* Nairobi, Kenya.

Biswas, J. C., Maniruzzaman, M. M. H. M., & Kabir, M. H. A. W. (2019). Natural hazards and livestock damage in Bangladesh. *Natural Hazards, 99*(2), 705–714. https://doi.org/10.1007/s11069-019-03768-0.

Bruwer, A., van Staden, M., Le Roux, A., & van Niekerk, W. (2017). Understanding the social and economic implication of global change. In J. Mambo & K. Faccer (Eds.), *South African Risk and Vulnerability Atlas* (2nd ed.). African Sun Media: Pretoria, South Africa.

Carrera, L., Standardi, G., Bosello, F., & Mysiak, J. (2015). Environmental Modelling & Software Assessing direct and indirect economic impacts of a fl ood event through the integration of spatial and computable general equilibrium modelling. *Environmental Modelling and Software, 63*, 109–122. https://doi.org/10.1016/j.envsoft.2014.09.016.

CIA World Fact Book. (2004). *South Africa livestock.* Retrieved June 5, 2017, from https://photius.com/countries/south_africa/economy/south_africa_economy_livestock.html

ComMark Trust. (2009). *Giving development a face [Eastern Cape red meat project].* Eastern Cape Province.

Conway-Smith, E. (2011, January 28). South Africa: Floods kill 120 and destroy crops. *Global Post.*

Daniel, L. (2019, April 23). *Port St Johns floods: Residents evacuated amid heavy rainfall.* The south African. Retrieved from https://www.thesouthafrican.com/news/port-st-johns-floods-evacuation-news-updates-april-2019/

Davis-Reddy, C. L., & Vincent, K. (2017). *Climate risk and vulnerability: A handbook for Southern Africa* (2nd ed.). Pretoria, South Africa: CSIR. https://doi.org/10.1201/b17047-5.

DCG. (2019). *Minister Zweli Mkhize visits area affected by floods in Eastern Cape.* Pretoria, South Africa. Retrieved from https://www.gov.za/speeches/minister-zweli-mkhize-see-first-hand-damages-caused-recent-rains-eastern-cape-27-apr-2019

Deen, S. (2015). Policy gaps in disaster preparedness and response. *International Journal of Disaster Risk Reduction, 12*, 341–349. International Journal of Disaster Risk Reduction Pakistan 2010 floods. https://doi.org/10.1016/j.ijdrr.2015.03.007.

Elum, Z. A., Modise, D. M., & Marr, A. (2017). Farmer's perception of climate change and responsive strategies in three selected provinces of South Africa. *Climate Risk Management, 16*, 246–257. https://doi.org/10.1016/j.crm.2016.11.001.

IDC. (2016). *A study on the potential product development for the commercialization and value add to beef products, tender issue report.* Johannesburg, South Africa.

IDP. (2016a). *Amatole District municipality integrated development plan 2016–2017.* Eastern Cape Province.

IDP. (2016b). *Chris Hani District municipality integrated development plan 2016–2017.* Eastern Cape Province.

IDP. (2016c). *OR Tambo District municipality integrated development plan 2016–2017.* Eastern Cape Province.

IPCC. (2014). *Climate change 2014: Impacts, adaptation and vulnerability, part A: Global and sectoral aspects. Contribution of working Group II to the fifth assessment report of the intergovernmental panel on climate change.* Cambridge: Cambridge University Press.

Jacob, M. K. (2011). *Flood risks and their impact on insurance companies: The case of George local municipal-*

ity, at the south coast of South Africa. University of the Free State.

Jacobs, T. L. (2017). *A meteorological analysis of extreme flood events in the southern parts of the eastern cape*. South Africa: Rhodes University.

Khandlhela, M., & May, Æ. J. (2006). Poverty, vulnerability and the impact of flooding in the Limpopo Province, South Africa. *Natural Hazards, 39*, 275–287. https://doi.org/10.1007/s11069-006-0028-4.

Lahiff, E., Maluleke, T., Manenzhe, T., & Wegerif, M. (2008). *Land redistribution and poverty reduction in South Africa: The livelihood impacts of smallholder agriculture under land reform*. Cape Town, South Africa: Programme for Land and Agrarian Studies.

Le Roux, A., & Van Huyssteen, E. (2010). Socio-economic landscape: The south African socioeconomic and settlement landscape. In *South Africa risk and vulnerability atlas* (pp. 15–21). Pretoria, South Africa: CPD Print.

Le Roux, A., van Huyssteen, E., Arnold, K., & Ludick, C. (2019). *Green Book. The vulnerabilities of South Africa's settlements*. Pretoria, South Africa. Retrieved from https://pta-gis-2-web1.csir.co.za/portal/apps/GBCascade/index.html?appid=280ff54e54c145a5a765f736ac5e68f8.

Mahanta, R., & Das, D. (2017). Flood induced vulnerability to poverty: Evidence from Brahmaputra Valley, India. *International Journal of Disaster Risk Reduction, 24*, 451–461. https://doi.org/10.1016/j.ijdrr.2017.04.014.

Mahmood, S., Shaker, A. K., & Mayo, M. (2016). Exploring underlying causes and assessing damages of 2010 flash flood in the upper zone of Panjkora River. *Natural Hazards, 83*(2), 1213–1227. https://doi.org/10.1007/s11069-016-2386-x.

Mandleni, B. (2011). *Impact on climate change and adaptation on cattle and sheep farming in the eastern Cape Province of South Africa*. University of South Africa. Retrieved from http://umkn-dsp01.unisa.ac.za/handle/10500/5477

Masifunde, & Zingisa. (2011). Threats to the food security and food sovereignty in the Eastern Cape: Impacts of the Massive Food Production Programme (MFPP), GMOs and cash crops in four villages in the Amathole District Municipality.

McElwee, P., Nghiem, T., Le, H., & Vu, H. (2017). Flood vulnerability among rural households in the red River Delta of Vietnam: Implications for future climate change risk and adaptation. *Natural Hazards, 86*(1), 465–492. https://doi.org/10.1007/s11069-016-2701-6.

Municipalities of South Africa. (2020). *Port ST Johns local municipality*. Retrieved October 27, 2020, from https://municipalities.co.za/map/1038/port-st-johns-local-municipality

Musyoki, A., Thifhulufhelwi, R., & Murungweni, F. M. (2016). The impact of and responses to flooding in Thulamela Municipality, Limpopo Province, South Africa. *Jàmbá: Journal of Disaster Risk Studies, 8*(2), 1–10. https://doi.org/10.4102/jamba.v8i2.166.

NAMC. (2013). *National red Meat Development Programme: 3rd beef cattle farming report*. Pretoria, South Africa.

Ndoro, J. T., Mudhara, M., & Chimonyo, M. (2014). Cattle commercialization in rural South Africa: Livelihood drivers and implications for livestock marketing extension. *Journal of Human Ecology, 45*(3), 207–221.

Nethengwe, N. S. (2007). *Integrating participatory GIS and political ecology to study flood vulnerability in the Limpopo Province of South Africa*. University of West Virginia.

News24. (2019, April 22). *Hundreds evacuated over flooding as heavy rainfall batters Port St Johns. News24*. Retrieved from https://www.news24.com/SouthAfrica/News/hundreds-evacuated-over-flooding-as-heavy-rain-batters-port-st-johns-20190422

Ngarava, S. (2019). *Evaluating livestock development Programmes through the production risk Interface: Case of the Kaonafatso ya Dikgomo (KyD) scheme in South Africa*. University of Fort Hare.

Niehaus, J. (2019). *Message from the NSTF Executive Director: No longer "safe as houses."* Pretoria, South Africa. Retrieved from http://www.nstf.org.za/wp-content/uploads/2019/05/EDs-May-2019.pdf

OR Tambo DM. (2016a). Agricultural industry analysis. In *OR Tambo master Agri-Park business plan* (pp. 51–82). Mthatha, Eastern Cape Province: OR Tambo District Municipality.

OR Tambo DM. (2016b). *Coastal management programme 2016–2021*. Mthatha, Eastern Cape Province.

OR Tambo DM. (2016c). District economic and demographic overview. In *OR Tambo master agri-park business plan* (pp. 38–50). Mthatha, Eastern Cape Province: OR Tambo District Municipality.

OR Tambo DM. (2016d). Red meat. In *OR Tambo master agri-park business plan* (pp. 83–117). Mthatha, Eastern Cape Province: OR Tambo District Municipality.

Pentz, E. (2009). *Marketing perceptions of the Drakensberger breed of cattle*. North West University.

Rojas-downing, M. M.. (2017). *Evaluating the impacts of climate change and variability on grazing dairy production*. Michigan State University.

Rojas-downing, M. M., Nejadhashemi, A. P., Harrigan, T., & Woznicki, S. A. (2017). Climate change and livestock: Impacts, adaptation and mitigation. *Climate Risk Management, 16*, 145–163. https://doi.org/10.1016/j.crm.2017.02.001.

SABC. (2019, April 23). *Heavy floods in Port St Johns*. SABC Online. Retrieved from https://www.sabcnews.com/sabcnews/heavy-floods-in-port-saint-johns/

Shisanya, S., & Mafongoya, P. (2016). Adaptation to climate change and the impacts on household food security among rural farmers in uMzinyathi District of Kwazulu-Natal, South Africa (pp. 597–608). doi: https://doi.org/10.1007/s12571-016-0569-7.

StatsSA. (2016). *Community survey 2016 agricultural households*. Pretoria. Retrieved from http://www.statssa.gov.za/?page_id=964

Theron, H. (2015). *Challenges and opportunities for beef production in developing countries of the Southern Hemisphere*. Retrieved April 5, 2018, from https://www.icar.org/Documents/Riga_2010/ppt/Theron.pdf

Turpie, J., & Visser, M. (2015). The impact of climate change on South Africa's rural areas. In *Financial and Fiscal Commission* (pp. 100–106). Johannesburg, South Africa: 2013/14 Division of Revenue.

UN. (2015). *Transforming our world: The 2030 Agenda for Sustainable Development* (Vol. 16301). Washington D. C.

UN. (2019). *The sustainable development goals report*. Geneva. doi: https://doi.org/10.29171/azu_acku_pamphlet_k3240_s878_2016.

Wikipedia. (2016). *Map of the Eastern Cape with Port St Johns highlighted*. Retrieved October 27, 2020, from https://en.m.wikipedia.org/wiki/File:Map_of_the_Eastern_Cape_with_Port_St_Johns_highlighted_(2016).svg

Williams, R., Malherbe, J., Weepener, H., Majiwa, P., & Swanepoel, R. (2016). Anomalous high rainfall and soil saturation as combined risk indicator of Rift Valley fever. *Emerging Infectious Diseases, 22*(12), 2054–2062.

Worklaw. (2019). *Chapter 7 of the constitution of the Republic of South Africa local government*. Retrieved March 17, 2020, from https://www.worklaw.co.za/SearchDirectory/Legislation/conchapter7.asp

Zuma, B. M., Luyt, C. D., Chirenda, T., & Tandlich, R. (2012). Flood disaster management in South Africa: Legislative framework and current challenges. In *International Conference on Applied Life Sciences (ICALS)* (pp. 127–132). Turkey, September 10–12: ISALS.

Part V

Increasing Risk of Tornadoes

14 Characterisation and Analysis of Emerging Localised Severe Storms in Malawi: How Common Are Tornadoes?

Sosten Staphiel Chiotha, Donald Reuben Kamdonyo, and Patrick Jeremy Likongwe

Abstract

Southern Africa has, of late, been on the receiving end of severe storms that have left behind serious trails of destruction from strong winds, heavy rainfall and floods. The most recent is Tropical Cyclone Idai, one of the most intense and most powerful storms to ever hit southern Africa. Idai brought death and destruction in Malawi, Mozambique and Zimbabwe in March 2019. While much attention is given to these severe storms that cover large areas and several countries, our attention has been drawn to the concurrent emergence of more localised storms that can be just as destructive as they are poorly understood, and not well integrated into early warning and response systems in the region. An analysis of these devastating storms was based on eyewitness accounts and post-event damage assessment from five sites between 2012 and 2020. Results show that some of the severe localised storms in Malawi are tornadoes. This first attempt in Malawi to characterise these emerging storms will emphasise the need for improved tornado forecasting capacity, documentation of occurrence and effects, and awareness raising on risk reduction.

Keywords

Disaster impacts · Thunderstorms · Rotating wind · Severe local storms · Tornados

14.1 Introduction and Background

Tornadoes are intense, rare, localised events which create large amounts of damage and destruction. There is need to increase the understanding of the causes of vulnerability to these severe phenomena which can also lead to injury and death (Fricker 2015). The definition of tornadoes has been evolving (Edwards 2012). More recent definitions consider a tornado to be a violently rotating column of air, in contact with the ground, either pendant from or underneath a cumuliform cloud and often (but not always) visible as a funnel cloud (Goliger et al. 1997; Glickman 2000; NOAA 2003; Agee 2014; The University of Rhode Island, 2020). Because of the violent rotation, tornadoes are also popularly called twisters (National Geographic 2019). The intense gradients of wind speed and direction associated with most tornadoes make them unique among windstorms (NOAA 2003).

S. S. Chiotha (✉) · P. J. Likongwe
Leadership for Environment and Development (LEAD), Zomba, Malawi
e-mail: schiotha@leadsea.mw; plikongwe@leadsea.mw

D. R. Kamdonyo
Kamdonyo & Associates, Lilongwe, Malawi

G. Nhamo, L. Chapungu (eds.), *The Increasing Risk of Floods and Tornadoes in Southern Africa*, Sustainable Development Goals Series, https://doi.org/10.1007/978-3-030-74192-1_14

Tornadoes are destructive, and this is attributed to the strong vertical acceleration of the upward wind motion near the surface, driven by the pressure field of the tornado itself, creating a suction effect through which objects and structural elements in their path are lifted into the air. The nature of debris in a tornado is a function of the wind speed and the type of objects encountered in the tornado path.

The evolving definition of the tornado over time has been reflected in the changing nature of frameworks for differentiating one tornado from another. The initial attempt to standardise assessment and characterisation of tornadoes was done by Theodore Fujita in the 1970s when he came up with a rating scale which was adopted at the time as a standard for estimating the intensity of tornadoes (Fujita 1971; NOAA 2003). The scale comprised six categories by wind speed and observed damage associated with each category from scale F0 being light damage to F5 as the highest (Fujita 1971). The F scale was initially assigned to levels of destruction to well-built homes, corresponding to wind speed estimates based on Beaufort and Mach scales (Edwards et al. 2013). The estimation of wind speed from damage assessment was justified because the extent of damage is related to wind speed, and damage assessment could even be applied to historical events through analysis of archival records of newspaper accounts and photographs.

Proxies are important because the chances of a tornado strike on a fixed, well-calibrated and sufficiently sturdy anemometer to withstand the enormous force of the tornado are quite rare (Edwards et al. 2013; National Geographic 2019). This means that post event damage surveying remains the most common method for estimations of tornado strength (National Geographic 2019). However, there is a challenge because the relationship between wind speed and damage is not a simple one but is rather influenced by the nature and distribution of structures and vegetation encountered along the path. For instance, a tornado passing through an open field experiences less resistance at its boundary layer compared to one passing through a built-up area, or an area with large trees. Therefore, a relatively weak tornado entering a community from an open space might do its worst damage early in its track and then seem to weaken as it encounters more obstacles.

The lack of simple relationship between wind speed and damage may be one of the factors contributing to lack of robust and well-researched information on tornadoes in many parts of the world. This is compounded by lack of capacity for timely assessment, especially in countries where actual and perceived tornado frequency is low. Nevertheless, the improvement in knowledge of structural integrity and how structures react to the stress imposed by windstorms such as tornadoes has significantly enhanced the validation of damage-wind speed relationship assumed by the Fujita scale. As argued by Van Niekerk and Sampson (1999), there is a need to carry out damage surveys within 2 days of a severe event to have as much information as possible for analysis later. Apart from wind speed, the other characteristics of tornadoes include size as defined by path length and width (Goliger et al. 1997). However, Van Niekerk and Sampson (1999) have argued that the size of a tornado is not necessarily an indication of its intensity because tornadoes exhibit dynamic variation in size and strength with time and as argued earlier this may also be influenced by the obstacles in the tornado path. Since February 2007, the National Weather Service in the USA has used the EF (Enhanced Fujita) scale, which takes more variables into account when assigning wind speeds to a tornado (NOAA, 2006; National Geographic 2019).

A critical challenge associated with tornadoes relates to the latitudinal position and forecasting techniques because it requires sophisticated monitoring tools which include Doppler radar, satellites, weather balloons and computer modelling to watch the skies for severe storms and tornadic activity by identifying wind speeds and areas of rotation within thunderstorms (National Geographic 2019). Malawi lies with the tropics with a barotropic atmosphere which makes it more chaotic thus contributing to forecasting and tracking challenges to this rare and microscale phenomenon. Malawi has a satellite-based ground receiving station coupled with 7 km resolution consortium of small-scale modelling (COSMO) which can be downscaled to 2.7 km. However, even where the capacity and conducive weather conditions exist like in the USA, the warning time for tornadoes, unlike hurricanes, is very limited. According to the National Geographic (2019), the Doppler radar technology has marginally improved the warning time for tornadoes from fewer than 5 min in the 1980s to an average of 13 min by the late 2000s. Conducive weather conditions for tornado formation include the collision of warm humid air and cold air creating wind shear and instability producing a fast change in wind speed or direction in a short period of time or distance.

While some countries in Africa have confirmed reports on tornado activity, Malawi did not have any such reports until more recently (Chiotha and Kamdonyo 2016). At the onset of every rainy season, between November and early December, Malawi faces severe local thunderstorms with strong winds that cause extensive damage to buildings and trees, but there had been no attempt to investigate if some of these tropical storms would, in fact, be tornadoes. The trend for the storms, however, has been that once the season gets established in mid-December, the damaging storms reduce in frequency and severity because by this time most of the heat energy generated into the atmosphere during the hot months of October and November have been expended (Kimambo 2018).

This chapter reports the investigations that were conducted to validate the suspicion that some of these tropical storms could indeed have been tornadoes. It further attempts to characterise these emerging storms to inform policy on early warning, response and rehabilitation, including managing the trauma experienced by affected communities. The findings are significant in that the impacts and adaptation actions to these phenomena are linked to Sustainable Development Goals (SDGs), especially SDG1 (impact on the poor and the vulnerable), SDG2 (impact on food security), SDG4 (education disruption as the school infrastructure gets damaged), SDG7 (disruption of energy infrastructure), SDG11 (need for resilient and safe infrastructure) and SDG13 (understanding impacts of climate change and need for appropriate action).

14.2 Materials and Methods

Climate variability is now recognised as a major influence on the performance of the economy and social well-being in Malawi and Southern Africa more widely (Clay et al. 2003), hence it is important that all emerging extreme weather events are characterised to inform early warning and response systems. Emerging extreme weather events include several catastrophic or unusual weather events taking place at an accelerated pace, notably floods, fire hazards, droughts and tropical storms (IPCC 2007; Stern 2007; IPCC 2014; Serdeczny et al. 2016). Malawi has two main seasons: the wet season is from October to April, while the dry season is from May to September. At the onset of every rainy season, especially between November and December, Malawi faces severe local thunderstorms with strong winds that cause extensive damage to buildings and trees, but there had been no attempt to investigate if some of these tropical storms would, in fact, be tornadoes. This study investigates five thunderstorm events of unusually strong winds.

The study was conducted in three districts across two regions, covering rural, peri-urban and urban settings. The study sites were selected based on reported 'unusual' thunderstorms through WhatsApp social media groups and confirmed by local government officials. The study period covers summer rainy seasons from 2012 to 2020. The Eastern and Central Regions of the country were picked. From the Eastern Region, Zomba District was selected, with Songani township (peri-urban), Sikinala township (urban) and Mulunguzi township (urban) being study sites. From the Central Region, Dowa District was selected, with Mtiti Village (rural) as the site. In addition, Lilongwe District's Mliritsa Village (rural) was the site.

The study targeted eyewitnesses who were first asked to narrate the occurrence of the storm in terms of time, duration, movement and any destruction caused. There were no credible eyewitnesses at Songani because 5 years had passed since the event occurred. A survey consisting of open-ended and closed responses was administered in person at four sites that had experienced these unusual storms.

In the four remaining sites, the eyewitness accounts were both individual key informant interviews and focus group discussions (FGD). The big test for the methodology was to show the respondents one of the top five tornado videos captured by tornado chasers in the USA. This was a violent storm at Katie, Oklahoma on ninth May 2016, captured by Brunin and McGowan (2015) (https://www.weather.gov/oun/events-20160509-ef4tornado). This was shown to the eyewitnesses to relate what they saw locally against what they saw in the video to make their independent comparison which was used to access if what they saw was indeed a tornado or just mere whirlwind. Permission was sought and granted for the interviews to be recorded. The first three questions on the questionnaire were for establishing the credentials of the interviewees (names, age and gender). The next five questions were to establish if indeed what they saw was an electric storm (thunderstorm) with lightning, thunder and hail which are often characteristic of storms that produce tornadoes. The next open-ended question was for the interviewees to explain in their own words what they saw. As they explained, questions were asked to ascertain if the wind was rotating like a whirlwind, and also in what direction it was moving. This was followed by a question that was to establish if the person had witnessed such a phenomenon before to ascertain that the occurrence was the first time or had happened before. In Mitundu, the FGD involved community members of Mliritsa Village including the Village Headman. Also present was Village Headman Chimphaninji as recommended by the members since the latter's village had also been affected by the same storm.

At a community workshop in southern Malawi, an opportunity availed itself to survey severe storms in the southern region of the country. The workshop had 25 participants who were rural farmers engaged in irrigation in the various districts of the region. These participants listed the impacts of climate change in their areas and severe storms with strong winds came up repeatedly. Ten of these participants were asked to complete the same questionnaires and shown videos of tornadoes that had been shown to the eyewitnesses above.

14.2.1 Damage Assessment

Damage assessment was conducted in three case studies, namely Sikinala and Mulunguzi in Zomba and Mliritsa in Lilongwe because the researchers were able to visit the affected locations within the recommended time period: within 2 days for Mliritsa and within hours for Mulunguzi and Sikinala. The damage assessment was based on elements of the F scale and EF scales. These included describing the damage patterns to buildings, trees and crops (maize) as well as the nature of the debris and drop-off pattern of debris.

14.3 Presentation and Discussion of Results

14.3.1 Eye-Witness Accounts

Eye-witness accounts for some damaging wind events covering the periods 2012/2013, 2016/17 and 2017/18 appeared unusual, with many descriptions of 'whirlwinds embedded into thunderstorms'. The eye-witness accounts mention rotating wind, rotating smoke-like whirlwind, and terrifying roaring. This is consistent with the characterisation of tornadoes as twisters (National Geographic 2019). All the eye-witness accounts were in agreement on unusually loud noise associated with the event, and this is consistent with tornado reports elsewhere (Hyndman and Hyndman 2006; Grams et al. 2011; National Geographic 2019) although the noise on its own does not define a tornado because any intense thunderstorm wind can produce a roaring sound (Wisconsin Bureau of Environmental and Occupational Health 2019). The noise may have been attributed to various sources, such as the sound associated with destruction or collision of flying debris, or even the thunder that accompanies the twister. The noise phenomenon has been an area of studies (Abdullah 1966; NOAA 2011). One eye-witness account mentioned that the rotating wind was hanging from black clouds, and this is consistent with the scientific definition of a tornado being a rotating column of air, in contact with the ground, either pendant from a cumuliform cloud or underneath a cumuliform cloud (Goliger et al. 1997; Glickman 2000; NOAA 2003; Agee 2014). The eye witnesses also characterised impact from shaking and breaking of trees, knocking down of perimeter brick fences, roofs being blown off and debris lifted into the rotating column of air. Iron sheets were not only blown off buildings but were blown longer distances and also twisted—damages that are synonymous with tornadoes.

These eyewitness observations were validated in the damage surveys but are also consistent with F and EF scales relating wind speed to associated damage (Fujita 1971; Edwards et al. 2013). All eyewitnesses alluded to rains on that particular day although some indicated that the rains came after the winds. This is normal in a severe storm because the winds come as downdrafts preceding the rains (Hyndman and Hyndman 2006). Some eyewitnesses reported about thunder and lightning, while others could not remember. Only the Mulunguzi group mentioned hail. Lightning and hail are common in thunderstorms that produce tornadoes (Wisconsin Bureau of Environmental and Occupational Health 2019), but some researchers have argued on the need for discrimination between significant tornadoes and significant hail and wind events in the absence of tornadoes (Grams et al. 2011). All eye-witness accounts indicated that the events occurred in the afternoon between 1500 and 1600 and that the strong winds were brief—in the range of 5–10 min. This is in agreement with Kimambo (2018) who quoted the South African weather service that in South Africa, most tornadoes typically occur in the late afternoon or early evening and that they are brief. The results were that seven districts in the Southern region of Malawi had experienced tornadoes, some districts every year since 2009.

An insight into some of the study sites is further outlined in the succeeding paragraphs. For instance, in Mitundu eye witnesses reported red dust being lifted high into the sky which got mixed with other debris creating this moving structure with exploding and roaring sound and at times looking like a column of smoke. This is validated by a picture (Fig. 14.1) taken by an amateur photographer some distance from the epicentre of the event who thought someone was burning tyres. This smoke-like and red dust characterisation of the tornado may be explained by Fricker (2015) who has argued that a tornado is not a purely rotational object but has both volume and density components to its mass.

In the Mliritsa case study, the area is relatively flat and the communities were able to spot the tornado from a distance, and because of the force it was lifting debris, they did not think a house would be a safe place to hide and so they sought shelter in the community cemetery because of the good tree vegetation cover. According to Ellis et al. (2019), perceptions of tornado risk are related to their physical environment, and in this case, the cemetery was perceived to be a safe place. Advice from the Wisconsin Bureau of Environmental and Occupational Health (2019) is to shelter in a low area under a thick growth of small trees in an emergency, rather than near large trees or powerlines that are easily felled by the wind. However, computer simulations and laboratory studies in tornado vortex chambers are inconclusive because they have shown that the surface roughness, being the measure of how disrupted the wind near the ground is by objects such as dirt, rocks, hills, trees, and even houses, can either increase or decrease the wind speeds in a tornado. In Mliritsa, an 80-year-old man escaped injury due to sheer luck because the roof and wall of his house collapsed except the corner where he was sleeping (Fig. 14.1). The younger people who saw it coming and ran to the cemetery to seek shelter left the old man behind. He was still asleep when they came back and was surprised to be rescued from a house that was partially collapsed. It is in Mliritsa where the eye-

Fig. 14.1 The Mitundu tornado, captured by an amateur photographer, showing a rotating wind with dust, rising from the ground into the base of a dark rain cloud. Source: Ellasy Chisale, 2018

witnesses were traumatised by the event, in keeping with the literature reporting that tornado exposure may cause post-traumatic stress disorder (PTSD) (Niederkrotenthaler et al. 2013). Reasons for this might be because of the flatness of the terrain they had a full view of the approaching tornado, something they had not experienced before and some of them reported having recurring nightmares. They also reported feeling guilty for leaving an elderly and sickly man behind. These reports are consistent with suggestions by Fricker (2015) who argues that middle-aged and elderly populations are at much greater risk than younger people during tornadoes.

In the Mtiti case study, one eyewitness indicated that the roof and household items (including clothes) were 'sucked' out of the house into the air and spun around. The suction is consistent with tornado dynamics because there is a strong vertical acceleration of the upward motion near the surface, driven by the pressure field of the tornado itself which acts to lift objects and structural elements into the air (NOAA 2003).

In the Mulunguzi case study, one eyewitness indicated that the house was shaking as if in an earthquake, and felt the safe place to hide was a maize field. This may have been wise in terms of reducing risk from falling trees, but there would still be a risk of injury from projectile debris carried in the wind (NOAA 2003). Another issue raised by the eyewitnesses was successive days of high afternoon temperatures leading up to the event. In all the case studies there were no injuries reported except one individual who had a minor cut to the sole of his left foot after stepping on a sharp object that was part of the debris that landed on a footpath.

The responses from some of the workshop participants were similar to the eyewitness accounts above. Based on such responses, it was established that seven districts in the Southern region of Malawi may have experienced tornadoes, with some districts having episodes every year.

14.3.2 Damage Assessment

14.3.2.1 Trees and Crops

The impact on maize crops exhibited damage patterns consistent with rotating wind because not only were the maize stalks flattened but they also portrayed circular pattern in different parts of the maize field. This observation is consistent with Hyndman and Hyndman (2006) who argue that tornadoes differ from downbursts because the former creates a damage pattern of trees and other debris lying at many angles while for the latter, they lie in a straight line away from the point of touch down. The damage to trees included broken branches and outright uprooting (Fig. 14.2). Interestingly, some maize and trees remained intact even though eyewitness account reported vigorous shaking of trees due to the strong wind. This observation is widely reported in the literature that severe damage adjacent to those with only minor damage was due to smaller vortices that rotate around a tornado causing intense damage while missing out others in their paths (Hyndman and Hyndman 2006). Some mango trees in one location were uprooted while in other locations they were not. The same was true for huge eucalyptus trees as well as billboard stands and power lines that fell while others remained intact. One of the trees whose branches and trunk was broken into so many small pieces was *Cussonia arborea*, the octopus cabbage tree, locally known as Mbwabwa.

14.3.2.2 Damage to Property

According to a preliminary assessment by Zomba City Council (2017) after a potential tornado event, a total of eight houses were damaged, affecting 36 people while 9 institutional and business structures were damaged (including a school, lodge, orphanage and guest house), affecting 111 people along the way. The direct storm damage to houses affected the roofs, house and perimeter walls. The simplest sign of the storm in some houses was ceiling trapdoors being pushed out of place with no other visible damage to the house infrastructure (Fig. 14.3). The next level minimal impact was a manifestation of upward warping deformation of roofing iron sheets at several points without a lift-off from the roof mountings. However, more serious damage involved whole iron sheets being blown off roof mountings on the whole or part of the buildings. Sometimes, the iron sheets were blown off together with roof mountings (rafters) and not the trusses.

Fig. 14.2 A tree uprooted along the path of the storm in Sikinala, Zomba. Source: Photo by Sosten Chiotha 2017

Fig. 14.3 Property damages. Source: Photos by Sosten Chiotha, 2017

Some of the iron sheets were circularly trapped in trees throughout the crown while others were stuck against tree trunks in a V shape showing the direction of the wind that was consistent with the direction of affected nearby maize stalks (Fig. 14.4). Some loose iron sheets manifested severe twisting deformations. The blown-off iron sheets were dropped from within metres up to a kilometre away from the houses, and one house owner simply picked a different iron sheet that had landed from elsewhere to fix his roof after being unable to locate all his iron sheets near his house. In urban centres, the Mulunguzi tornado resulted in some perimeter brick fences collapsing. House brick wall collapses were only observed in the rural areas and not in the urban centres. According to Goliger and Retief (2007) information on damage to informal settlements and possibly also the rural areas is grossly underestimated, owing to poor communication and reporting systems, including inadequate coverage by the media.

The variation in the extent of damage to infrastructure and other objects may be explained as alluded to earlier by Hyndman and Hyndman (2006) as a function of the nature of the wind hazard (i.e. risk) but also as a function of nature of the impacted object which constitutes resistance (i.e. vulnerability). This suggests that the damage potential is the product of risk and vulnerability (Davenport 1997; Goliger and Retief 2007) where high-risk areas are bare grounds as compared to those with a good tree cover (Fig. 14.5). Potential risk factors for the damage include effects of climate change manifested in increased extreme weather events, as echoed by IPCC (2014), loss of windbreaks due to reduced tree cover and habitation in fragile areas (Zomba City Council 2017). Almost all structures affected were in planned and con-

Fig. 14.4 Severe twisting deformations of iron sheets and a different iron sheet picked from around and fixed to roof. Source: Photos by Sosten Chiotha, 2017

Fig. 14.5 High-risk, bare ground areas. Source: Photo by Sosten Chiotha, 2017

trolled places where risk as a result of having substandard buildings was ruled out.

14.3.3 Weather Patterns—Synoptic Analysis

The Mulunguzi case study benefited from the nearby automated weather station at LEAD, located about 300 m from the epicentre of this storm, which covered an area approximately 200 m wide by 700 m long. On 10 February 2017, rain clouds were seen forming to the east of Zomba in the early afternoon, and a picture of the cloud formation was captured at 1507 out of curiosity (Fig. 14.6). Because of the windy weather conditions outside that were also being monitored on the automated weather station console, it was noted that at 1510 the wind speed had quickly risen from about 5–30 km/h and kept on rising, reaching 70 km/h in just 6 min at 1516, at which tree branches started breaking off and one tree fell and knocked down part of the perimeter brick wall for the LEAD office. While the wind speed was picking up, it had also started raining heavily. The weather monitor showed that the rainfall intensity was at 245 mm/h at 1516 h when the wind speed was at 69.2 km/h (Fig. 14.7), the highest in the 5 years LEAD has been recording weather using the Davis automated station. While the strong winds quickly subsided in about 5 min, the rainfall intensity continued to rise to reach the peak of 340 mm/h at 1527 h but quickly eased and stopped by 1600 h. The 24-h rainfall reading was 83.3 mm for that day. The average wind speed for that day (0.8 km/h) was not very different from the other days indicating that the extreme wind was just a brief unusual episode.

14.3.4 National Weather Picture

During the period 3–12 February 2017, the Congo Air Mass got established, and the Inter-Tropical Convergence Zone (ITCZ) frequently oscillated across the country. The combined effects of these two rain-bearing systems resulted in scattered to widespread rains and thunderstorms over Malawi which were severe in places. This is a conducive atmosphere for tornado-bearing storms (Figs. 14.8, 14.9, 14.10). Box 14.1 gives a sample of reports of extreme weather events that occurred in February 2017 reported through Malawi Weather Chasers Forum on WhatsApp.

Fig. 14.6 Cloud cover a few minutes before the storm that caused damage in Zomba. Source: Photo by Sosten Chiotha, 2017

Box 14.1 Sample of Reports on the Extreme Weather Episodes in 2017 Submitted Through the Malawi Weather Chasers Forum on WhatsApp

- Lilongwe area 15 recorded 64.8 mm of rainfall while Kamuzu international airport recorded 108.0 mm on 3 February.
- Houses collapsed due to heavy rains in Mchesi township in Lilongwe on 4 February.
- Nankhaka River in area 18, and Lumbadzi River, both in Lilongwe, burst their banks on 4 February.
- Malosa agriculture recorded 94.0 mm of rainfall on 4 February and Domasi River burst its banks on the same day resulting in floods.
- Lifidzi River in Salima burst its banks on 10 February.
- Nankhaka and Lingadzi rivers burst their banks resulting in floods in Lilongwe's areas 47 and 49 on the morning of 10 February

During the period 24–26 January 2018, the Inter-Tropical Convergence Zone (ITCZ), which is one of the main rain-bearing systems for Malawi, was active over the Central and Northern areas of the country. The weather forecast by DCCMS for 25 January talked of a broad rain-belt being active over Malawi resulting in thunderstorms over central and northern areas. Picture of rapidly developing thunderstorms (RDT) on 25 January 2018 shows a series of thunderstorms strewn across the central and northern areas at 1345- and 1515-h local time, respectively. Over the Mitundu area, there was a mature thunderstorm (purple colour) visible (Fig. 14.11).

14.4 Conclusions

Based on the observed characteristics of the winds that have been analysed, the type of destruction in the storm path and the identification of a tornado through the videos by all eyewitnesses, it can be inferred that tornadoes touched down from severe storms at Mtiti in Dowa District on 8 February

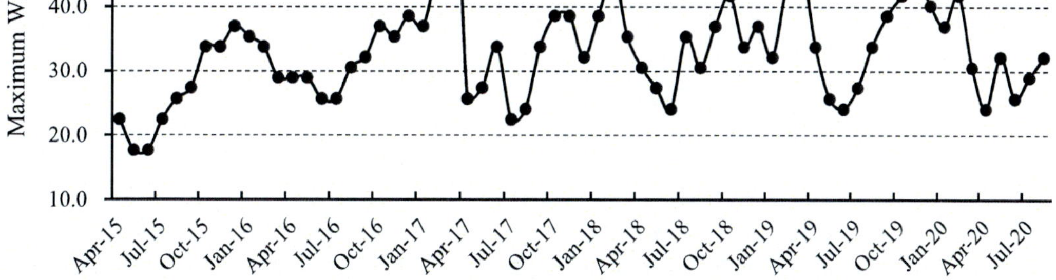

Fig. 14.7 A trend of the highest wind speed recordings in the month. Source: Authors, 2020

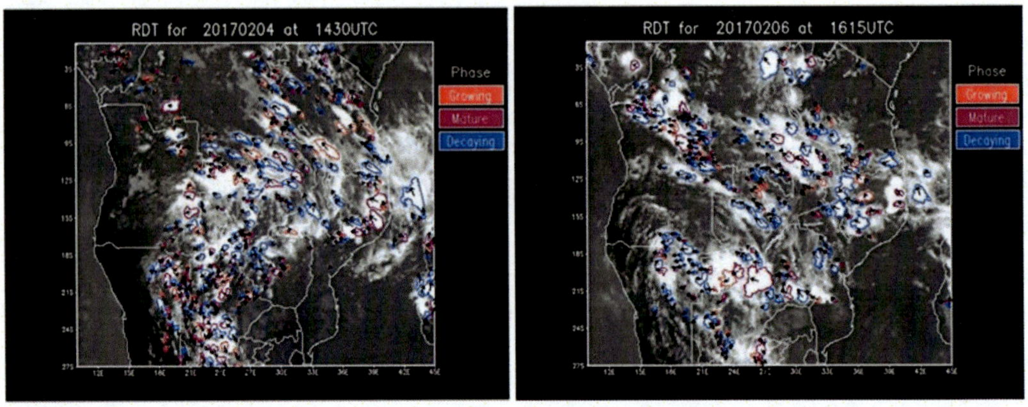

Fig. 14.8 Picture of Rapidly Developing Thunderstorms (RDT) on 4 and 6 February 2017. Source: Severe Weather Forecast Demonstration Project—Regional Specialised Meteorological Centre, Pretoria, South Africa 2017

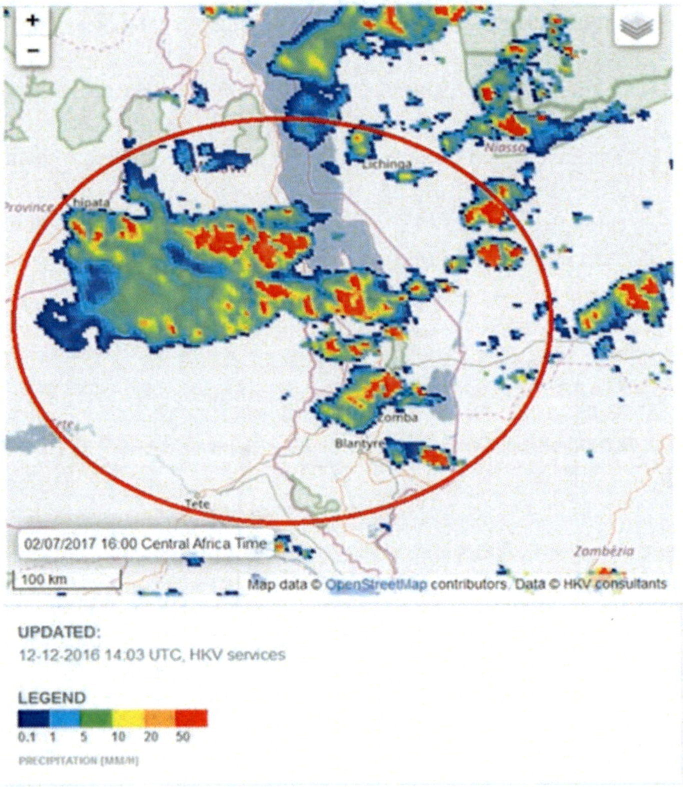

Fig. 14.9 Forecast rainfall for 7 February 2017 over Central and Southern Malawi. Source: HKV Services 2017

2017, in the Mulunguzi area of Zomba City on 10 February 2017 and at Mitundu in Lilongwe District on 25 January 2018. The first two tornadoes occurred during the period the country was under active weather brought about by the simultaneous prevalence of both the Congo Air Mass and the ITCZ, which are the main rain-bearing systems in this part of the continent, while the third tornado occurred during the period the ITCZ alone was active.

This chapter is the first to formally report on tornadoes in Malawi, but as reported by Goliger et al. (1997), the most significant wind damage to human development over the inland areas of the

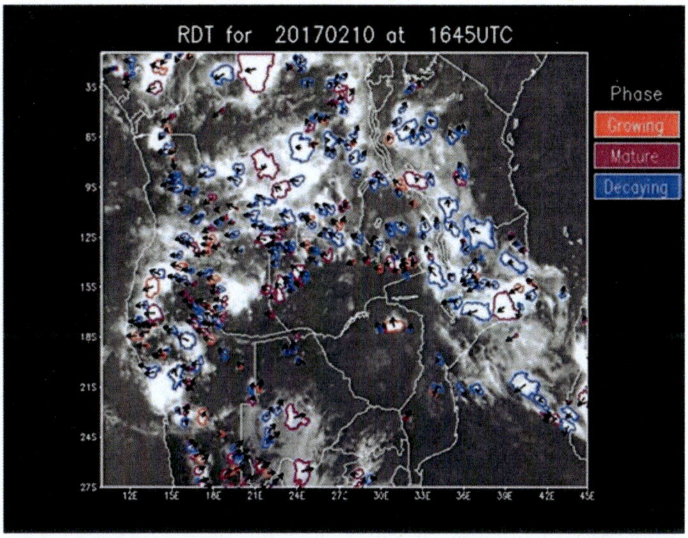

Fig. 14.10 Picture of Rapidly Developing Thunderstorms (RDT) on 10 February 2017. Source: Severe Weather Forecast Demonstration Project—Regional Specialised Meteorological Centre, Pretoria, South Africa 2017

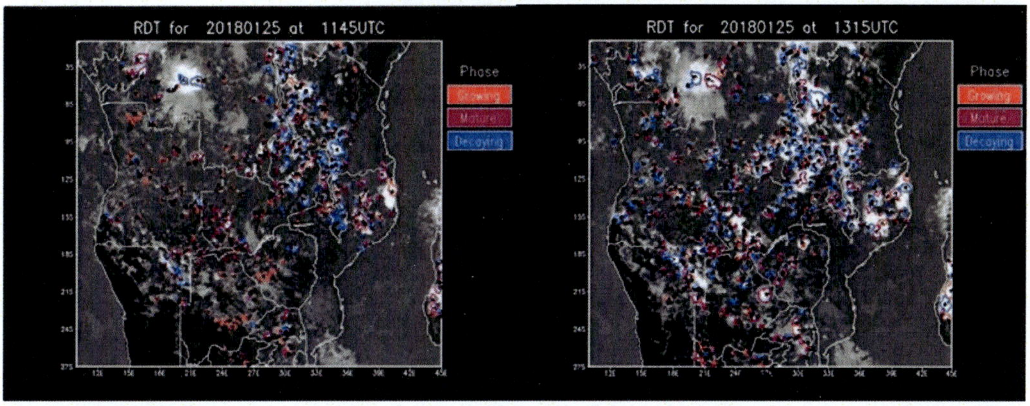

Fig. 14.11 Pictures of Rapidly Developing Thunderstorms (RDT) on 25 January 2018 at 1145- and 1315-h UTC (1345 and 1515 h's local time, respectively). Source: Severe Weather Forecast Demonstration Project—Regional Specialised Meteorological Centre, Pretoria, South Africa 2018

southern African region is caused by tornadoes and downbursts. The engagement with stakeholders at a climate change meeting showed from their recollections that such events have been occurring as early as 2009. The naming of the tornado at Mtiti as *namondwe* in the vernacular Chichewa language by a local eyewitness may suggest that these events existed even much earlier than the current analysis especially that they can differentiate *namondwe* as rotating destructive winds during the rainy season as opposed to *kavuluvulu* (whirlwind) that occurs during the dry months.

There is, therefore, need for tornadoes to be given greater attention in Malawi to get a true picture of their occurrence and impact. This is in agreement with NOAA (2011) on the need for developing capacity for rapid updates on tornado location, movement and damage. In Malawi, the Weather Chasers WhatsApp group and others help to draw attention to these events, and, as reported by NOAA (2011), these social media platforms can support communication of damage path information after the event. As argued by NOAA (2011) documenting and sharing photos and videos of tornado damage help people to appreciate the destructive potential of major tornadoes and that where this has been done it has influenced the change of behaviour from perceiving a tornado as unlikely and therefore not requiring action, to being motivated to take action despite tornadoes being low probability occurrences.

References

Abdullah, A. J. (1966). The "musical" sound emitted by a tornado. *Monthly Weather Review, 94*(4), 213–220.

Agee, E. M. (2014). A revised tornado definition and changes in tornado taxonomy. *Weather and Forecasting, 29*, 1256–1258. https://doi.org/10.1175/WAF-D-14-00058.1.

Brunin, D. & McGowan, D. (2015). Tornado videos: Copyright by korry@severestudios.com, Retrieved April 6, 2017.

Chiotha, S. & Kamdonyo, D. (2016). *Tornadoes Affecting Malawi?* Paper presented at the First Stakeholders' Workshop on Enhancing Early Warning Systems in Malawi; LEAD SEA Publications, Blantyre, Malawi, 10–11 August 2016.

Clay, E., Bohn, L., de Armas, E. B., Kabambe, S., & Tchale, H. (2003). *Malawi and Southern Africa-climatic variability and economic performance,* Disaster Risk Management Working Paper Series No. 7, World Bank Group, Washington DC, USA.

Davenport, A. G. (1997). The vulnerability of structures in catastrophic windstorms. In: Goliger, A.M., Retief, J.V. (2007). Severe wind phenomena in southern Africa and the related damage. *Journal of Wind Engineering and Industrial Aerodynamics, 95*, 1065–1078.

Edwards, R. (2012). Tropical cyclone tornadoes: A review of knowledge in research and prediction. *Electronic Journal of Severe Storms Meteorology, 7*(6), 1–61.

Edwards, R., LaDue, J. G., Ferree, J. T., Scharfenberg, K., Maier, C., & Coulbourne, W. L. (2013). Tornado intensity estimation–Past, present and future. *American Meteorological Society, 94*(5), 641–653.

Ellis, K. N., Mason, L. R., & Gassert, K. N. (2019). Public understanding of local tornado characteristics and perceived protection from land-surface features in Tennessee, USA. *PLoS One, 14*(7), e0219897.

Fricker, T. A. (2015). *Estimating kinetic energy of US tornadoes*. Dissertation, The Graduate School, Florida State University. Retrieved June 10, 2020.

Fujita, T. T. (1971). *Proposed characterization of tornadoes and hurricanes by area and intensity. Satellite and Mesometeorology Research Project Report 91*, Univ. of Chicago, 42 pp. Retrieved May, 14, 2020.

Glickman, T. (Ed.) (2000). *Tornado. Glossary of meteorology* (p. 585). 2nd ed. American Meteorological Society, 585.

Goliger, A. M., Milford, R. V., Adam, B. F., & Edwards, M. (1997). *Inkanyamba—Tornadoes in South Africa.* United Litho

Goliger, A. M., & Retief, J. V. (2007). Severe wind phenomena in southern Africa and the related damage. *Journal of Wind Engineering and Industrial Aerodynamics, 95*, 1065–1078.

Grams, J. S., Thompson, R. L., Snively, D. V., Prentice, J. A., Hodges, G. M., & Reames, L. J. (2011). A climatology and comparison of parameters for significant Tornado events in the United States. *Weather and Forecasting, 27*, 106–123.

Hyndman, D., & Hyndman, D. (2006). Hurricanes and nor Easters: The big winds and hurricane Katrina, a case study of the costliest disaster in US history. In D. Hyndman & D. Hyndman (Eds.), *Natural hazards and disasters*. USA: Cengage Learning.

IPCC. (2007). *Climate change 2007: Synthesis report. Contribution of working groups I, II and III to the fourth assessment report of the intergovernmental panel on climate change Core writing team: R.K. Pachauri, A. Reisinger, (Eds.)*. Geneva, Switzerland: IPCC. (104 pp). Retrieved April 20, 2020.

IPCC. (2014). *Climate change 2014: Synthesis report. Contribution of working groups I, II and III to the fifth assessment report of the intergovernmental panel on climate change core writing team: R.K. Pachauri, L.a. Meyer (Eds.)*. Geneva, Switzerland: IPCC, (151 pp). Retrieved June 7, 2020.

Kimambo, O. N. (2018). Tornadoes in South Africa: An overview with a case study. doi:https://doi.org/10.20944/preprints201811.0547.v2. Retrieved July 22, 2020.

National Geographic. (2019). *Tornadoes, explained.* Retrieved May, 17, 2020 from https://www.nationalgeographic.com/environment/natural-disasters/tornadoes/.

National Oceanic and Atmospheric Administration (NOAA) National Weather Service. (2003). *A guide to F-Scale damage assessment.* Maryland: US Department of Commerce.

National Oceanic and Atmospheric Administration (NOAA) National Weather Service. (2011). *The historic tornadoes of April 2011*. Maryland: US Department of Commerce.

National Oceanic and Atmospheric Administration (NOAA) National Weather Service. (2006). *The enhanced Fujita scale (EF Scale)*. Retrieved July 14, 2020, from http://www.spc.noaa.gov/efscale. Accessed 15 July 2020.

Niederkrotenthaler, T., Parker, E. M., Ovalle, F., Neo, R. E., Bell, J., Xu, L., Morrison, M. A., Mertzlufft, C. E., & Sugerman, D. E. (2013). Injuries and post-traumatic stress following historic tornadoes: Alabama, April 2011. *PLoS One, 8*(12), e83038.

Serdeczny, O., Adams, S., Baarsch, F., Coumou, D., Robinson, A., Hare, W., Schaeffer, M., Perrette, M., & Reinhardt, J. (2016). Climate change impacts in sub-Saharan Africa: From physical changes to their social repercussions. *Regional Environmental Change, 17*(6), 1585–1600.

Stern, N. (2007). Climate Stern review: The economics of climate change. *New England Journal of Public Policy, 21*(4). Retrieved September 15, 2020, from http://scholarworks.umb.edu/nejpp/vol21/iss2/4.

The University of Rhode Island. (2020). *Hurricanes: Science and society. The hurricanes: Science and society register*. Retrieved May 14, 2020, from http://www.hurricanescience.org/society/impacts/tornadoes/.

Van Niekerk, H., & Sampson, G. (1999). *Hell season or par for the course: Tornadoes over the eastern cape 1998/99 season*. Internal report number OB/17, South African Weather Bureau.

Wisconsin Bureau of Environmental and Occupational Health. (2019). *Severe thunderstorms and Tornadoes toolkit*. State of Wisconsin, USA: Wisconsin Climate and Health Program.

Zomba City Council. (2017). *Zomba city disaster report February 2017*. Malawi: Zomba City Council.

15 Characterisation and Impacts of Tornado-Induced Flooding and Windstorms in Mpumalanga Province, South Africa

Kaitano Dube and David Chikodzi

Abstract

Extreme weather events have significantly affected human civilisation and development in recent periods. While COVID-19 has been the top news headline in 2020 and 2021, it is important to note that the additional impact of extreme weather events is catastrophic to human development. Tornadoes rank amongst the deadliest extreme weather events of global concern. This study seeks to assess the trend and impacts of tornadoes on the highveld area of Mpumalanga province, South Africa. The Mkhondo Local Municipality area was used in a case study to research primary and archival data from key informants and weather bureaus. The study found that high-impact tornadoes are on the increase in Mpumalanga. Tornadoes that occur mainly in summer months with a peak between November and January have led to loss of property, timber plantations, homes, schools, and health institutions and threaten the progress made in achieving the Sustainable Development Goals. The study noted a statistically significant ($p = 0.015$) increase in the frequency and severity of tornadoes in Mpumalanga, particularly in the last 5 years that coincide with probably the worst El Niño event experienced in Southern Africa. The study recommends a risk-adjusted approach to infrastructure development considering the increased frequency of tornadoes due to global warming and climate change. The risk-adjusted approach must encompass smart climate initiatives for rural and urban development to ensure climate resilience if sustainability is to be achieved.

Keywords

Tornadoes · Climate change · Extreme weather · Mpumalanga · South Africa · Timber plantations

15.1 Introduction and Background

With changes in climate-disrupting weather and climatic components occurring across the world, there are increasing concerns of how such changes will alter extreme weather events such as tornadoes. Given the fact that tornadoes are not large weather systems, they are often not cap-

K. Dube (✉)
Ecotourism Management, Vaal University of Technology, Andries Potgieter BlvD, Vanderbijlpark, South Africa
e-mail: kaitanod@vut.ac.za

D. Chikodzi
Exxaro Chair in Business and Climate Change, Institute for Corporate Citizenship, University of South Africa (UNISA), Pretoria, South Africa
e-mail: chikod@unisa.ac.za

G. Nhamo, L. Chapungu (eds.), *The Increasing Risk of Floods and Tornadoes in Southern Africa*, Sustainable Development Goals Series, https://doi.org/10.1007/978-3-030-74192-1_15

tured by meteorological stations. Consequently, the study and understanding of tornadoes are largely unexplored in various parts of the world (Yao et al. 2015). There have been efforts to capture and understand this weather phenomenon using various technologies such as Doppler weather radar and analysing features of tornadoes and radar echoes to determine the occurrence and impacts of tornadoes in other countries (Houfu et al. 2014). In the United States of America (USA), digital beamforming, or radar imaging, is often used to observe local storms and tornadoes (Kurdzo et al. 2017).

As climate change increasingly leaves its mark on the natural and human environment, studies indicate that changes in climate are altering the frequency and occurrence of tornadoes differently across the world. While some regions are reporting a decline in the occurrence of tornadoes due to climatic conditions that promote their development, other regions are reporting an increase in tornadoes due to conditions that are favourable to their development. In China, the average number of tornadoes occurring per year between the 1960s and 1970s was 7.5 times higher than the present count (Yao et al. 2015). There has been significant interest among scholars and the general public in the USA to understand the impact of climate change on tornadoes. According to Brooks et al. (2014), the evidence seems to suggest that the mean annual number of tornadoes in the USA has remained relatively constant over the years, but variability has increased since the 1970s. Changes in the timing of the starting of tornadoes have also been observed. Molina and Allen (2020) have observed regional differences in the USA's parameters and the occurrence of tornadoes.

Tornado formation usually requires large vertical wind shear and adequate humidity near the ground. Some studies have highlighted the importance of a vertical wind shear of about 0–1 and 0–6 km (Yao et al. 2015). Molina and Allen (2019) have studied the influence and source of low-level moisture for tornadoes and found that for the USA various potential sources critical for the formation of tornadoes include the Gulf of Mexico, which contributes about 50%. The same study also found that the Caribbean Sea, the North Atlantic Ocean, and the ground and vegetation contribute to tornadic thunderstorms with warmer sea surface temperatures helping to enhance fluxes of overlaying air parcels. Lee et al. (2016) also argue that the North Atlantic sea surface temperature tripole variability plays a crucial role in determining the occurrence of tornadoes. Other studies have also indicated that the moisture source for tornadoes can be modulated by external factors which can also influence their frequency, such as the El Niño–Southern Oscillation (Allen et al. 2018; Allen et al. 2015). Most tornadoes in countries, including the USA and China, originate and occur in plains (Molina and Allen 2020; Yao et al. 2015).

Tornadoes occur in various parts of the world, including Africa. However, most tornadoes occur in the USA with more than 1200 tornadoes reported every year (NOAA 2021). In as much as tornadoes have a long history, their causes are largely a subject of research. The World Meteorological Organization (WMO) (2021) defines tornadoes as a "rotating column of air, extending from the base of a cumuliform cloud, and often visible as a condensation funnel in contact with the ground, and/or attendant circulating dust or debris cloud at the ground". It is not known why tornadoes are formed, but they are a consequence of a supercell thunderstorm. This intense thunderstorm has a persistent updraft that can reach up to about 161 km/h. Inside the storm, strong vertical wind shear causes a rotational cylinder of air. According to NOAA (2021), "the rotating cylinder of air narrows, becoming stretched, and spinning faster and faster forming a tornado. The rotation within the thunderstorm gives the supercell its classic 'hook' appearance". These tornadoes can be classified into three groups as shown in Box 15.1.

Box 15.1 Categories of Tornadoes

Type I—in association with supercells;

Type II—in association with quasi-linear convective systems; and

Type III—localised convective and shear vortices. These comprise landspouts, waterspouts and cold-air funnels.

Source: Authors' data from WMO

The different categories of tornadoes have various socio-economic and environmental impacts, given their range of intensities. Tornadoes have been of interest to both geographers and society mainly because of their impacts on human activities as they are rather similar to hurricanes and tropical cyclones. A team of meteorologists and wind engineers developed the F-scale to estimate the intensity of tornadoes based on damage imposed by the tornado. The scale uses 3-s gusts estimated at the point of damage based on a judgement of 8 levels of damage to develop 28 indicators. Table 15.1 shows the revised enhanced Fujita (F) scale for tornado damage that is often used.

Given the wind and rainfall that can be dumped as a consequence of tornadoes, they can cause extensive damage to the environment and the economy. Between 2005 and 2014, tornadoes costed the economy of the USA a staggering US$ 21.7 billion due to property and crop damage with as many as 1100 lives lost. However, such losses vary across the northern states, and southern states of the USA (Biddle et al. 2020). Tornadoes are very disruptive to sea, ground and air travel and outdoor activities (Bluestein 2013). Evidence also shows that tornadoes pose a significant threat to wind farms as turbines have proven to be vulnerable to the high-intensity wind, which often characterises F2 tornadoes (AbuGazia et al. 2020). Studies conducted in Ontario, Canada and Harper County in Kansas in the USA reveal a high number of wind turbine failures due to tornadoes.

Even though the USA started issuing tornado warnings in 1952, here and across the world these predictions are often criticised for their inaccuracy and the rudimentary ability to predict tornadoes (Bluestein 2006). Tornadoes remain a mystery with very little known about them; consequently they are called one of the last frontiers of atmospheric science. This pattern also applies in Southern and South Africa where tornadoes are barely understood and hardly researched with many theories evolving around their occurrence and causes. According to Simpson and Dyson (2018), tornadoes categorised as severe weather are under-researched in South Africa. There has been very little generated over the years to document the trend, impact and occurrence of tornadoes in the country, apart from Goliger and Retief (2002) who mapped areas with strong winds in South Africa where tornadoes occur.

Goliger and Milford (1998) built on Goliger (1993) who drew a map of tornado hotspots in South Africa. The two studies reveal that the areas most affected by tornadoes lie along South Africa's highlands located predominantly in the KwaZulu-Natal and Mpumalanga provinces—both areas are located off the coast on the western side of the Indian Ocean. Simpson and Dyson (2018) report that low-level moisture from the Mozambique Channel favours the development of tornadoes on South Africa's highveld region which have a devastating impact on socio-economic life.

Understanding tornadoes requires a regional stratification approach to avoid the masking of regional details (Molina and Allen 2019). In that regard, this study adopts a regional approach in the analysis of the trends, patterns and impacts of tornadoes in South Africa by focusing on one particular area in Mpumalanga. The study seeks to respond to the following critical research questions: (1) What has been the trend and occurrence of tornadoes in the Mpumalanga highveld area? (2) What has been some of the observed impacts of tornadoes in the province?

15.2 Materials and Methods

The study adopts the case study research approach that focuses on Mpumalanga province as a whole (Fig. 15.1) using the Mkhondo Local Municipality as a sample point to employ a convenient sampling technique. Mkhondo Local Municipality (Piet Retief) is in the Gert Sibande District Municipality, Mpumalanga province in South Africa (Fig. 15.1). The study area was chosen as it lies along the tornado hotspot zones as observed by Goliger (1993) and Goliger and Retief (2002). Piet Retief is a small town located close to the border of Eswatini, Mpumalanga and KwaZulu-Natal about 300 km southeast of Johannesburg. The main economic activities

Table 15.1 Enhanced F-scale for tornado damage

Fujita scale			Derived EF scale		Operational EF scale		Observable impact
EF number	Fastest 1/4-mile (mph)	Three-second gust (mph)	EF number	Three-second gust (mph)	EF number	Three-second gust (mph)	
0	40–72	45–78	0	65–58	0	65–85	Some light damage: Chimneys, road signs, branches broken off trees
1	73–112	79–117	1	86–109	1	86–110	Moderate damage: Peeled surface off roofs, temporary houses pushed off foundations or overturned, moving cars pushed off the road
2	113–157	118–161	2	110–137	2	111–135	Considerable damage: Roofs torn off frame houses, shacks demolished, boxcars pushed over, large trees snapped or uprooted, light object missiles generated
3	158–207	162–209	3	138–167	3	136–165	Severe damage: Roofs and some walls torn off well-constructed houses, trains overturned, most trees in forest uprooted, heavy cars and similar objects lifted off the ground and thrown into the air
4	208–260	210–261	4	168–199	4	166–200	Devastating damage: Well-constructed houses levelled, structures with weak foundations blown off some distance, cars thrown and large missiles generated
5	261–318	262–317	5	200–234	5	Over 200	Incredible damage: Strong frame houses lifted off foundations and carried considerable distance to disintegrate, automobile-sized missiles flown through the air in excess of 100 m, trees debarked

Source: Authors' data from NOAA and Disaster Center (2021)

include timber plantations mostly owned by four different companies Mondi, Sappi, TWK and Komati Land Forests. The plantations are a vital component of the socio-economic livelihoods of people in the area (Ofoegbu 2014). Outside the town are rural and farming areas, and some mining activities are also conducted in the area.

A case study research approach allowed the study to take an in-depth look at one geographic area, which allowed for in-depth analysis and understanding of the phenomenon as advanced by Cousin (2005). Case studies have been used to document and better understand various climatic events in Southern Africa and abroad by various scholars such as Dube et al. (2020, 2021) and Simpson and Dyson (2018).

The study utilised archival data from South African Weather Services (SAWS). The website has a complete archived history of tornadoes in South Africa which include their categorisation. The website also provides a whole range of weather and climatic data. Primary data was collected from key informant interviews with relevant employees from Mkhondo Local Municipality, including the environmental manager, disaster management team members, traditional leadership and managers for human settlement. Other participants included managers from timber plantation companies in the area. The key-informant in-depth interviews included questions on trends, impacts, costs and mitigation and adaptation measures, among other critical questions, and interviews lasted from 45 min

Fig. 15.1 Study area—Mpumalanga provincial map. Source: Authors

to 1 h. The fieldwork was undertaken between November and December 2020. This was complemented by field observations where the researchers took a tour of the district's affected areas to understand the context and some of the observed impacts of tornadoes in the area.

Data analysis was done using a geographic information system (GIS) to produce the provincial heatmap, whereas the Microsoft Excel Tool Pak was used to analyse quantitative data. Qualitative data from interviews were transcribed, and content and thematic analysis were done using the research questions.

15.3 Results and Discussions

The study analysed archival data on tornadoes from 1905 to 2020. The study showed that during this period Mpumalanga province recorded a total of 65 observed tornadoes. Of the various types of tornadoes that were observed in Mpumalanga province, a huge percentage (42%) were not given a category. Of the 65 tornadoes, almost a third were categorised as ranging between category F1 and F3 tornadoes (Fig. 15.2). Very few of these, however, were categorised as F3 tornadoes. Category F3 tornadoes were reported in 1986 in Piet Retief, and in 2002 and 2020 in Panbult (located between Ermelo and Piet Retief). It is important to note that all the category F3 tornadoes have taken place within the last 40 years, which could indicate an increase in the intensity and frequency of tornadoes in the area. It emerged from the study that the bulk of tornadoes occurs between November and December and generally during the summer months as opposed to winter months when there are generally few tornadoes (Fig. 15.2). The long-term average occurrence is about three tornadoes per year.

Figure 15.3 shows the frequency of tornadoes in the Mpumalanga province of South Africa. The hardest-hit areas are areas around Middelburg, Witbank, Bethal and Piet Retief. Middelburg is the most tornado-prone in the province with a count of eight incidences over the study period.

No category F4 and F5 tornadoes were reported in Mpumalanga during the period of

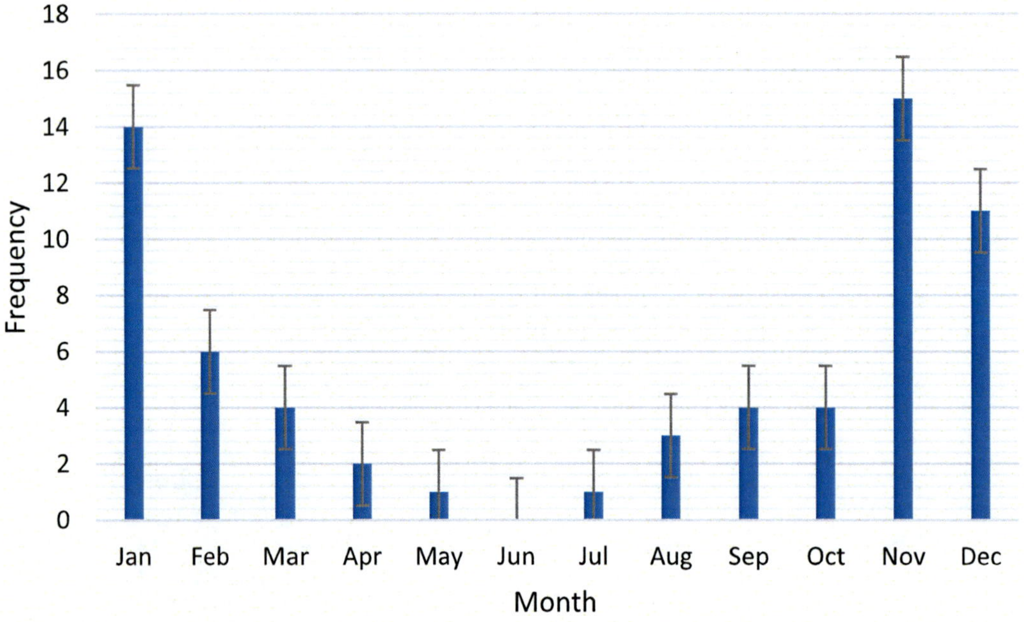

Fig. 15.2 Monthly distribution and occurrence of tornadoes in Mpumalanga province, South Africa. Source: Authors

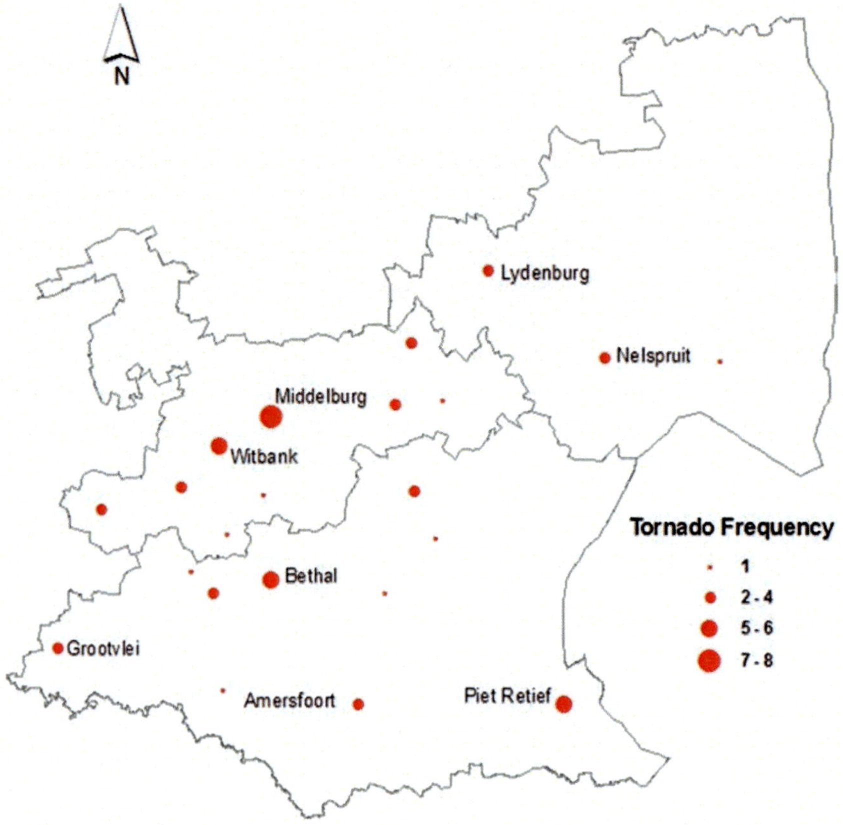

Fig. 15.3 Tornado hotspots and frequency in Mpumalanga province. Source: Authors

study. If the Fujita-Pearson (FP) scale of tornado intensity were taken into consideration, several areas where tornadoes were experienced witnessed light to severe damage from the impact of tornadoes which could have resulted in property damage to housing by close to a third of tornadoes that were categorised F1 to F3 tornadoes (Fig. 15.4). The high number of uncategorised tornadoes could be attributed to the lack of technical expertise and or capacity to properly document or assess the damage to categorise it as the province is largely rural with meteorological offices being far apart.

The trend of tornado occurrence in the province was not static or uniform across the period of study but varied across the entire time spectrum (Fig. 15.5). Between 1905 and 1947, no tornadoes were observed and reported in the area, with the first tornado only reported in 1948. The highest number of tornadoes in the province was reported in the last decade of the study between 2010 and 2020. The period between 2014 and 2020 saw a record number of tornadoes in the Mpumalanga province. This period coincides with the period in which Southern Africa observed one of the worst El Niño events in the Southern African region, which led to record high temperatures in one of the longest drought periods in the region (Dube and Nhamo 2018). The trend shows that there was an increase in the frequency and occurrence of tornadoes, which is considered an extreme weather event. The understanding of the impact of the El Niño–Southern Oscillation (ENSO) phenomenon has been of interest to academics. A study by Allen et al. (2015) examined the influence of La Niña on tornadoes in the USA and established an increase in La Niña episodes and a decline in tornadoes during the El Niño episodes. In a separate study, Allen et al. (2018) observed an annual peak probability of 1.5–2 weeks in tornadoes during a La Niña period. On the contrary, in Mpumalanga, the last episodes of El Niño have been characterised by an increase in frequency and occurrence of tornadoes—a factor that can be attributed to the role of climate variability in shaping tornado occurrence in the province.

Tornadoes that occur over the South African highveld are influenced in part by low-level moisture from the Mozambique Channel (Simpson and Dyson 2018). In other parts of the world, an increase in the occurrence of tornadoes has been attributed to increased sea surface temperature (Chu et al. 2019; Cook et al. 2017). The increase in tornado activity can be attributed, in part, to an increase in sea surface temperature in the Indian Ocean's Mozambique Channel, which could have influenced tornadoes on the South African highveld as this provided much needed low-level moisture crucial for the development of tornadoes. Reports indicate that sea surface anomalies have recently peaked over the Indian Ocean, and in the Mozambique Channel, due to global warming and climate change (Marine Copernicus 2019). The India Meteorological Department (2020) reports that sea surface temperature anomalies as high as +0.29 °C to 0.71 °C were observed in the Indian Ocean between 2010 and 2020.

As mentioned earlier, there has been an increase in the number of category three (F3) tornadoes. Nonetheless, a decline in F1 tornadoes has been observed in the last three decades in the Mpumalanga province with the last F1 tornadoes reported in 1987. F2 tornadoes have also been on the decline with the last F2 tornado reported in 2006 (Fig. 15.6). At the same time, there has been an increase in other extreme weather events such as dust storms, dust devils, land spouts and funnel clouds in the province. This is in line with the global narrative, where extreme weather events are increasing globally and are mainly driven by global warming. Climate change is believed to be one of the key drivers of extreme weather events across the world (D'Andrea et al. 2020; Otto et al. 2017; Stott 2016).

15.3.1 Socio-Economic Impact of Tornadoes in Mpumalanga

The increased occurrence of tornadoes in the Mpumalanga province is a growing concern as tornadoes are associated with serious damages to the environment and infrastructure. Tornadoes,

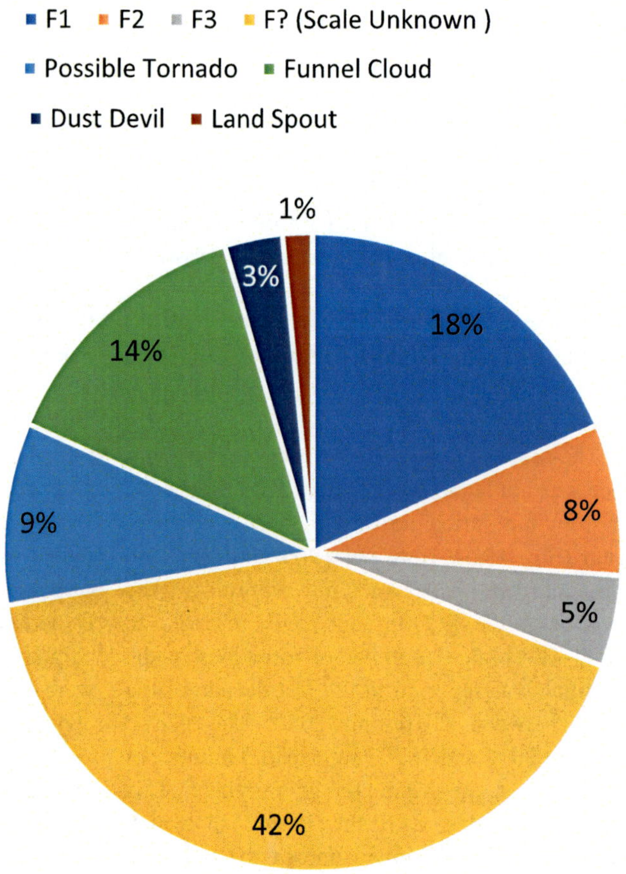

Fig. 15.4 Categories of various types of tornadoes that affected Mpumalanga province 1905–2020 ($n = 65$). Source: Authors (2021)

which are often associated with hailstorms, are one of the major causes of property and infrastructure damage in the province. This calls for better disaster preparedness and management and control to reduce both human and socio-economic costs associated with tornadoes. The last EF3 tornado that took place in the evening of 3 January 2020 is a reminder of the devastating impact of tornadoes in the province. The tornado that affected Gert Sibande District left a trail of destruction, as shown in Fig. 15.7. The tornado caused extensive damage to property, houses and forestry plantations in the area resulting in massive financial costs.

It emerged from interviews that residents of Piet Retief experience occasional tornadoes and have different beliefs about the cause of these natural disasters. According to key informants, tornadoes have been on the increase as observed in climate data presented earlier. The history of tornadoes in the area shows that they have a long-lasting impact on the community—a traditional leader's homestead, and a number of residential areas in and around the town have suffered extensive damage as a consequence of tornadoes in the area. According to tornado archival records, about 500 homes and the classrooms and offices of several schools have been destroyed. The wing of the old provincial hospital was damaged at one point compromising health delivery systems in the area. Instances were reported of destruction of livestock and human lives lost as a consequence of tornadoes.

The destruction of homes, schools and health infrastructure is a serious concern as it hampers service delivery and progress towards achieving the Sustainable Development Goals. There is a strong focus and desire to deliver quality education in pursuit of ensuring education for all—a distant dream to achieve for poor communities if

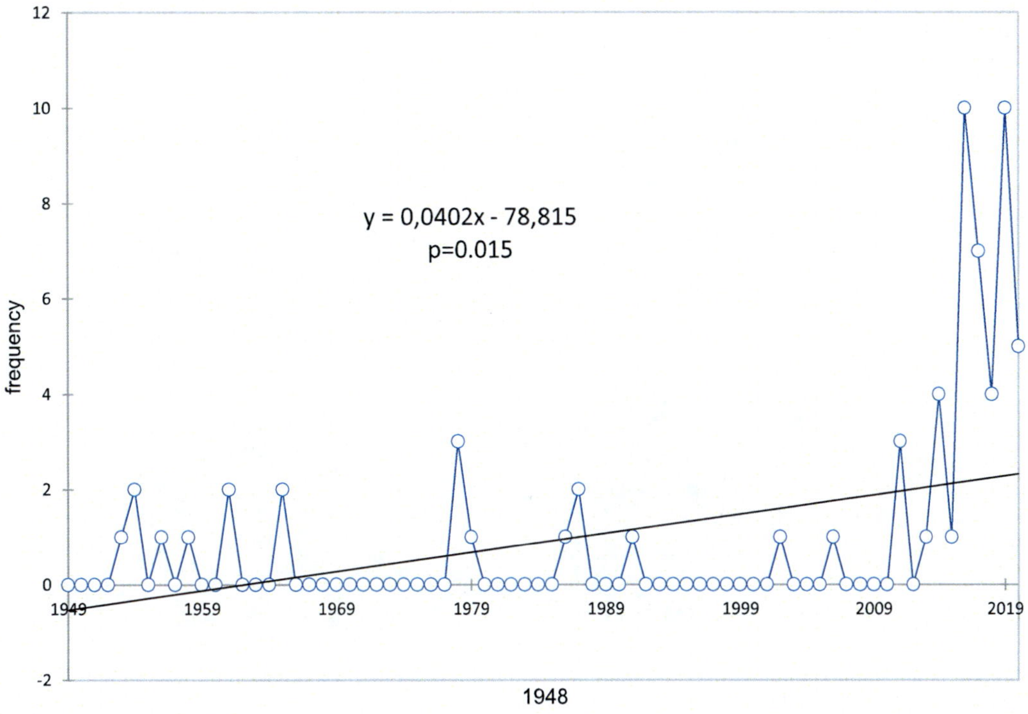

Fig. 15.5 Trend in the total number of tornadoes in Mpumalanga province ($n = 65$). Source: Authors

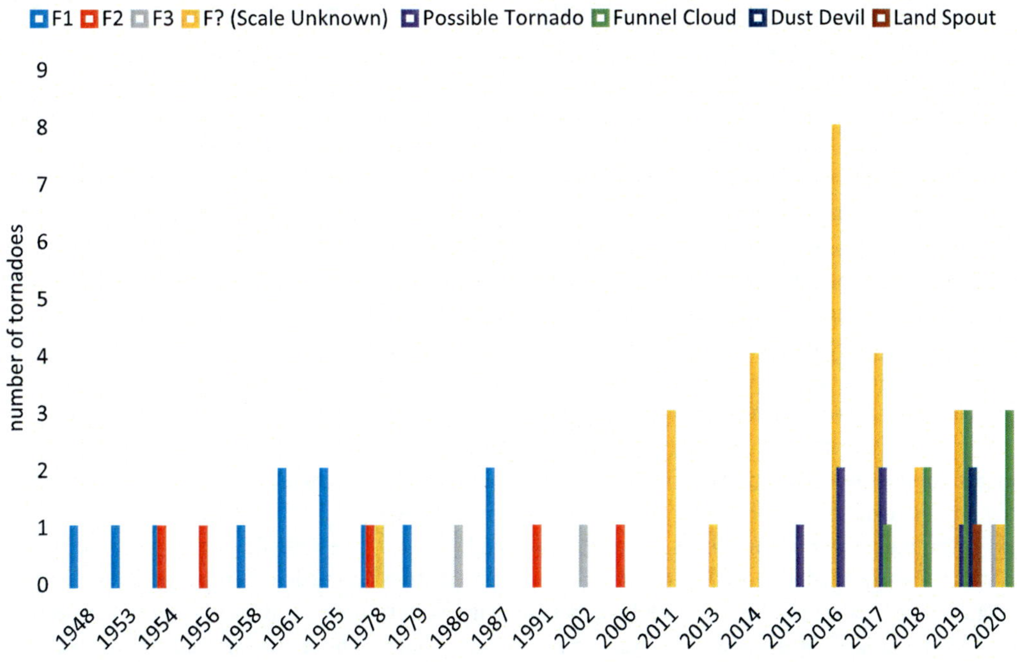

Fig. 15.6 Trend in tornadoes and their intensity in the Mpumalanga province ($n = 65$). Source: Authors

Fig. 15.7 Impact of an EF3 tornado that affected an area between Ermelo and Piet Retief in the Mpumalanga province. Source: Pictures courtesy of Kurt Paul

tornadoes in the area constantly disrupt the infrastructure that supports education delivery. The destruction of homes is a wake-up call for the area to develop a robust housing plan that considers the extreme weather events prevailing in the province to reduce both the chances and impact of weather risks such as tornadoes as they become a regular occurrence. Livestock destruction is also problematic as livestock is a source of economic empowerment and also provides the necessary tillage power for smallholder farmers in the area. Destruction of livestock is likely to have an adverse impact on crop production and, by proxy, food security for those affected. Achieving food security is a crucial aspect of achieving Zero Hunger in line with Sustainable Development Goal 2. This cannot be done without efforts by both government and private individuals to secure insurance from tornadoes to allow communities to better cope and respond to the shocks experienced in the area. Education and awareness are also critical to ensure that community members have an understanding of proper community response mechanisms using both conventional and traditional means.

15.3.2 Impact of Tornadoes on Forest Plantations in Mpumalanga

The forest plantation industry is an important source of employment in Mpumalanga. Damage to plantations due to tornadoes in the area takes a heavy financial toll on the province. Tornadoes have led to massive damage to sections of forest plantations in their path, mainly due to the associated strong winds that cause wind throw, which uproot and break trees (Fig. 15.8).

Information from key informants suggests that not many specific plant species are damaged when tornadoes strike. As shown in Fig. 15.7, mature and taller trees are the most vulnerable to damage from tornadoes as they are less likely to bend in the wind as younger and smaller trees would. From a conservation perspective, tornadoes enhance the speed at which environmental degradation occurs on forestry estates. Of particular concern are soil erosion and landslides that may result in mountainous terrain. Erosion may lead to the formation of gullies which require ample investment to control. Damage to operations and road networks to and from forest plantations can be damaged as a result of tornadoes, restricting salvage operations and temporarily rendering areas inaccessible. Even though fire is rated the biggest threat to forest plantations in Mkhondo Local District, an increase in the frequency of tornadoes damaging these forests calls for the urgent consideration of adaptation measures to mitigate this risk.

The destruction of plantations is of concern as plantations are a source of livelihood in the area and provide employment for both skilled and semi-skilled individuals in the province. Plantations also provide raw materials for several other industries that operate in the area. In addition, the disruption of plantations is also a concern from a climate change perspective as trees act as carbon sinks in this carbon-intensive province which produces carbon emissions from a number of coal-fired power stations.

15.3.3 Community Response and Adaptation to Tornadoes in Mpumalanga

Community response to the emerging problem of tornadoes in the study area is still haphazard and mainly reactionary. Within the local communities in Mkhondo, there is both a scientific and traditional understanding of the occurrence of flooding. The scientific understanding explains their occurrence as being influenced by the changes in climate and as such response to these is guided by the district municipality's draft climate change strategy which was in the final stages of adoption during the fieldwork for this study. The climate change policy acknowledges climate change as a threat to the district and details the way in which the district can respond to this perceived threat and build resilience. The municipality has adopted building standards and codes that are

Fig. 15.8 Damage to forest plantations after the January 2020 tornado (Mkhondo Local Municipality). Source: Authors

robust and produces structures that are, to a large extent, resilient to damage imposed by tornadoes and other extreme weather events.

The traditional explanation for the increased occurrence of tornadoes in Mkhondo is the growing disregard of local communities to traditional leadership and lack of respect for sacred sites. According to one traditional leader as a result of traditional belief disregard, mining and communication towers have been placed on sacred sites and people are building on land which, in the past, they were forbidden to do. Accordingly, in the traditional leadership's view the tornadoes are a form of punishment for those who disregard traditional leadership. In-depth interviews with several key informants revealed that there are traditional ways and methods of stopping the occurrence of these hazards, but they can only work if people submit to traditional leadership and respect cultural sites.

15.4 Conclusions and Recommendations

The study found that Mpumalanga province is one of the areas experiencing a backlash from nature in the form of increasingly frequent and intense tornadoes. Tornadoes in the area occur at a huge cost to the community's socio-economic well-being. Tornadoes have proven to be devastating to the plantations and rural communities. They have destroyed houses, health facilities, and loss of livestock and human lives. Disruptions caused by tornadoes take a heavy toll on socio-economic activities and livelihood security. Most tornadoes were observed to occur in the summer months, especially between November and January with November as the peak month. Tornado activity during the winter period was much less frequent, with no tornadoes recorded in June during the period under investigation. Given the increased frequency of tornado occurrence, there is a need to provide concrete relief to the vulnerable community through a number of possible interventions, namely proper recording and forecasting of tornadoes, to ensure that reliable early warnings are issued to the community. Recognising the importance of climate and weather data in human settlement planning and development is crucial in a changing climate era that is not given pre-eminence by the province's local government. A climate risk-adjusted approach is imperative in developing and constructing private and public infrastructure to ensure climate resilience. This includes revision of housing infrastructure plans to ensure that such houses and other developments are climate-proof. In as much as indigenous knowledge is crucial in human development, there is a need for traditional communities to acknowledge the role of science in adopting a robust approach to human settlement that ensures human safety. The role of traditional leadership in building a climate-proofed future can never be underestimated; nonetheless, there is a need to ensure a hybrid approach to build a climate-reliant rural community.

References

AbuGazia, M., et al. (2020). Numerical model for analysis of wind turbines under tornadoes. *Engineering Structures, 223*, 111,157.

Allen, J. T., Molina, M. J., & Gensini, V. A. (2018). Modulation of annual cycle of tornadoes by El Niño–southern oscillation. *Geophysics Research Letters, 45*, 5708–5717.

Allen, J. T., Tippett, M. K., & Sobel, A. H. (2015). Influence of the El Niño/southern oscillation on tornado and hail frequency in the United States. *Nature Geoscience, 8*(4), 278–283.

Biddle, M. D., Brown, R. P., Doswell, C. A., III, & Legates, D. R. (2020). Regional differences in the human toll from tornadoes: A new look at an old idea. *Weather, Climate, and Society, 12*(4), 815–825.

Bluestein, H. (2013). Introduction. In *Severe convective storms and tornadoes. Springer Praxis Books* (pp. 1–25). Berlin, Heidelberg: Springer.

Bluestein, H. B. (2006). *Tornado alley: Monster storms of the Great Plains*. Oxford: Oxford University Press.

Brooks, H. E., Carbin, G. W., & Marsh, P. T. (2014). Increased variability of tornado occurrence in the United States. *Science, 346*(6207), 349–352.

Chu, J. E., Timmermann, A., & Lee, J. Y. (2019). North American April tornado occurrences linked to global sea surface temperature anomalies. *Science Advances, 5*(8), eaaw9950.

Cook, A. R., Leslie, L. M., Parsons, D. B., & Schaefer, J. T. (2017). The impact of El Niño–southern oscillation (ENSO) on winter and early spring US tornado outbreaks. *Journal of Applied Meteorology and Climatology, 56*(9), 2455–2478.

Cousin, G. (2005). Case study research. *Journal of Geography in Higher Education, 29*(3), 421–427.

D'Andrea, E., et al. (2020). Frost and drought: Effects of extreme weather events on stem carbon dynamics in a Mediterranean beech forest. *Plant, Cell & Environment, 43*(10), 2365–2379.

Disaster Center. (2021). *Fujita-Pearson Scale of Tornado Intensity.* [Online]. Retrieved January 10, 2021, from http://www.disastercenter.com/tornado/fujipear.htm.

Dube, K., & Nhamo, G. (2018). Climate variability, change and potential impacts on tourism: Evidence from the Zambian side of the Victoria Falls. *Environmental Science & Policy, 84*, 113–123.

Dube, K., Nhamo, G., & Chikodzi, D. (2020). Climate change-induced droughts and tourism: Impacts and responses of Western cape province, South Africa. *Journal of Outdoor Recreation and Tourism*, 100,319.

Dube, K., Nhamo, G., & Chikodzi, D. (2021). Rising sea level and its implications on coastal tourism development in Cape Town, South Africa. *Journal of Outdoor Recreation and Tourism, 33*, 100,346.

Goliger, A. M. (1993). Tornado activity in South Africa. *Civil Engineering Siviele Ingenieurswese, 1*(2), 13–14.

Goliger, A. M., & Milford, R. V. (1998). A review of worldwide occurrence of tornadoes. *Journal of Wind Engineering and Industrial Aerodynamics, 74*, 111–121.

Goliger, A. M., & Retief, J. V. (2002). Identification of zones of strong wind events in South Africa. *Journal of Wind Engineering and Industrial Aerodynamics, 90*(11), 1227–1235.

Houfu, Z., Xiuguang, D., & Wenmei, X. (2014). Analysis of the tornado supercell storm and its environmental parameters in the Yangtze-Huaihe region. *Acta Meteorologica Sinica, 72*, 306–317.

India Meteorological Department. (2020). *State of Climate in India 2020.* [Online]. Retrieved January 11, 2021, from https://mausam.imd.gov.in/Forecast/marquee_data/Statement%20of%20Climate%20of%20India2020.pdf?fbclid=IwAR0CUU6jz5qb83vWg7_c9FKEdIbWVIihJqdnc3kGEqQbu_PF8YMRDYlyzU0.

Kurdzo, J. M., et al. (2017). Observations of severe local storms and tornadoes with the atmospheric imaging radar. *Bulletin of the American Meteorological Society, 98*(5), 915–935.

Lee, S. K., et al. (2016). US regional tornado outbreaks and their links to spring ENSO phases and North Atlantic SST variability. *Environmental Research Letters, 11*(4), 044008.

Marine Copernicus. (2019). *July 2019, hottest month on record: A look at warming ocean*. Retrieved January 11, 2020, from https://marine.copernicus.eu/july-2019-hottest-month-on-record-warming-ocean/#:~:text=The%20Copernicus%20Climate%20Service%20(C3S,hardest%20hit%20by%20this%20warming.

Molina, M. J., & Allen, J. T. (2019). On the moisture origins of tornadic thunderstorms. *Journal of Climate, 32*(14), 4321–4346.

Molina, M. J., & Allen, J. T. (2020). Regionally-stratified tornadoes: Moisture source physical reasoning and climate trends. *Weather and Climate Extremes, 28*, 100,244.

NOAA. (2021). *Tornadoes.* [Online]. Retrieved January 9, 2021, from https://www.noaa.gov/education/resource-collections/weather-atmosphere/tornadoes.

Ofoegbu, C. (2014). The contribution of tree plantations to household welfare: Case study of Piet Retief and Iswepe communities. *International Forestry Review, 16*(2), 172–179.

Otto, F. E., et al. (2017). Assigning historic responsibility for extreme weather events. *Nature Climate Change, 7*(11), 757–759.

Simpson, L. A., & Dyson, L. L. (2018). Severe weather over the Highveld of South Africa during November 2016. *Water SA, 44*(1), 75–85.

Stott, P. (2016). How climate change affects extreme weather events. *Science, 352*(6293), 1517–1518.

World Meteorological Organization (WMO). (2021). *Tornado.* [Online]. Retrieved January 9, 2021, from https://cloudatlas.wmo.int/en/tornado.html.

Yao, Y., et al. (2015). Climate analysis of tornadoes in China. *Journal of Meteorological Research, 29*(3), 359–369.

Part VI

The Policy Space

16 Emerging Key Findings, Conclusions and Recommendations

Lazarus Chapungu and Godwell Nhamo

ABSTRACT

Climate-induced hazards, including floods and tornadoes, are ravaging communities the world over. Floods and tornadoes have immensely contributed to the disasters, which have costed lives and imposed a strain on financial resources for several economies, including those regarded as economically developed. It is fear-provoking to observe that the frequency of both phenomena has increased over time and it is projected that the trend will remain upward into the foreseeable future. Hence, the proliferation of major flood and tornado-related calamities is expected to perpetuate in the coming decades. In southern Africa, the increasing risk of flooding is a subject that is worth investigating given the widespread occurrence of socio-economic depravities associated with recent flood events spawned by successive tropical cyclones. On the other hand, tornado occurrences have received little attention, yet they are causing deaths and destroying infrastructure, especially in South Africa. This chapter focuses on the policy space, profiling the key findings, conclusions and recommendations emanating from this book. Generally, the book shows that floods and tornadoes are atrocious hazards seriously threatening societal progress and economic development at various spatial and temporal levels in southern Africa. The tornadoes and floods have contributed to the orchestration of slow progress towards the achievement of sustainable development goals. Recent observations of tornadic incidences in some countries, the increasing understanding of their risks and the evident increase in flooding risks should galvanise policymakers into action. The need for the policies and actions to strengthen the early warning systems is emphasised as an important step towards building community resilience to hazards related to floods and tornadoes.

Keywords

Floods · Tornadoes · Southern Africa · Disaster Risk Reduction

16.1 Introduction

The ramifications of climate change have garnered academic and public attention for decades (von Uexkull and Halvard 2021) although influential policymakers have only become more concerned in the recent years (Ide et al. 2020).

L. Chapungu · G. Nhamo (✉)
Exxaro Chair in Business and Climate Change, Institute for Corporate Citizenship, University of South Africa (UNISA), Pretoria, South Africa
e-mail: nhamog@unisa.ac.za

G. Nhamo, L. Chapungu (eds.), *The Increasing Risk of Floods and Tornadoes in Southern Africa*, Sustainable Development Goals Series, https://doi.org/10.1007/978-3-030-74192-1_16

Disasters now occupy a significant amount of space in policy debates and discussions (Peters and Peters 2018). For example, the third United Nations Security Council in 2018 saw Sweden initiating a discussion on climate change and security. German upheld the disasters theme throughout its presents as a member of the council (Auswärtiges Amt 2019). Disasters have exposed socio-economic systems, while their frequency and intensity have increased over the past three decades (Guiseppe and Feyen 2019). There is considerable scientific evidence that these hazards, including floods and tornadoes, are driven mostly by the changes in the earth-atmospheric system. However, for tornadoes, Howard (2019) argues that their formation and link to climate change remain obscured. This is besides the fact that, since they are largely influenced by moisture updrafts, the atmospheric thermodynamics play a significant role in their formation. Hence, a change in the thermodynamics of the earth atmospheric system will influence a change in the character and form of tornadoes.

The world over, the probability of flooding has increased due to climate change (Lemperiere 2017). The rise in sea level driven by the melting of ice caps and the expansion of oceans due to global warming is threatening coastal areas (O'Donnell and Thorne 2020). It is predicted that between 2030 and 2040, some coastal areas will be partially or completely submerged (Saintilan et al. 2020). Recent flooding events have been associated with increased rainfall intensity, extratropical cyclones and tropical cyclones among other factors (Martinez et al. 2020). Furthermore, it is predicted that climate change will modify the hydrological cycle resulting in more extreme precipitation, which will eventually amplify the magnitude and frequency of floods (Stagl et al. 2014).

Historical flood events have led to infrastructural damages and economic losses. Svetlanaa et al. 2015 concur with the observation that insured and economic losses due to extreme floods have increased abruptly in the last two decades. For example, each year, the United Kingdom spends 1.3 billion pounds in remedial work after flood damages (Svetlanaa et al. (2015)). The increase in the cost of flood damage is driven by social, economic and spatial planning policies. Munich RE (2005) argues that wealth and exposure have increased in areas vulnerable to flooding. Populations located along rivers have been on the increase recently due to targeted economic opportunities. Areas prone to flooding remain attractive for economic activities. Thus, there is a lot of investment in these areas, and when a flood occurs, economic losses are huge. However, on the other hand, Rolfe et al. (2020) and Lee and Jung (2014) aver that exposure to floods is greater among socially vulnerable individuals and families. Vulnerable populations usually inhabit flood-exposed areas because of the society barriers linked to social stratification (Qiang 2019).

The frequency and severity of extreme hydrological events such as floods are predicted to be on the rise as a consequence of the changing climate (Mysiak et al. 2018; He et al. 2021). These events will affect several facets of life, including human health, social well-being, disease proliferation as well as environmental quality. To this end, it is expected that the projected increase in the risk of floods will be associated with a plethora of impacts on diverse facets of life. This will require well-coordinated policies and implementation plans to reduce the risks

Tornadoes, just like floods, contribute to the world's total number of fatalities each year. Even in cases where there are no instant fatalities, the massive damages inflicted by tornadoes on specific regions affect several facets of life including public health and economic stability (Molloy and Mihaltcheva 2013). The flying debris in tornadoes causes serious injuries to humans and the natural ecosystems. Tornado occurrences, depending on their magnitude as measured on the Fujita scale, may result in a surfeit of health impacts including psychological trauma, fractures, cervical spine trauma and wound infections among other complications.

Tornadoes (also known as twisters in other regions) continue to cause the loss of life and property and their increasing frequency may worsen the situation. Although their highest frequencies of occurrences have been recorded in north America, from this work, it emerged that southern Africa has received its fair share of tornadic events, most of which have been recorded in South Africa with a significant upward trend showing that their frequency is increasing. Hence, tornadoes may be rampant in several southern African countries but the capacity to detect them and ascertain their impacts is still lacking. The occurrence of tropical cyclones is also viewed as increasing the possibility of tornado occurrences as the "land falling tropical cyclones sometimes spawn tornadoes within bands of convective storms, especially just to the right of the track of the centre of the cyclones" (Ide et al. 2020, p. 2). Disasters have been shown to increase the risk of conflict by current recent studies although the link has generally been weak and dependent on conditions (Ide et al. 2020; von Uexkull 2016).

There is now consensus that the risk of floods and tornadoes is increasing across the globe and southern Africa will continue to receive its fair share of challenges associated with these phenomena. It is also expected that the predicted proliferation of the risk of floods and tornadoes will be accompanied by attendant catastrophes that will affect diverse facets of life including urban and rural living, socio-economic development and environmental sustainability among others. Such a melange of problems requires proactivity in the form of policies that build the adaptive capacity of flood and tornado-prone communities as well as reduce the proliferation of the attendant risks. It is in this context that this book unpacked issues surrounding the increasing risk of floods and tornadoes in southern Africa. The purpose of this chapter is to summarise the book's emerging key findings, conclusions and policy recommendations.

16.2 Emerging Key Findings

16.2.1 Flood Modelling, Early Warning and Partnerships

Part II focuses on flood modelling, early warning systems and partnerships. The section comprises work that interrogates satellite-based approaches in the detection and monitoring of selected hydro-meteorological disasters. Chapter 2 provides an overview of the state-of-the-art remote sensing techniques and tools for the detection and monitoring of hydro-meteorological disasters, focusing on Cyclones Idai and Kenneth. It shows that, from a perspective of disaster preparedness, rainfall measurements provided by satellites can enable decision-makers to take urgent measures in the pre-event phase and can be used as an input for early warning systems. Furthermore, in Chap. 3, there is an examination of ICT readiness for flood risk reduction and management, drawing lessons from Eastern Cape Province's Port St Johns Local Municipality, South Africa. In this work, findings show that mobile phones are the most dominant ICT device used during flood disaster reduction efforts although there is a deluge of barriers to the effective use of ICT for flood risk reduction, including the lack of capacity to buy data, the unavailability of a reliable network and low levels of literacy of the elderly. The need for capacity enhancement in that regard cannot be overemphasised.

Chapter 4 focuses on hydrological and other emerging perspectives of Lake Kariba's Operation Noah. It investigates the hydrological perspectives of Operation Noah, taking another look at why the floodwaters rose so rapidly within a short period, catching many by surprise. Drawn from the Biblical Noah's flood times, Operation Noah was a global rescue plan to save the animals marooned in over 500 islands that formed as the Kariba Dam waters rose. It was observed that the filling of the dam coincided with the all-time high 1:110 years water flow regime of the Zambezi River and its tributaries. The flood waters rose so rapidly that it took everybody by surprise and unprepared resulting in the loss of life. In addition, Chap. 5 discusses water flow regimes focusing on Victoria Falls and comparing two half-centuries. A time-series analysis was done for the Victoria Falls water flow regimes, with specific reference to two half-centuries marked by 1919/1920–1968/1969 (Half Century I) and 1969/1970–2018/2019 (Half Century II), thereby using 100 years of available data. The key finding is that Half Century I had higher flows compared to Half Century II. In addition, daily mean flow regimes for Half Century II were declining, with marked negative trends for

January, August, September, October, November and December. From these months, several Mann-Kendall tests confirmed significant declines for October, November and December.

16.2.2 Urban Flooding and Insurance Matters

Part 3 of the book investigates urban flooding and insurance matters. Chapter 6 is a thorough assessment of hotel management under increasing and more intense floods, focusing on the Centurion Hotel, South Africa. Through repeated direct observations, interviews and Geographical Information Systems, the chapter investigates the extent, and potential causes, of repeated flooding at the Centurion Hotel, South Africa. It further examines the disaster risk reduction and management protocols put in place by the hotel and their effectiveness. The authors note that the Centurion Hotel has been a victim of repeated flooding, with the most recent event in December 2019 being the worst, forcing it to close twice in 2020, affecting 84 employees. The flooding is aggravated by the heavily silted Centurion Lake and land use changes in the upper catchment of the Hennops River stretching across the three municipal jurisdictions of Ekurhuleni, Johannesburg and the City of Tshwane metropoles. The results of this study further revealed that the hotel's design originally took into account the possibility of flooding, although the flood lines have since shifted and high-impact floods have become more frequent. The chapter also notes that the hotel has put in place measures to counter the effects of flooding although these were overwhelmed by the 2019 floods. The hotel has a sound standard operating procedure for the evacuation of guests in the event of a disaster. Given that such extreme weather events will only increase in the future, the authors recommend continuous adaptation and building back better and smarter with lessons from each flood event, including carefully planning for possible seasonal operations in the future.

Following the above is Chap. 7, which noted the increase in concern over the state of ocean pollution resulting from extreme hydro-meteorological events. The challenges include deposition of solid waste into harbours, beaches and estuaries by floods. A thought on harbours and beaches takes one to the blue (ocean) economy space, a subject that is embedded in the United Nations' Sustainable Development Goal 14. The Durban Harbour is one of Africa's most prominent and busiest ports handling cargo for South Africa and several southern African countries. In addition, its beaches rank among the world's best. Through the use of events study, in-depth interviews and document analysis, the chapter determines how the April 2019 floods revealed the challenges of waste management practices in Durban, South Africa, and how the concerned stakeholders, among them government, the private sector and non-governmental organisations, responded to the challenge. The results showed that extreme floods caused a significant challenge at the Durban harbour along beaches and estuaries as plastic waste mainly comprising shopping bags heavily polluted these facilities. The heavy plastic pollution at these places is a clear symptom of poor waste management in the upper catchment areas. Different stakeholders have put forward some innovative solutions to the problem but largely remain inadequate because of the complexity of the problem. During the decade of action, there is a need for the municipality and stakeholders to increase the efforts of ensuring environmental law enforcement so as to ensure the protection of the oceans and blue ocean economy. The problem needs to be confronted from multiple dimensions, using mass education coupled with intensive consultations as prerequisites to success in dealing with plastic litter.

Chapter 8 focuses on building flood resilient communities through storm water management in Port St Johns, South Africa. In this chapter, the authors observed that globally, urbanisation attracts informal settlements, which often do not comply with official spatial planning guidelines and are hardly provided with basic services, including drainage infrastructure. Storm water management is particularly important to build resilience in cities, especially with the increasing impact of climate change and the efforts to achieve the Sustainable Development Goals (SDGs) by 2030. It is against this background that this chapter explores the factors influencing the poor provision and maintenance of storm water management systems in different areas located in Port St Johns, South Africa. Using a household survey, key informant interviews and observations, the study points to the importance of storm water management as a key infrastructure component that requires sufficient funding to reduce flood risks. In addition, the study recommends community-led initiatives to enable storm water intervention strategies. It further proposes that the local government develops prevention measures against the building of houses in flood-prone areas.

Marking the end of part 3 is Chap. 9, which investigates the impacts of cyclones Idai and Kenneth and the 2019 floods on the insurance sector in South Africa and Mozambique. The authors argue that the insurance industry plays an important role in the management of risks, including risks from extreme weather events such as tropical cyclones and floods. They examined the impacts of the aftermath of the 2019 cyclones Idai and Kenneth and their associated floods on the insurance sector in Mozambique and the impacts of the 2019 floods in South Africa. Data was collected through an online questionnaire survey in Mozambique and a desktop-based analysis in South Africa. Results show that the insurance sector was susceptible to the cyclone and flood events in both countries. The study notes that the insurance sector was ill-prepared for such risks, with substantial losses to the sector. Recommendations are made for sector improvements in responding to the predicted increase in frequency and intensity of cyclone and flood events in southern Africa.

16.2.3 Challenges from Rural Flooding

Part 4 focuses on the assessment of a plethora of flooding challenges bedevilling rural settlements. The first chapter (Chap. 10) explores the migration effects of cyclones and floods in southern Africa focusing on the 2019 floods and cyclones Idai and Kenneth. Using the case study research design, this chapter investigates the impacts of floods on migration patterns of people in Zimbabwe and South Africa focusing on Chimanimani, Zimbabwe, and Port St Johns, South Africa. The chapter further explores whether cyclones and floods and related environmental catastrophes will be the next drivers of intra- and international migration within the southern African region. The results of this study indicate that despite the severity of effects (transient and permanent), cyclones and floods do not push individuals to move outside their countries of origin. Instead, migration to other countries is typically triggered by household economic constraints and tends to be voluntary.

In Chap. 11, the second chapter of part 4, the authors assess the climate-related effects on access to drinking water in wetlands and/or flood-prone areas with a focus on Magalasi Village, Chikwawa District in Malawi. The authors noted that access to drinking water remains a challenge in many rural areas in sub-Saharan Africa. This situation is further compounded by frequent disasters such as floods, which affect water supply infrastructure. However, few scholarly works have examined access to drinking water for rural households vulnerable to floods in southern Africa. This study analysed the extent to which the 2019 flooding induced by Cyclone Idai affected households' access to water in Magalasi Village in Chikwawa District in Malawi. Data was collected and analysed following the mixed methods research design in which both quantitative and qualitative approaches were used. The results show that during the floods, households relocated to a nearby school that has solar-powered tap water. However, the increased numbers at the camp resulted in a decline in the daily minimum water per capita of 20l/c/d in about 60% of the households, as each tap is meant to cater for 250 people only. Furthermore, the use of the taps was restricted to 8 h (6 a.m. to 2 p.m. during weekdays and 8 a.m. to 5 p.m. during weekends), which coincided with times for farming activities. Consequently, households had to fetch water from a village about 3kms away. However, the alternative water sources were also inaccessible during severe flooding conditions, with intermittent and contaminated water. Considering the health challenges associated with the consumption of insufficient and contaminated water, this study proposes the construction of more resilient deep boreholes for continuous water supply.

Chapter 12 deals with settlements' vulnerability to floods in South Africa. The objective of the study was to highlight the flood vulnerabilities exhibited in different settlements in South Africa, using the case of Port St Johns. The results showed that the causes of flood vulnerabilities included environmental degradation, poor drainage, poor settlement planning, location of settlements on wetlands, slopes and waterways. This was exhibited at varying degrees in the high density, medium density, low density, rural, rural (urban) like and informal settlements. Informal settlements identified were located on waterways, while the rural settlements identified were on degraded land, as well as on wetlands and slopes. An FVI of 0.39 was obtained indicating that the study area was vulnerable to flooding. Furthermore, physical FVI had the highest loading of 0.47, followed by environmental FVI (0.40), social FVI (0.35) and economic FVI (0.33), respectively. Households in informal settlements showed high physical and social vulnerabilities, and were characterised by blocked drainages, inadequate storm water infrastructure and a high degree of social differentiation, which increased their vulnerability. Rural settlements exhibited high economic and environmental vulnerabilities, and were characterised by a high unemployment rate, low incomes, lack of alternative livelihoods, and location on slopes and flood-prone areas, increasing their vulnerability to flooding. The chapter concludes that Port St Johns is vulnerable to flooding, with high physical and environmental vulnerabilities.

In Chap. 13, the book examines the impacts of floods on livestock production using cross-sectional survey data from 195 purposively selected respondents in Port St Johns Local Municipality. The results indicate that among all livestock, small stock was the most affected by flooding, especially among households settled in flood-prone areas. This resulted in reduced agricultural production for sheep, poultry, pig, goat and cattle producers. The flooding affected the availability of meat, with beef, mutton and chicken experiencing reduced availability, while the availability of goat and pork meat increased. Flooding also caused mutton to be relatively expensive, whereas the other livestock meat products were relatively cheaper. The availability and affordability of dairy products decreased. The flooding caused partial damage to agricultural and grazing land, taking between one and two weeks to restore vegetation after the flooding. It was perceived that environmental degradation was the leading cause of flooding that affected cattle production. Climate change and excessive rainfall were the leading causes of flooding affecting sheep and chickens, respectively. The packaging of the flood warning message was explicit and timeously communicated for sheep producers. The study concludes that flooding affected sheep supply and demand more disproportionally than other livestock species. It recommends the promotion of flood-tolerant vegetation to ease shocks in livestock production. There is also a need to strengthen infrastructure against flooding.

16.2.4 Increasing Risk of Tornados

Part 5 comprises Chaps. 14 and 15, which focus on the increasing risk of tornados. In Chap. 14, the authors note that southern Africa has, of late, been on the receiving end of severe storms that have left behind serious trails of destruction from strong winds, heavy rainfall and floods. The most recent being Tropical Cyclone Idai, one of the most intense and most powerful storms to ever hit southern Africa. Idai brought death and destruction in Malawi, Mozambique and Zimbabwe in March 2019. While much attention is given to severe storms that cover large areas and several countries, this study focused on the concurrent emergence of more localised storms that can be just as destructive as they are poorly understood, and not well integrated into early warning and response systems in the region. The analysis was based on eyewitness accounts and post-event damage assessment from five sites between 2012 and 2017. Results show that some of the severe localised storms in Malawi are tornadoes. The authors identify the need for improved tornado forecasting capacity, documentation of occurrence and effects, and awareness-raising on risk reduction.

"Characterisation and impacts of tornado-induced flooding and windstorms in Mpumalanga province, South Africa" are the matters discussed in Chap. 15. In this chapter, the authors note that extreme weather events have significantly affected human civilisations and development in recent years. While COVID-19 has been the top news headline in 2020 and 2021, it is important to note that the additional impact of extreme weather events is catastrophic to human development. Tornadoes rank among the deadliest extreme weather events of global concern. This study seeks to assess the trend and impacts of tornadoes on the highveld area of Mpumalanga province, South Africa. The Mkhondo Local Municipality was used in a case study to collect primary and archival data from key informants and weather bureaus. The study found that tornadoes have increased in Mpumalanga. Tornadoes that occur mainly in summer months with a peak between November and January have led to the loss of property, timber plantations, homes, schools, and health institutions and threaten the progress made in achieving the Sustainable Development Goals. The study noted a statistically significant increase in the frequency and severity of tornadoes in Mpumalanga, particularly in the last five years that coincide with probably the worst El Niño event experienced in southern Africa. The study recommends a risk-adjusted approach to infrastructure development considering the increased frequency of tornadoes due to global warming and climate change. The risk-adjusted approach must encompass smart climate initiatives for rural and urban development to ensure climate resilience if sustainability is to be achieved.

16.3 Policy Recommendations

Findings in this book have shown that earth observation systems play an instrumental role in early warning systems, life-saving activities and impact assessments in the phases immediately following a hydro-meteorological disaster. It has proven useful in providing near real-time products and collecting trusted information from other services that can be used effectively in disaster reduction. It is recommended that earth observation systems be integrated with social media streams to help verify areas affected or pinpoint target regions for satellite image acquisitions. The ability of communities to understand and use this information is also an important aspect. Matching capabilities and needs are also a key aspect of SEM mechanisms. The IWG-SEM role in the definition of both mapping guidelines and standard formats/products could foster better agreement between maps produced by different organisations involved in the mechanism. Solving restrictive data licensing could further increase data availability in the aftermath of disasters. The role of satellites is not limited to the emergency phase. Activities in the recovery phase, especially in areas where the agricultural sector is dominant, can take advantage of the possibility of performing reliable analysis with wide spatial coverage and frequent updates from satellites. Short-term damage assessment can take advantage of the information produced by SEM mechanisms without necessarily planning further satellite acquisitions whereas a similar analysis performed with in situ observation would require a more massive resource-intensive effort.

It is evident from the current work that ICT in the form of mobile phones play an important function in flood disaster management and risk reduction. However, it is noted that there are several barriers to effective use of ICT for flood risk reduction and management, including the lack of capacity to buy data, the availability of a reliable network and low levels of literacy in the elderly. The world is moving swiftly in scaling up the use of mobile phones, harnessing such for flood risk reduction and management can be very useful. At the policy level, there is a need to improve network coverage and subsidising data for flood-prone regions to ensure that it is affordable to most of the households and individuals. The need to enhance literacy levels through workshops tailor-made for flood risk reduction among people residing in vulnerable areas cannot be overemphasised.

The book has highlighted unusual hydrological events on the Zambezi River during Operation Noah, which influenced the fast rate at which the levels of the lake were rising. A large number of people and animals needing to be resettled and rescued drastically increased the total amount of resources required. Many animals drowned and/or were marooned on islands leading to biodiversity loss and ecosystem disturbance. It is recommended that for mega projects such as dam construction, comprehensive planning, which takes into consideration direct and indirect impacts before, during and after project implementation, needs to be done and this should include ecological, social and physical aspects of the environment. Adequate resource mobilisation should be carried out before project commencement.

This work indicates the possibility of Victoria Falls drying up particularly in October, November and December if the population increases, industrialisation takes place, irrigation schemes expanded and domestic consumption of water upstream increases. There is a need for authorities and responsible stakeholders to accept this reality and take appropriate resilience-building measures such as population control, urban land use planning to ensure there is control on agricultural activities and water consumption patterns.

The repeated flooding of The Centurion Hotel over six years as shown in this book has been attributed to the shifting flood line. The magnitude of flooding has intensified owing to the land cover and land use changes occurring on the upper parts of the Hennops River catchment, as well as the amount of rainfall received in the catchment. The catchment has witnessed increases in urbanisation, urban informal agriculture, wetland destruction and sand poaching along the tributary streams. All these factors, among several others, contributed to influencing flooding at the hotel. Despite instituting several control measures, the floods still occurred and destroyed. It is recommended that an integrated catchment management system be implemented in the Hennops River catchment to deal with the problems of flooding. Stakeholders need to be involved in the management of the catchment in general and water resources and must be cognisant of the impact of their actions on the hydrological cycle. Discussion forums such as the Hennops River Forum, which include local governments, government environmental departments, members of academia, civic organisations as well as other experts, are a good starting point. Lake Centurion needs to be managed and monitored in terms of the level of sediment, as well as the outlet valves that are in most cases blocked with sediment or various types of plastics. This will help to speed up the rate at which water will flow from the lake, thus preventing backflows that may end up flooding the hotel. There is also a need to give special consideration to foreign nationals, especially those from non-English-speaking countries, during flooding or other emergencies because they have limited capacity to react swiftly to perceived threats as a result of the language barrier. Given the trend towards increased flood incidences, this work recommends hotels to explore new avenues to take an insurance cover, which includes flood losses. Building back better and smarter is recommended too to ensure the structures and all infrastructures are resilient to the whims of nature.

Investigation of the impacts of flooding on the pollution of harbours, beaches and urban estuaries in this book has shown that the beaches, harbours and urban estuaries were hit hard by plastic litter after the floods, an indication of poor waste management systems in the upper catchments. Pollution is a threat to marine ecosystems and the benefits they offer to humanity. Initiatives put in place to manage the pollution did not yield good results due to the complexity of the pollution problem. At the policy and operational levels, the problem needs to be tackled from a multidisciplinary and multi-stakeholder front. This should include tourism, retail, communities, industry and government. The solutions must be suited to regions and communities with limited resources, and societal challenges. Such solutions can only be arrived at through intensive consultation processes. There is a need for more investment in waste management and the environment, including both the hardware and software components. Awareness raising and mass environmental education relating to environmental issues associated with the use and disposal of plastics should be promoted at schools, social media, radios and newspapers. These and many more relevant media should be used more wisely to inform and draw attention to key environmental issues.

The book also critically assesses the factors influencing flood resilience, drawing lessons from St Port Johns. The work establishes that there are a plethora of factors affecting the resilience of urban settlements. In fact, flooding has compounded the negative impacts on the livelihoods and lives of those living in informal settlements. The under-funding of proper storm water drainage systems coupled with the lack of basic service delivery, and proper urban planning further worsen the consequences on the poor and the most vulnerable. From this deduction, this study recommends the establishment of an integrated institutional structure that comprises the roads and transport department and other departments within local government such as planning and environment to manage storm water. There is a need for the proactive management of storm water, rather than treating it as an issue that is only dealt with in times of floods; that should minimise the impact of flooding. The lack of investment in drainage also results in the rapid deterioration and ageing of the infrastructure. The study also advocates for community-led initiatives towards flood risk reduction. Partnerships between local government and community members build formidable coalitions to respond to the flooding events. This can be through the formation of community-led initiatives to reduce flood risk vulnerabilities within the community. An example of this initiative could include constructing storm water channels that will direct water during rainy seasons. Another example could be the establishment of a local team of volunteers to collect waste in the area on a weekly basis and dispose of it at a designated spot. This can prevent waste from getting into and blocking storm water drains. These interventions certainly require some sort of incentives for the work done from the local government's side. Sufficient funding mechanisms for storm water management and prevention measures against settling households in flood-prone areas are some of the recommendations in this regard. If storm water management is prioritised, the damage from flooding could be drastically reduced in relation to the already impoverished households.

The impacts of cyclones Idai and Kenneth on the insurance sector were examined in this book. The study argues that research data-driven insurance models for predicting and mapping cyclone and flood risks can assist in sending early warnings to clientele and in estimating exposure as well as pricing insurance products. This can help increase people's understanding of localised cyclone and flood activities through raising the necessary awareness, encouraging adaptive avoidance practices and adoption of insurance products, thereby developing a culture that promotes loss prevention, reduction and risk management. Similarly, innovative insurance products that incentivise the establishment of the cyclone and flood-resistant infrastructure through providing economic incentives (lower premiums) can contribute significantly to socio-economic and ecological adaptation, thereby helping to develop relevant cyclone and flood resilience practices in areas prone to such disasters. The insurance sector also plays an important role in cushioning clients from economic shocks resulting from cyclone and flood damage through assessing damage and providing timely finance in the form of claim pay-outs (risk transfer). The insurance industry also contributes to climate change mitigation through promoting carbon emissions reduction, such as refusing to insure carbon-heavy industrial practices such as coal mining and coal-powered electricity generation. Insurance companies are therefore in a powerful position to play an important role in promoting the transition to a sustainable global future. Southern Africa has limited cyclone and flood risk insurance coverage, which provides an opportunity for the development of a catastrophic risk insurance market in this area.

The book also explored the migration effects of cyclones and floods in southern Africa, focusing on the 2019 floods and cyclones Idai and Kenneth. It indicates cyclones and floods have the propensity to induce migration in southern Africa. This work recommends that response measures that address the problem of internal or external migration need to be tailored to fit the socio-economic realities of the affected communities. Community-specific issues such as rural settings, limited access to necessities and information need to be taken into consideration. Southern African governments should strengthen their disaster management policies on response measures to cyclones and floods by allocating financial resources and building human resource capabilities that can respond rapidly when disasters strike. When disasters have taken place, governments should consider revisiting economic development plans and reallocating budgets promptly in order to prioritise developmental projects as a way of utilising limited resources. This should facilitate the recovery process in the aftermath of environmental, social and economic shocks. As a communication strategy and for public participation purposes, governments should utilise existing structures in their respective communities, such as traditional authorities and local government, to reach people at the grass-roots level, taking into account local languages and communication methods used in specific areas.

Tropical cyclones affect community livelihoods and vital infrastructure such as water because of the floods that they are normally associated with. In this book, the effects of the 2019 Cyclone Idai-induced flooding on drinking water for households affected by flooding were assessed in Chikwawa District in Southern Malawi. The study shows the scarcity of water resources during a flood event. Although a deluge of challenges inhibits access to clean water during non-flood periods, their impacts worsened during Cyclone Idai-induced flooding conditions. Considering the health challenges associated with the consumption of insufficient and contaminated water, this study proposes the construction of secured deep boreholes which are less affected by the climate extremes and are likely to provide continuous water supply. Further, in cases where existing water infrastructure fails to cater for the water needs of all those affected by floods, relief interventions should include drilling of boreholes within the camp. In addition, there is a need to provide basic water treatment kits to flood-affected communities depending on alternative water sources to reduce the risk of water-borne diseases.

Worldwide, various settlements are shown to be vulnerable to flooding. In this book, the vulnerability of settlements has been studied with a focus on St Port Johns in South Africa. The book notes that Port St Johns is vulnerable to flooding, with high physical and environmental vulnerabilities. Furthermore, informal settlements were found to be the most vulnerable to flooding, followed by rural settlements. Informal settlements had high physical and social vulnerabilities, while rural settlements had high economic and environmental vulnerabilities. Therefore, informal settlements are characterised by blocked drainages; impermanent, neglected and poorly designed houses; lack of connectivity; inadequate storm water infrastructure; high poverty; overcrowding; social problems; low levels of service provision; limited networks; and a high degree of social differentiation, which tend to increase their physical and social vulnerability to flooding. Rural settlements are characterised by a high unemployment rate; low incomes; lack of alternative livelihoods; and location on slopes and flood-prone areas such as flood plains, which tend to increase their economic and environmental vulnerability to flooding. There is a need to increase adaptive capacity to reduce vulnerability. Building the capacity of communities to migrate before a flood event is recommended as one way of reducing vulnerability.

In a study that assessed the impacts of floods on livestock production, this book argues that small livestock are the most affected by flooding especially when people are settled in flood-prone areas. The study recommends that waterlogged-tolerant vegetation should be promoted to circumvent the reduction in vegetation and grazing land that occurs due to flooding. This could ease the shocks in livestock production. However, a balance between flood tolerance of the plants as well as its palatability to the various livestock species needs to be reached. There is also a need to strengthen the infrastructure against flooding such as the communication infrastructure and thus maintaining market access and supporting farmers to cope with changes to agricultural and grazing land conditions.

The book assessed the increasing risk of tornadoes, looking at the characterisation and analysis of emerging localised severe storms and tornadoes in Malawi. The study concluded that the first two tornadoes occurred during the period the country was under active weather brought about by the simultaneous prevalence of both the Congo Air Mass and the Inter-tropical Convergence Zone (ITCZ), which are the main rain-bearing systems in this part of the continent, while the third tornado occurred during the period the ITCZ alone was active. This work is the first to formally report on tornadoes in Malawi. There is, therefore, need for tornadoes to be given greater attention in Malawi to get a true picture of their occurrence and impact. This agrees with NOAA (2011) on the need for developing capacity for rapid updates on tornado location, movement and damage. In Malawi, the Weather Chasers WhatsApp group and others help to draw attention to these events, and these social media platforms can support communication of damage path information after the event. As argued by NOAA (2011), documenting and sharing photos and videos of tornado damage helps people to appreciate the destructive potential of major tornadoes and that where this has been done it has influenced the change of behaviour from perceiving a tornado as unlikely and therefore not requiring action, to being motivated to take action despite tornadoes being low probability occurrences.

As observed in Chap. 15, focusing on the characterisation and impacts of tornado-induced flooding and windstorms in Mpumalanga province, South Africa, it emerged that there was an increase in the frequency and intensity of tornadoes. These have negatively affected the community's socio-economic well-being. Most tornadoes were observed to occur in the summer months, especially between November and January with November as the peak month. Given the increased frequency of tornado occurrence, there is a need to provide concrete relief to the vulnerable community through a number of possible interventions, namely proper recording and forecasting of tornadoes, to ensure that reliable early warnings are issued to the community. Recognising the importance of climate and weather data in human settlement planning and development is crucial in a changing climate. A climate risk-adjusted approach is also recommended as it is imperative in developing and constructing private and public infrastructure to ensure climate resilience. This includes the revision of housing infrastructure plans to ensure that such houses and other developments are climate-proof. In as much as indigenous knowledge is crucial in human development, there is a need for traditional communities to acknowledge the role of science in adopting a robust approach to human settlement that ensures human safety. The role of traditional leadership in building a climate-proofed future can never be underestimated; nonetheless, there is a need to ensure a hybrid approach to build a climate-reliant rural community.

16.4 Conclusion

This chapter has shown that the risk of floods and tornados is increasing across southern Africa. It is therefore expected that the occurrence of the risk of floods and tornadoes will be accompanied by attendant disasters that will affect diverse facets of life. The resultant problems require proactivity in the form of policies that build the adaptive capacity of flood and tornado-prone communities as well as reduce the proliferation of the attendant risks.

References

Auswärtiges Amt. (2019). *Statement by Federal Minister for Foreign Affairs Heiko Maas at the open debate of the UN Security Council addressing the Impacts of Climate-related Disasters on International Peace and Security*. Retrieved June 14, 2019, from https://www.auswaertigesamt.de/en/newsroom/news/maas-climate-un/2182052.

Guiseppe, F., & Feyen, L. (2019). Empirical evidence of declining global vulnerability to climate-related hazards. *Global Environmental Change, 57*(1), 1–9.

He, C.-Y., Tung, C.-P., & Lin, Y.-J. (2021). Applying the DRCA risk template on the flood-prone disaster prevention community due to climate change. *Sustainability, 13*, 891. https://doi.org/10.3390/su13020891.

Howard, B. C. (2019). *What we know—and what we don't—about the science of tornadoes*. Scientists probe the mysteries of violent twisters. Retrieved from https://www.nationalgeographic.com/science/article/140430-tornadoes-meteorology-atmospheric-science-disasters

Ide, T., Brzoska, M., Donges, J. F., & Schleussner, C. F. (2020). Multi-method evidence for when and how climate-related disasters contribute to armed conflict risk. *Global Environmental Change, 62*(May). https://doi.org/10.1016/j.gloenvcha.2020.102063.

Lee, D., & Jung, J. (2014). The growth of low-income population in food plains: A case study of Austin, TX. *KSCE Journal of Civil Engineering, 18*, 683–693. https://doi.org/10.1007/s12205-014-0205-z.

Lemperiere, F. (2017). Dams and floods. *Engineering, 3*, 144–149.

Martinez, M., Bakheet, R., & Akib, S. (2020). Innovative techniques in the context of actions for flood risk management: A review. *Engineering, 2*, 1–11. https://doi.org/10.3390/eng2010001.

Molloy, S. L., & Mihaltcheva, S. (2013). Vulnerability of human health to climate. *Climate Vulnerability*.

Munich RE. (2005). *Hurricanes, typhoons and cyclones*. Retrieved from February 4, 2020, from https://www.munichre.com/en/risks/natural-disasters-losses-are-trending-upwards/hurricanes-typhoons-cyclones.html#-1979426458.

Mysiak, J., Castellari, S., Kurnik, B., Swart, R., Pringle, P., Schwarze, R., Wolters, H., Jeuken, A., & van der Linden, P. (2018). Brief communication: Strengthening coherence between climate change adaptation and disaster risk reduction. *Natural Hazards and Earth System Sciences, 2018*(18), 3137–3143.

National Oceanic and Atmospheric Administration (NOAA) National Weather Service. (2011). *The historic tornadoes of April 2011*. Maryland: US Department of Commerce.

O'Donnell, E. C., & Thorne, C. (2020). Drivers of future urban flood risk. *Philosophical Transactions of the Royal Society A - Mathematical Physical and Engineering Sciences, 378*(20190216), 2020.

Peters, K. & Peters L. E. R. (2018). Disaster risk reduction and violent conflict in Africa and Arab States. Implication for the Sendai Framework priorities, Briefing note, ODI, London.

Qiang, Y. (2019). Disparities of population exposed to food hazards in the United States. *Journal of Environmental Management, 232*, 295–304.

Rolfe, M. I., Pit, S. W., & McKenzie, J. W. (2020). Social vulnerability in a high-risk food-affected rural region of NSW Australia. *Natural Hazards*. https://doi.org/10.1007/s11069-020-03887-z.

Saintilan, N., Khan, N. S., Ashe, E., Kelleway, J. J., Kerrylee, R., Colin, D. W., & Horton, B. (2020). Thresholds of mangrove survival under rapid sea level rise. *Science, 368*, 1118–1121.

Stagl, J., Mayr, E., Koch, H., Hattermann, F. F., & Huang, S. (2014). Effects of climate change on the hydrological cycle in Central and Eastern Europe. In *Managing protected areas in central and eastern Europe under climate change* (pp. 31–43). Berlin/Heidelberg, Germany: Springer.

Svetlanaa, D., Radovana, D. & Ján, D. (2015). The Economic impact of floods and their importance in different Regions of the World with Emphasis on Europe, *Business Economics and Management 2015 Conference, BEM2015, Procedia Economics and Finance 34* (pp. 649–655).

von Uexkull, N. (2016). Climate, Conflict and Coping Capacity. The Impact of Climate Variability on Organized Violence. Report / Department of Peace and Conflict Research 111. 44 pp. Uppsala: Department of Peace and Conflict Research. ISBN 978-91-506-2585-1.

Von Uexkull, N., & Halvard, B. (2021). Security implications of climate change. *Journal of Peace Research, 58*(1), 3–17.

Index

G. Nhamo, L. Chapungu (eds.), *The Increasing Risk of Floods and Tornadoes in Southern Africa*, Sustainable Development Goals Series, https://doi.org/10.1007/978-3-030-74192-1